WIDE BANDGAP SEMICONDUCTOR ELECTRONICS AND DEVICES

SELECTED TOPICS IN ELECTRONICS AND SYSTEMS

Editor-in-Chief: **M. S. Shur** ISSN: 1793-1274

---

*Published**

Vol. 63: *Wide Bandgap Semiconductor Electronics and Devices*
eds. Uttam Singisetti, Towhidur Razzak and Yuewei Zhang

Vol. 62: *High Performance Logic and Circuits for High-Speed Electronic Systems*
eds. F. Jain, C. Broadbridge, M. Gherasimova and H. Tang

Vol. 61: *High Performance Materials and Devices for High-Speed Electronic Systems*
eds. F. Jain, C. Broadbridge, H. Tang and M. Gherasimova

Vol. 60: *Microelectronics and Optoelectronics*
The 25th Annual Symposium of Connecticut Microelectronics and Optoelectronics Consortium (CMOC 2016)
eds. F. Jain, C. Broadbridge and H. Tang

Vol. 59: *Scaling and Integration of High Speed Electronics and Optomechanical Systems*
eds. Magnus Willander and Håkan Pettersson

Vol. 58: *Fundamental and Applied Problems of Terahertz Devices and Technologies*
Selected Papers from the Russia-Japan-USA-Europe Symposium (RJUSE-TeraTech 2016)
by Maxim Ryzhii, Akira Satou and Taiichi Otsuji

Vol. 57: *Frontiers in Electronics*
Selected Papers from the Workshop on Frontiers in Electronics 2015 (WOFE-15)
eds. Sorin Cristoloveanu and Michael S. Shur

Vol. 56: *Fundamental & Applied Problems of Terahertz Devices and Technologies*
Selected Papers from the Russia–Japan–USA Symposium (RJUS TeraTech-2014)
ed. Michael S. Shur

Vol. 55: *Frontiers in Electronics*
Selected Papers from the Workshop on Frontiers in Electronics 2013 (WOFE-2013)
eds. Sorin Cristoloveanu and Michael S. Shur

Vol. 54: *Frontiers in Electronics*
Advanced Modeling of Nanoscale Electron Devices
eds. Benjamin Iñiguez and Tor A. Fjeldly

Vol. 53: *Frontiers in Electronics*
Selected Papers from the Workshop on Frontiers in Electronics 2011 (WOFE-2011)
eds. Sorin Cristoloveanu and Michael S. Shur

Vol. 52: *Frontiers in Electronics*
Proceedings of the Workshop on Frontiers in Electronics 2009 (WOFE-2009)
eds. Sorin Cristoloveanu and Michael S. Shur

*The complete list of the published volumes in the series can be found at
https://www.worldscientific.com/series/stes

Selected Topics in Electronics and Systems – Vol. 63

WIDE BANDGAP SEMICONDUCTOR ELECTRONICS AND DEVICES

Editors

Uttam Singisetti
University of Buffalo, USA

Towhidur Razzak
The Ohio State University, USA

Yuewei Zhang
University of California Santa Barbara, USA

World Scientific

NEW JERSEY • LONDON • SINGAPORE • BEIJING • SHANGHAI • HONG KONG • TAIPEI • CHENNAI • TOKYO

Published by

World Scientific Publishing Co. Pte. Ltd.
5 Toh Tuck Link, Singapore 596224
USA office: 27 Warren Street, Suite 401-402, Hackensack, NJ 07601
UK office: 57 Shelton Street, Covent Garden, London WC2H 9HE

British Library Cataloguing-in-Publication Data
A catalogue record for this book is available from the British Library.

Selected Topics in Electronics and Systems — Vol. 63
WIDE BANDGAP SEMICONDUCTOR ELECTRONICS AND DEVICES

ISBN 978-981-121-647-3

For any available supplementary material, please visit
https://www.worldscientific.com/worldscibooks/10.1142/11719#t=suppl

Preface

With the dawn of Gallium Oxide (Ga_2O_3) and Aluminum Gallium Nitride (AlGaN) electronics and the commercialization of Gallium Nitride (GaN) and Silicon Carbide (SiC) based devices, the field of wide bandgap materials and electronics has never been more vibrant and exciting than it is now. Wide bandgap semiconductors have had a strong presence in the research and development arena for many years. Recently, the increasing demand for high efficiency power electronics and high speed communication electronics, together with the maturity of the synthesis and fabrication of wide bandgap semiconductors, has catapulted wide bandgap electronics and optoelectronics into the mainstream.

Wide bandgap semiconductors exhibit excellent material properties, which can potentially enable power device operation at higher efficiency, higher temperatures, voltages, and higher switching speeds than current Si technology. GaN-based electronics have already made a big impact in the performance of several applications including RF cell phone base stations and military radar. GaN and SiC based devices have also made significant contributions in the reduction of worldwide energy consumption. On the other hand, emergent materials like Ga_2O_3, an ultra-wide bandgap semiconductor that can be grown from a melt using methods including the Czochralski process, high Al-composition AlGaN, and diamond hold great promise in the field of power electronics and high-speed electronics by virtue of their high critical breakdown field strengths and high electron saturation velocity. Thus, wide bandgap electronics is a rapidly evolving area with research being actively conducted worldwide. It is therefore timely for us to put forth this volume as a platform to share the most up-to-date research in this area.

We hope that this edited volume will serve as a useful reference for researchers in this field — newcomers and experienced alike. This book discusses a broad range of topics including fundamental transport studies, growth of high-quality films, advanced materials characterization, device modeling, high frequency, high voltage electronic devices and optical devices written by the experts in their respective fields. The topics covered in this volume span over a large spectrum of wide bandgap materials including AlGaN, Ga_2O_3 and diamond.

We would like to thank Prof. Michael Shur for his encouragement to put together this edited volume in *Wide Bandgap Semiconductor Electronics and Devices* as volume 63 in his book series on *Selected Topics in Electronics and Systems*. Here, we would like to thank all the contributing authors for the excellent chapters. We would also like to thank all the reviewers for their diligent assessment of the manuscripts. Last but not the least, we would like to express our gratitude to the desk editors' Dr. Ng Yan Hong and Ms. Nurul Affiah, for tirelessly helping us with the preparation of this edited volume.

Editors
Uttam Singisetti (University at Buffalo)
Towhidur Razzak (The Ohio State University, Columbus)
Yuewei Zhang (University of California, Santa Barbara)
November, 2019

Contents

Substrate Effects in GaN-on-Silicon RF Device Technology

Hareesh Chandrasekar

Department of Electrical and Computer Engineering,
The Ohio State University,
2015 Neil Avenue, 205 Dreese Labs, Columbus OH 43210, USA
chandrasekar.28@osu.edu

The influence of the semiconducting Si substrate on the performance of GaN-on-Si RF technology is reviewed. Firstly, the formation of a parasitic conduction channel at the substrate-epitaxy interface is discussed in terms of its physical mechanism and its influence on RF loss, followed by schemes to minimize this effect. Secondly, it is shown that the presence of the parallel channel serves to back-bias the III-nitride epitaxial stack and lead to current collapse even on the highly-resistive Si substrates used for RF device fabrication, analogous to GaN-on-doped Si power devices. Strategies to mitigate this issue are also presented and critically compared. Thirdly, thermal generation of carriers in Si at elevated operating temperatures leading to increased substrate loss is quantified, also followed by a discussion of possible techniques to reduce its influence on RF loss.

Keywords: GaN devices; RF loss; parasitic channel; highly-resistive Si substrate; dielectric loss; GaN-on-Silicon; temperature-dependent loss; back-biasing, current collapse.

1. Introduction

Gallium Nitride high electron mobility transistors (HEMTs) are currently the mainstream RF technology of choice for discrete and MMIC-based high-power, high-frequency power amplifiers in a range of applications ranging from radars, CATVs, and satellite communications, to wireless broadband in the 4G and the up-and-coming 5G spectrums [1–3]. Despite the stellar performance of GaN devices on silicon carbide (SiC) substrates [1], the inherently high cost of this technology due to the limited areas and high starting substrate costs, have spurred the development of GaN devices on alternate, more inexpensive substrates in order to achieve cost-competitiveness with existing Si LDMOS technology for sub-3.6 GHz power amplifiers and for future applications at higher frequency bands. Silicon (111) has been widely used as a substrate for GaN growth despite the multifarious challenges involved in obtaining good quality III-nitride films such as the very large lattice (17%) and thermal mismatch which in turn leads to higher dislocation densities and film cracking [4–6]. These challenges have been successfully addressed by a variety of stress management schemes and present day epitaxy on Si is of comparable quality and repeatability to that on SiC substrates albeit with a higher complexity of the epitaxial process [6–9]. Nevertheless, the large size of Si wafers (up to

12" diameter currently), their ultra-low cost, and possibility of co-processing in existing CMOS foundries promises to make GaN-on-Si devices the cost-effective gallium nitride solution for power amplification and power electronic applications. Indeed the bulk of R&D efforts in GaN power electronics has focused on material grown on Si substrates in view of the above advantages [10, 11].

Although the development of GaN-on-Si HEMTs for RF applications far predates that for power electronics [12–15], this technology is becoming more mainstream thanks to its adoption currently in wireless broadband applications, and device & circuit level metrics of reliable GaN-on-Si devices comparable to GaN-on-SiC have been reported in literature [12, 16–21]. The continued development of GaN-on-Si devices goes hand in hand with advances in hetero-epitaxy in terms of defect reduction and stress management schemes, maturity of CMOS-compatible process flows for device fabrication, and informed device design and modeling based on accurate device characterization data – both electrical and thermal – as well as reliability implications. Such an interplay between device design-growth-processing-characterization is key to realizing the full potential of this technology.

In this context, it is important to thoroughly understand the role of the Si substrate itself on the intrinsic performance of GaN-on-Si RF electronics from a device standpoint. Here we review recent work on three issues stemming from the use of semiconducting Si as the substrate of choice for GaN RF devices. Firstly, the formation of a parasitic/parallel conduction channel at the Si-epitaxy interface is discussed in terms of the physical mechanisms involved and its effect on RF losses, with solutions to address the same. Secondly, we discuss the increased susceptibility of GaN-on-HR-Si HEMTs to dispersion/trapping effects in the GaN buffer due to the presence of such a parallel channel and small amounts of leakage through the substrate backplane. We will then briefly present and evaluate approaches to minimize such buffer-induced current collapse for GaN-on-Si devices. Lastly, thermal generation of carriers in the semiconducting Si substrates, either due to heat flow from the active device or elevated ambient operating temperatures is considered. This causes a drop in the resistivity of the highly-resistive Si substrates (HR-Si) used for GaN devices, leading to RF substrate losses which are quantified, and we conclude with presenting approaches to minimize this effect.

2. Parasitic Channel Formation and RF Loss in GaN-on-Si Technology

2.1. *Physical Origin of Parasitic Channel Formation*

The formation of a parasitic channel, i.e. a conductive interfacial/near-interfacial layer at the Si/III-nitride epitaxy interface giving rise to parasitic loss effects in GaN-on-Si active devices and passives has been well documented at this point. Apart from increased losses/parasitic loading effects for RF devices, the presence of such a parallel channel has also been attributed to be the cause of the high vertical leakage and pre-mature breakdown of GaN-on-Si power devices [22–27]. However, the mechanisms involved in the formation of such a parasitic channel are multifarious and complex with published

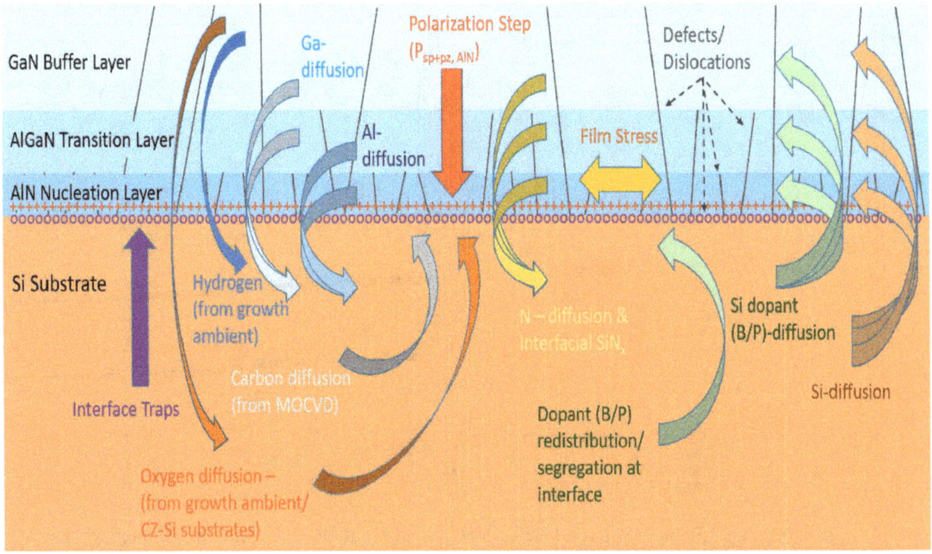

Fig. 1. A summary of potential factors affecting formation of the parasitic channel at or near the epitaxy – Si substrate interface.

reports offering differing explanations of its origin. Figure 1 attempts to pictorially summarize the various factors that could potentially contribute to the formation of such a parasitic/parallel channel.

For instance, several reports have identified the diffusion of Ga-and Al-species into the Si substrate as the cause of a conductive p-type layer [23, 28, 29], while yet others have attributed this to a polarization driven n-type inversion layer of electrons at the interface [25–27, 30], both of which are quite distinct mechanisms by themselves. We now examine these and several other factors in more detail.

The diffusion of Al and Ga-containing species into the Si substrate has been shown to form a p-type conductive layer of holes with peak densities ranging from 10^{18}-10^{15} cm^{-3} at the interface and a roll-off ranging from <100 nm – 3 μm away from the interface as characterized by SIMS and spreading resistance measurements [23, 28]. Ga and Al, both p-type dopants in Si, could diffuse into the substrate from –

1) the vapor phase directly, most notably in the case of Al as an AlN nucleation layer is the first layer deposited on Si prior to GaN growth, and from reactor residues containing Ga- and Al-species, and/or
2) out-diffusion from the deposited films as the growth temperatures used are typically very high (900-1200°C for MOCVD).

Figure 2 shows the 1-D equilibrium band diagram for the case of a Ga-profile with a peak concentration of 10^{18} cm^{-3} at the interface and a roll-off of 3 μm into highly resistive n-type Si (10 kΩ.cm), without accounting for any interfacial polarization charge.

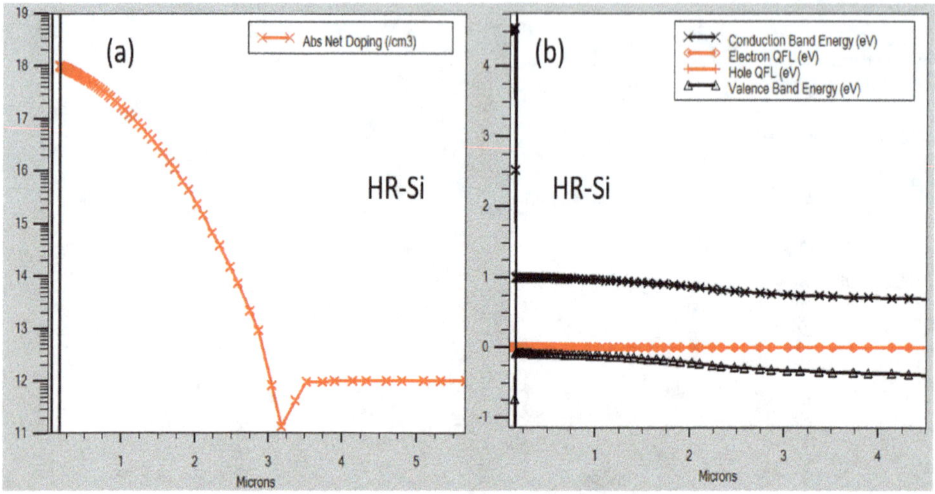

Fig. 2. (a) Ga-diffusion profile in the Si substrate with peak concentration of 10^{18} cm^{-3} at the interface and roll-off of 3 μm after Hanson *et al.* [23] (b) Band diagram of near-interface HR-Si region at the AlN/Si hetero-interface showing the accumulation of holes close to the interface, generated using the Silvaco ATLAS device simulator.

It is important to note that the III-nitride films grown on Si are not only highly defective but also have large film stresses (in the GPa range) even at the growth temperatures [6]. The initial layers such as the AlN nucleation layer and some part of the subsequent AlGaN transition layers are typically grown under tensile strain and have dislocation densities ranging from 10^{13} cm^{-2} in the AlN layer to 10^{11} cm^{-2} in the AlGaN layers. The GaN buffers are typically engineered to have a compressive strain in order to obtain crack-free films on cool-down and typically have dislocation densities of 10^{9}-10^{10} cm^{-2} at the active device regions [31, 32]. These large dislocation densities could act as local diffusion pathways for Ga- and Al-species towards the substrate with the large film stresses further aiding such diffusion. It is also noteworthy that only Al and Ga in substitutional Si sites act as dopants. While such substitutional doping would be aided by the high growth temperatures typically used, the total density of these species as reported by SIMS measurement would also include interstitial Al/Ga. Furthermore, the parasitic channel is expected to be composed of holes (p-type) in this case.

Another commonly proposed explanation for the parasitic channel is the formation of an inversion layer of electrons at the Si/AlN interface [25–27]. AlN is a polar material with a high spontaneous (P_{sp}) and piezoelectric polarization (P_{pz}) while Si is a non-polar material. This gives rise to a "polarization step" at this polar/non-polar hetero-interface. Since AlN on Si is almost always Al-polar, a positive polarization charge would be manifest at this interface. Given the large spontaneous polarization of the AlN (0.081μC/cm^2), this should correspond to a sheet charge of ~5x10^{13} cm^{-2} at the interface. Furthermore, the AlN films are typically strained in tension (>1 GPa) and hence the piezoelectric polarization should add to the spontaneous polarization and increase the

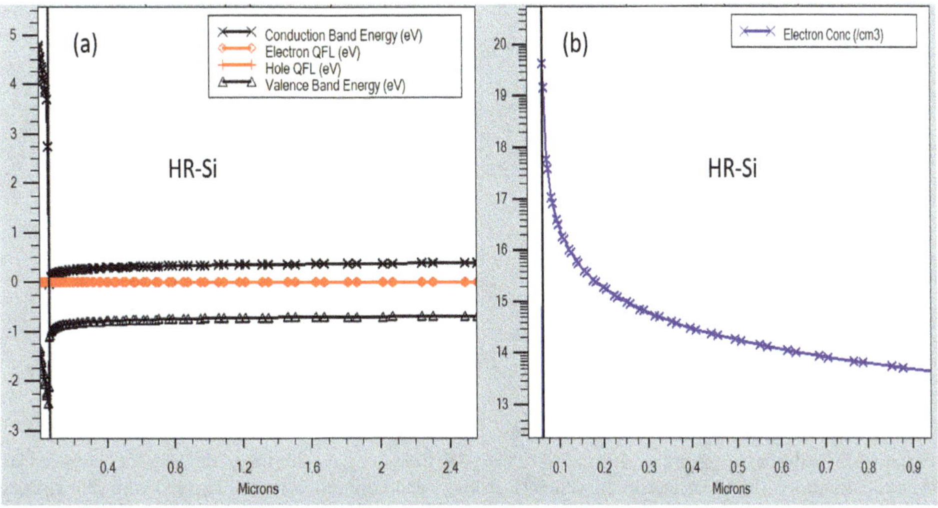

Fig. 3. (a) Band diagram at the AlN/Si interface assuming spontaneous polarization only in relaxed AlN films on Si showing formation of an inversion layer of electrons at the interface. (b) Plot of electron concentration expected due to the polarization step at an ideal AlN/Si interface, generated using the Silvaco ATLAS device simulator.

interfacial sheet charge and expected electron density at this interface. Figure 3 shows the equilibrium band diagram for this scenario assuming a relaxed 100 nm AlN layer ($P_{pz} = 0$) on a highly resistive p-type Si (10 kΩ.cm). The parasitic channel in this case is composed of an inversion layer of electrons (n-type), the origin of which would most likely be thermal generation in the silicon substrate or surface donors in case of only AlN layers on Si [27].

The interface between Si and AlN is expected to be highly defective due to the large lattice and thermal mismatch between these materials. Based on the lattice mismatch, it has been shown that 4 lattice segments of (111) Si correspond to 5 planes of (1-100) AlN with one misfit dislocation created for strain relaxation [33]. In addition, AlN grows in the 3D Volmer-Weber growth mode with 50-100 nm grains coalescing to form grain boundaries. This leads to threading dislocation densities of 10^{13}-10^{12} cm^{-2}, comparable to the expected electron density, in these layers. All of the above factors are also expected to give rise to a large trap density at this highly disordered interface. This in turn should lead to a sharp reduction in effective carrier mobility and hence the interfacial conductivity with the interfacial charge determined by the electrostatics of such trap states. Indeed AlN layers have been explored as a dielectric for Si MIS structures and both negative and positive shifts in flat-band voltages have been observed, corresponding to negative and positive interfacial charges [30, 34–36]. Reported interface trap densities vary widely from 10^{11}-10^{13} cm^{-2} depending on the deposition conditions [37]. For instance, Fig. 4 shows the extracted D_{it} from conductance spectra for MIS structures of standard MOCVD grown 50 nm AlN layers on p-Si substrates from our earlier work [29].

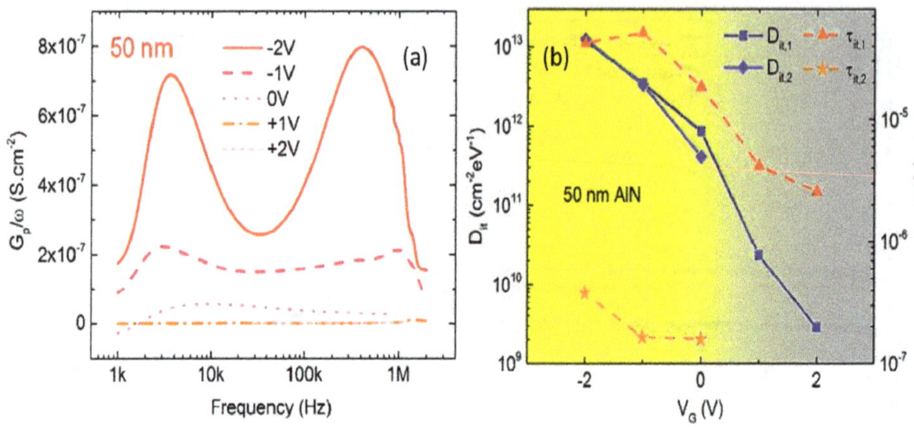

Fig. 4. (a) Conductance spectra of 50 nm MOCVD AlN films on p-Si substrates showing the evolution of slow and fast traps. (b) Interface trap densities (D_{it} in cm^{-2}) and trap time constants (τ_{it} in s) with gate voltage swept from depletion to weak inversion. (Reprinted with permission from *Thickness-dependent Parasitic Channel Formation at AlN/Si Interface*, by *H. Chandrasekar et al.*, in *Scientific Reports, Volume 7, Article number: 15749, 2017*).

Both slow and fast trap states were observed with D_{it} ranging from 10^{13}-10^{9} cm^{-2} as the capacitor was swept from depletion to weak inversion.

Similarly, Simoen *et al.* have undertaken detailed DLTS and ESR investigations of the AlN/p^{++}-Si interface and found no evidence of an inversion layer. Instead a large concentration of hole traps at the interface due to trivalent P_b centres in Si at 0.3-0.4eV above the Si valence band as well as defects associated with amorphous Si or extended defects in Si with a total interface state density of $>10^{12}$ cm^{-2} were observed [38, 39].

Additionally, the Si/III-nitride interface itself is typically composed of SiN_x layers due to nitridation of the Si surface by the N-containing species (NH_3 in case of MOCVD) during growth or the diffusion of nitrogen from the films (thus leading to nitrogen vacancies in III-N films) [40]. In some cases such nitridation is also intentional in order to reduce the dislocation density using SiN_x inclusions as in-situ masking layers [41]. These SiN_x layers are typically amorphous, thin, and/or discontinuous and may act as diffusion barriers for the movement of other species to the Si surface. But, the formation of SiN_x layers is very much dependent of growth conditions used and various reports can be found in literature to make the case for growth using Al-first, N-first or a simultaneous introduction of both species [42–44]. Similarly, TEM images show a plethora of interfacial configurations ranging from clean and abrupt AlN/Si hetero-interfaces to amorphous SiN_x inclusions to $AlSiN_x$ and Al-rich regions [9, 33, 39, 43, 45].

Furthermore, other chemical species such as oxygen, hydrogen and carbon are present in not insignificant proportions during the growth of III-nitride films. Residual oxygen is typically seen as a contaminant in precursor materials and gases used and can also originate from growth chambers with quartz components [46]. Some amount of oxygen observed at the epitaxy-substrate interfaces can also be due to the presence of oxygen in the silicon wafers themselves (>10^{17} cm^{-3} in Czochralski grown Si and <10^{16} cm^{-3} in float-zone silicon wafers) [47] which re-distributes at the high growth temperatures. Near-interface oxygen concentrations >10^{19} cm^{-3} have been reported for AlN and GaN growth on Si (111) using MOCVD [29, 46]. Hydrogen is a commonly used carrier gas in MOCVD growth of III-nitrides and significant incorporation and diffusion of hydrogen within these films have been shown to occur [48]. The diffusion of hydrogen towards the interface would also serve to passivate dangling bonds at the Si surface and thus reduce the trap density, but could also lead to the formation of complexes with other atomic species such as C, N and O at the interface. The interface trap density at AlN/Si interfaces has been shown to decrease both upon further annealing in NH_3/H_2, and the growth of AlGaN layers which indicates that further Ga and hydrogen diffusion passivates Si dangling bonds and reduces the extended defects in the Si [38, 39]. Carbon can be similarly traced back to the metal-organic precursors used for CVD growth of III-nitride films and additionally to the use of a carbon-doping source for obtaining highly-resistive GaN buffer and AlGaN transition layers [49, 50]. The presence of these atomic species has been shown by us earlier to form complexes such as the acceptor-like Si-O-N complex at high temperatures thus giving a p-type parasitic channel even for bare AlN/Si interfaces, without any GaN growth [29].

Moreover, in case doped Si substrates are used, there can also be significant dopant re-distribution within the substrate itself for growth temperatures >1000°C as has been shown for AlN growth on Si [29]. Such dopant pile-up at Si hetero-interfaces mediated by stress and defects has been well studied in the silicon literature [51, 52]. The increased presence of Boron, in case of typical p-type substrates, and Phosphorous, in case of typical n-type substrates, would lead to higher near-interfacial conductivity of the respective carrier polarity thus giving rise to parasitic conduction. This is expected to be much less of a problem for highly-resistive, uncompensated float-zone Si substrates (but not for compensated substrates).

Lastly, it is also of interest to note that significant diffusion of both Si and Si-dopants such as boron have been shown to occur upwards into the nitride films, presumably through some of the same dislocation and stress-mediated pathways discussed above [28, 29]. However, given the wide band gap nature and poor electrical conductivity of III-nitride layers, such diffusion is not expected to significantly contribute to parasitic channel effects. However, the presence of electrically active leakage paths, such as screw dislocations within the AlN layer itself has been recently shown by Berber *et al.* to be a major contributor to parasitic loss for MOCVD AlN films on Si ($\sigma > 15$ S/m) as opposed to any p-type layer ($\sigma < 1$ S/m) [53].

Given the diversity of observations listed above it is reasonable to conclude that the exact mechanism of parasitic channel formation is highly dependent on the starting Si substrate, pre-growth substrate treatments, and reactor & growth conditions used for the hetero-epitaxial growth. Changes to any one of these factors favors a particular interface/ near-interfacial configuration that in turn determines the origin and extent of the parasitic channel. We next discuss some of the implications of the parasitic channel on RF loss in GaN-on-Si devices.

2.2. *Effect of Parasitic Channel on RF Performance of GaN-on-Si Devices*

The presence of the parasitic channel has been shown to impact RF loss in GaN-on-Si devices which is evident in both small-signal and large-signal performance. The diversity of observations for the parasitic channel as discussed above manifests most clearly in terms of measured RF losses. In order to undertake a comparison of reported losses due to substrate parasitic conduction, we make use of reported small-signal CPW line losses in literature and this data is summarized in Table 1 below. The measured line loss on these structures depends on the conductivity of the parasitic channel as well as the epitaxial thickness and line dimensions, all of which affect the extent of capacitive coupling. The total loss reported typically also has conductor loss contributions which need to be isolated for one-to-one comparison of substrate losses. Hence, where available, these details are also included in the Table below. We see that for GaN buffers on Si, the transmission loss measured at 40 GHz varies from as high as >3.5 dB/mm for un-optimized epitaxial stacks to as low as 0.27 dB/mm for epitaxy with parasitic channel suppression and low conductor losses.

In terms of GaN-on-Si HEMTs, the presence of the conductive channel has been demonstrated to cause parasitic loading effects leading to a reduction in output power, efficiencies – both drain efficiency and PAE. For instance, Xiao *et al.* have projected improvements of 1.4dBm, 13.8% and 6 dB in output power, PAE and linear gain at 6dB gain compression point using large-signal modeling for devices without a parasitic conduction layer over those with. Similarly, Pattison *et al.*, demonstrated improvements in both drain efficiency (78% vs 71%) and class B PA efficiency (63% vs 71%) for devices with reduced parasitic channel effects as compared to non-optimized buffers [61]. Also, the f_t to f_{max} ratio of devices with a parasitic channel drops sharply to values ~1 as compared to those without, due to an increase in the output conductance ($f_{max}=f_t/2\sqrt{(R_i/R_{ds})}$ for the small-signal case). In addition to f_{max}, the intrinsic f_t itself has been shown to decrease by ~2x by Chumbes *et al.* as compared to the intrinsic f_t for GaN HEMTs on conductive Si substrates due to parasitic loading effects [12]. In this context, it is also important to mention that several approaches to construct RF small-signal models have been reported that explicitly account for substrate parasitic elements, by using bias-dependent S-parameter measurements and de-embedding the extrinsic circuit elements; and large-signal models constructed using these de-embedded intrinsic elements also show good agreement with measurements [24, 58–60].

Table 1. Comparison of reported CPW line losses in literature for GaN-on-Si at 40 GHz with available epitaxial and line dimensions and corresponding parasitic channel mechanism attributions.

Epitaxial Stack	**Line Dimensions**	**Line Loss @ 40 GHz (dB/mm)**	**Parasitic Channel Mechanism**	**Reference**
GaN/AlGaN/AlN/ HR-Si	unspecified	3.5 (@15 GHz)	Ga-diffusion - 10^{18} cm^{-3} peak conc. with 3 μm roll-off	Hanson *et al.* [23]
GaN/AlGaN/AlN/ HR-Si	unspecified	0.35 (@15 GHz)	Ga-diffusion - 10^{16} cm^{-3} peak conc. with 2 μm roll-off	Hanson *et al.* [23]
GaN/AlGaN/AlN/ HR-Si	W_c=80 μm, S=39 μm, t_{metal}=1.15 μm	0.4	Ga-diffusion 10^{16} cm^{-3} peak conc., conductivity=8 S/m	Marti *et al.* [54]
GaN (1μm)/AlGaN (500 nm)/HR-Si	unspecified	0.4	Ga/Al-diffusion into Si	Marcon *et al.* [55]
GaN(150 nm)/AlGaN (1.5 μm)/Transition layers/HR-Si	unspecified	1.05	unspecified	Menegehesso *et al.* [20]
GaN/AlGaN/AlN/ HR-Si	W_c=35 μm, S=20 μm, t_{metal}=0.195 μm	0.8	p-type layer/ conductive loss in leaky AlN	Cao *et al.* [56]
GaN/AlGaN/AlN/ HR-Si	W_c=35 μm, S=20 μm, t_{metal}=0.75 μm	0.27	p-type layer/ conductive loss in leaky AlN	Cao *et al.* [57]
100 nm AlN/HR-Si	W_c=17 μm, S=12 μm	0.65	Inversion layer of electrons at AlN/Si interface	Luong *et al.* [27]
200 nm AlN/HR-Si	W_c=17 μm, S=12 μm	2.6	Inversion layer of electrons at AlN/Si interface	Luong *et al.* [27]
GaN (1.2 μm)/AlGaN/ AlN (100 nm)/HR-Si	W_c=17 μm, S=12 μm	0.6	Inversion layer of electrons at AlN/Si interface	Luong *et al.* [27]
219 nm AlN/HR-Si	W_c=97 μm, S=57 μm, t_{metal}=0.70 μm	0.215	AlN conductivity due to dislocations, conductivity > 15 S/m, p-type layer < 1 S/m	Berber *et al.* [53]
345 nm AlN/HR-Si	W_c=97 μm, S=57 μm, t_{metal}=0.78 μm	0.25	AlN conductivity due to dislocations, conductivity > 15 S/m, p-type layer < 1 S/m	Berber *et al.* [53]
AlN/GaN(150 nm)/ AlGaN (1.5 μm)/ HR-Si	W_c=38 μm, S=32 μm, t_{metal}=0.5 μm	1.65	Substrate conductivity 0.25 S/m	Sahoo *et al.* [58]

2.3. *Minimizing the Influence of the Substrate Parasitic Channel*

We now discuss a few techniques that have been adopted to suppress the influence of the parasitic channel in GaN-on-Si devices. The first broad category relates to changes in the epitaxial growth techniques. The use of a low temperature AlN nucleation layer as the first step of the epitaxy is one such method, which reduces the thermal exposure of the Si top surface to the precursor chemicals while depositing a good quality AlN diffusion barrier at the start of growth [23]. Similarly, the use of other diffusion barriers such as SiN_x and SiC have also been proposed [56]. It is important to evaluate the film quality and thermal conductivity of these diffusion barriers as there will be a trade-off between the thickness of the diffusion barrier and the defect density in the III-nitride films in case of amorphous or non-lattice matched barrier materials, as well as affecting heat extraction from active channel layers. The diffusion barriers layers are expected to reduce the influx of species from the films/growth ambient into the substrate thus reducing the peak concentrations and depths of shallow acceptors in Si. It is important to note that these approaches are particularly sensitive to the growth of the initial III-nitride layers. Another approach that has been proposed to decrease the conductivity of the parasitic layer is to ion-implant H^+ or O^+ species post-epitaxy in order to create defects in the near-surface conductive region or to use counter dopants such as phosphorous to compensate the higher Al/Ga-acceptors in Si [56].

The other promising technique that has been employed to minimize the effect of the parasitic channel is the complete or partial removal of the substrate itself. Pattison *et al.* have shown improvements in the class B efficiency of GaN-on-Si devices from ~45% to 65% going up to 71% for devices with a parasitic channel (equivalent resistance of 400Ω), with back-side substrate removal under the channel region (R=3500 Ω) and back-side substrate removal under the channel and drain pad regions (R=11500 Ω) respectively [61]. Similarly, Chumbes et al have demonstrated that implementing a substrate removal scheme has increased the f_t/f_{max} ratio for GaN HEMTs on conductive Si substrates by 2.5x [12]. While the comparative ease of the Si etching process makes this a desirable option for parasitic channel suppression, an effective implementation of the substrate removal scheme also needs to consider the design of effective thermal management schemes for heat removal from the back-side etched III-nitride layers.

3. Back-bias Effects and Buffer-induced Current Collapse in GaN-on-Si

3.1. *Current Collapse in GaN HEMTs due to Buffer Traps*

Apart from the parasitic substrate loss issue discussed previously, the other major concern with GaN HEMTs in general is current collapse – whether due to surface or buffer traps. Effective passivation schemes and field plate designs have been shown to reduce surface-related current collapse in GaN HEMTs [62–67]. However, the optimal design of GaN buffers, which are made highly-resistive by doping them with deep-level impurities such as Fe or C in conjunction with growth defects, is also required to reduce buffer-induced current collapse in these devices [68–72]. This problem is exacerbated in the case of GaN-on-Si as compared to GaN-on-SiC as the presence of the parasitic conduction channel at the substrate interface serves to back-bias the epitaxy. We now discuss this effect in more detail.

The necessity to reduce parallel buffer conduction compels the use of highly-resistive GaN buffers which is typically achieved by the use of either Fe-doping or C-doping or intrinsic growth defects. Both Fe and C are deep level acceptors in GaN but with significantly different energy levels within the band gap. The Fe-acceptor is located at 0.6-0.7 eV below the conduction band whereas the C-acceptor level (C_N) is 0.8-0.9 eV above the valence band [73, 74]. Therefore, the preponderance of either Fe or C dopants pins the Fermi level at their respective trap levels in the upper and lower half of the band gap respectively, thus making the buffers either mildly n-type or p-type, albeit highly resistive for device characterization purposes [72, 75]. The exact location of the Fermi level in the GaN buffers is also determined by the amount of compensating donors – either due to self-compensation in case of Carbon doping (C-atoms in the Ga site C_{Ga}), unintentional donor states due to Oxygen incorporation or N-vacancies, or intentional doping with donors such as Si [76, 77]. If the Fermi level is indeed pinned to the lower half of the band gap, as in the case of higher C_N acceptor concentrations, this leads to a p-n junction between the n-type GaN channel and C:GaN buffer, thus giving rise to a floating buffer which can store charge under high applied drain bias [78–80]. It is this stored negative charge, due to the occupancy of deep-acceptor traps in the p-type buffer, which causes a strong voltage-dependent buffer-induced current collapse in C:doped epitaxy. In comparison Fe:doped GaN buffers, which are mildly n-type, have been shown to cause only mild current collapse due to the lack of such floating regions [68]. Furthermore, the presence of leakage paths in the unintentionally doped GaN channel regions has been shown by Uren *et al.* to be necessary to reduce current collapse by bringing the floating region to a similar potential as the 2DEG by using a back-bias technique to probe vertical charge transport in GaN-on-Si stacks [81]. Such back-biasing of the epitaxy due to the conductive substrate is more apparent for GaN on doped Si wafers typically used for power devices and the choice of Si substrate doping – n-type, p-type, n^+ and p^+ – also affects substrate depletion, vertical leakage and breakdown in GaN-on-Si power devices [82]. GaN-on-Si for RF, in contrast, is grown on highly-resistive Si as it is expected to act as more of an insulator, drop a larger proportion of applied drain voltage across it and hence have a reduced influence on the electric field distribution in the GaN epitaxial layers themselves. However, we now discuss that this is not the case, as even the resistive Si substrate has net vertical leakage across its thickness which contributes carriers to the interfacial channel at the AlN/Si interface and serves to back-bias the epitaxial stack leading to current collapse [83].

3.2. *Substrate Depletion and Back-biasing in Highly Resistive Si Substrates*

Buffer-induced current collapse effects in GaN-on-doped Si power devices have been accurately characterized and modeled using a substrate ramp technique, where the substrate voltage is slowly ramped from zero to more negative voltages (analogous to positive applied drain bias) and then swept back while the 2DEG conductivity is simultaneously monitored between two ohmic pads using a very small probe voltage [78]. This allows the isolation of buffer contribution to current collapse as the 2DEG serves to screen the applied back potential from the surface traps which hence do not contribute to current collapse until the 2DEG is significantly depleted. It also permits easier lumped element modeling (see Fig. 5) as the substrate plane makes this a 1-D problem as opposed to more involved 2-D field distributions in case of other methods such as pulsed-IV used to evaluate current collapse. In case of negative stored charge in the GaN buffer, the 2DEG conductivity is reduced when the substrate voltage is swept back to zero from negative voltages corresponding to current collapse [78]. The applicability of such a back-bias technique to highly-resistive Si substrates is not, however, obvious. Since the substrate potential is swept between rates of 0.1-100 V/s, we would also expect the HR-Si to be driven into deep-depletion even for modest sweep rates, which would widen the depletion layer in the substrate and hence support a higher voltage drop across it. The 2DEG conductivity for a perfectly capacitive epitaxial stack with a conductive substrate should lie along the dotted line for an ideal capacitor shown in Fig. 5. In case of additional voltage drop across the substrate, the effective voltage across the epitaxy is reduced and hence the 2DEG conductivity curve would lie above the capacitive line. It is also important to distinguish between substrate depletion effects and positive charge storage observed in current-collapse free GaN-on-Si stacks, which also manifest as 2DEG conductivity traces above the ideal capacitive line, but with the direction of the trace reversed, i.e. clock-wise direction [78].

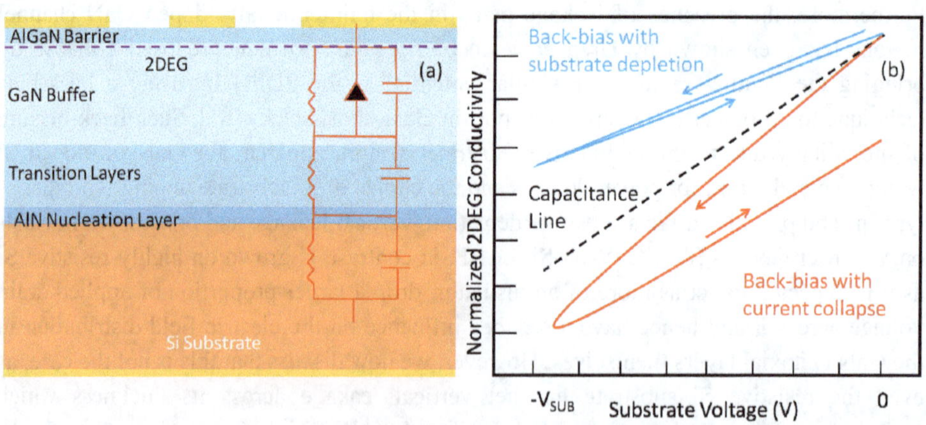

Fig. 5. (a) GaN-on-Si HEMT stack with 1-D lumped element equivalent circuit for charge storage during substrate ramp measurements. (b) Representative curves for normalized 2DEG conductivity for negative substrate ramps for a GaN HEMT with buffer-induced current collapse due to charge re-distribution and for the case where substrate depletion effects are observed. Also shown is the capacitance line for the entire epitaxy behaving as an ideal capacitor.

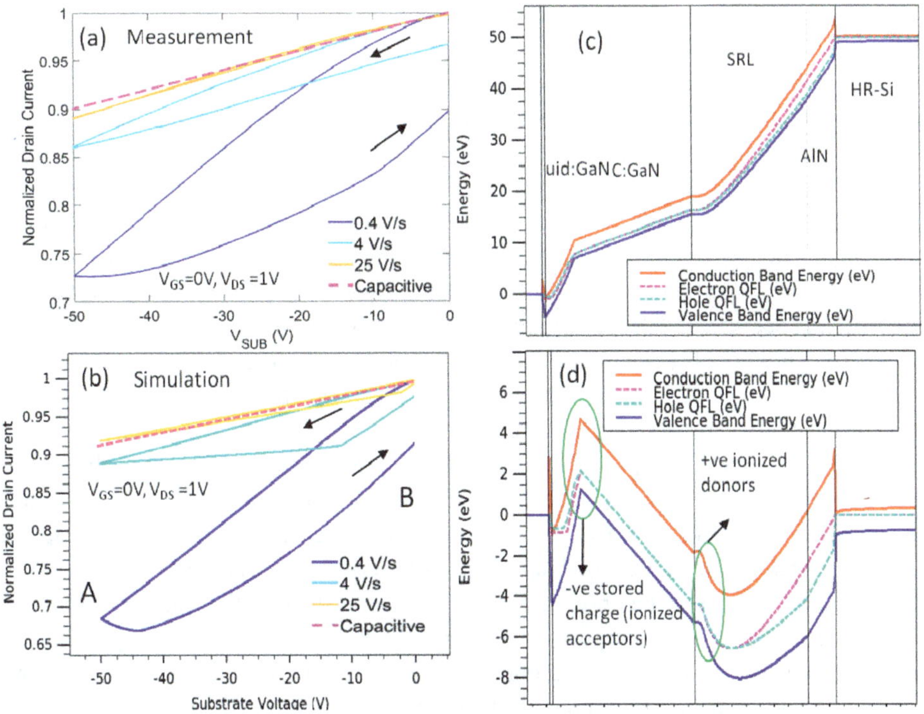

Fig. 6. (a) Measured substrate ramp traces for the C:doped GaN-on-HR Si HEMT stack shown in Fig. 5 for ramp rates of 0.4, 4 and 25 V/s with V_d of 1V and V_g of 0V. (b) Simulated substrate ramp traces for an identical stack using Silvaco ATLAS. (c) Band diagram at -50V (point A in (b)). The applied voltage drops entirely in the epitaxy and not in the HR-Si at all. (d) Charge re-distribution in the C:doped region when the applied substrate bias is removed (point B in (b)) (© IEEE 2018, Reprinted with permission from *Buffer-induced Current Collapse in GaN HEMTs on Highly Resistive Si Substrates* by *H. Chandrasekar et al.*, in *IEEE Electron Device Letters 39(10), pg. 1556-1559, 2018*) [83].

The substrate sweep rates for which the HR-Si is expected to be driven into deep-depletion can be estimated through a standard MOS capacitor analysis by treating the epitaxy as a dielectric, as the resistive III-nitride layers are typically highly-resistive (10^{10}-10^{14} Ω.cm). For an applied positive drain voltage, electron accumulation at the substrate-epitaxy interface should occur with the electrons in the Si being thermally generated. Since uncompensated, float-zone, highly-resistive Si has carrier lifetimes >1 ms [84], the diffusion length of carriers which can reach the interface is given by $L=\sqrt{(D_n\tau_n)}$=1.9 mm, which is comparable or greater than the thickness of most commonly used Si substrates for GaN growth. The sweep rates for the onset of deep-depletion can be estimated as [85],

$$\frac{dV}{dt} = \frac{qG_{th}x}{C_{III-N}} = 4.4\ V/s\,, \tag{1}$$

where G_{th} is the thermal generation rate given by n_i/τ_n and $C_{III\text{-}N}$ is the areal capacitance (F/cm^2) of the total III-nitride epitaxial stack. Substrate ramps above this rate should widen the depletion region and hence increase the voltage dropped across the Si. However, it has been shown that the measured 2DEG conductivity traces fall below the ideal capacitive limit even for very high sweep rates of 25 V/s for a proto-typical Carbon-doped GaN stack on HR-Si, as seen in Fig. 6(a) [83]. This shows that deep-depletion in silicon is not achieved even for such high sweep rates hence pointing to the inadequacy of thermal generation as the only carrier contributing mechanism to the inversion channel at the interface. Interestingly, the experimental findings could be reproduced by device simulations once carrier injection from the bottom substrate contact was considered (see Fig. 6(b)). Thus, small amounts of vertical current flow through the highly-resistive Si substrate, due to its semiconducting nature, coupled with the conductive layer at the epitaxy-substrate interface serves to back-bias the GaN layers and leads to current collapse. This current collapse in turn is due to charge-redistribution within the carbon doped buffer layer with ionized acceptors and donors at the GaN buffer/strain-relief layers (AlGaN) interface. Since the negative stored acceptor charge is closer to the 2DEG, this causes the 2DEG conductivity to drop leading to collapse. This can also be seen in band diagrams of Fig. 6(c) and (d) which shows the negative charge close to the 2DEG due to the ionized acceptors at the top of the C:doped layers leading to current collapse, with positive compensating donor charge at the top of the strain relief layer. We see that this phenomenon is identical to that for GaN-on-doped Si substrates for power electronic applications [78]. In comparison these effects would not arise in case of the insulating SiC substrates used for standard GaN-on-SiC RF devices and hence GaN-on-Si RF devices are inherently more susceptible to buffer-induced current collapse in comparison to those on SiC.

3.3. *Suppression of Buffer-induced Current Collapse for GaN-on-Si RF Devices*

The approaches for better buffer design with reduced current collapse developed for GaN-on-doped Si power stacks can be leveraged for GaN-on-HR-Si as the back-biasing effects are identical in these two configurations. This typically involves storing positive charge in the epitaxy, which screens the back-gated electric field due to positive applied drain bias (equivalent negative substrate bias) and reduces current collapse. Such positive charge storage in turn has been achieved by injecting holes into the buffer explicitly either using p^+ GaN layers connected to the drain electrode in the hybrid-drain GIT architecture proposed by Kaneko *et al.* [86], or by using a photonic ohmic drain as proposed by Tang *et al.* [87], which uses light to de-trap carriers from the deep states induced by carbon–doping; or more traditionally by introducing vertical leakage paths during the growth of the top-most GaN buffers themselves which prevents the formation of a floating buffer situation by injecting holes underneath the drain [78]. An alternate approach to drive the HR-Si into deep depletion for GaN-on-Si RF stacks would be to engineer a mildly leaky epitaxy throughout the III-nitride stack, instead of just the top-most layers, which would prevent the pile-up of electrons at the substrate interface. This would lead to added benefits in terms of increasing vertical breakdown by dropping more voltage across the substrate depletion region, quite unlike the more straightforward breakdown voltage/current-collapse trade-offs seen for GaN-on-doped Si stacks [82]. Another potential solution to suppress the kind of back-biasing effects observed due to current injection from the substrate contact is the use of a very thin insulating layer prior to back-side metallization in micro-strip or conductor-backed CPW architectures to form a blocking-back contact. The benefits of such an approach can be seen in Fig. 7 where the introduction of a 10 nm SiO_2 layer clearly cuts down on the injection of carriers into the substrate thus pushing it into deep-depletion. Therefore, the normalized 2DEG conductivity for the substrate ramp simulation now falls above the ideal capacitive line, for the same configuration discussed earlier in Fig. 6. Each of these schemes would cause a reduction in the current-collapse effects for GaN-on-Si RF HEMTs leading to improved power amplifier performance.

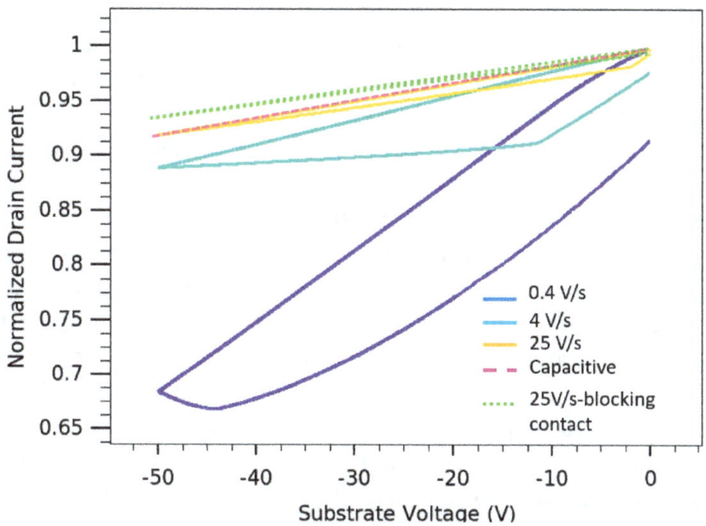

Fig. 7. Device simulations showing the effect of a blocking substrate contact in inducing substrate depletion for GaN-on-HR Si stack. The normalized drain current trace falls above the ideal capacitive line as expected when compared to a regular back-bias configuration.

4. Temperature-dependent Substrate Loss in GaN-on-Si RF Technology

4.1. *Background – Substrate Conductivity and Temperature*

The high power handling capabilities of lateral AlGaN/GaN HEMTs also lead to elevated channel temperatures in these devices which has been studied extensively [88–93]. Values for actual channel temperature depend on a wide range of factors such as choice of substrate, epitaxial design, device layout, die attach and heat spreader configuration, power density and device operating conditions among others [7, 94–96]. These high temperatures affect the electrical performance of the active device itself and constitute a major reliability concern [19, 97]. As a result, many heat removal strategies have been proposed and implemented both at the package and device levels, notably the use of high thermal conductivity substrates such as GaN-on-Diamond devices which have already shown power densities >3x higher than GaN-on-SiC devices with comparable electrical performance [98–101]. More specifically for GaN-on-Si RF devices, typical channel temperatures >150°C have been reported at power densities of ~4-6 W/mm under continuous wave (CW) operation [7, 102]. Thermal management strategies to reduce operating temperatures are out of the scope of the present review and will not be discussed further. Instead, we focus on the related problem of the device impact of heat flow from the active device regions down to the substrate in terms of increased RF losses. In comparison to the commonly used SiC or sapphire substrates for GaN growth, which can be made insulating/semi-insulating, Si is a narrow band gap semiconductor. Even though highly-resistive Si substrates (ρ>5 kΩ.cm) are almost exclusively used in GaN-on-Si RF device fabrication, the effect of substrate heating in Si serves to induce a significant number of thermally generated carriers in the substrate due to its small band gap. We also note here that it has been rightly pointed out in literature that the thermal conductivity difference between Si and SiC substrates reduces from a factor of 3:1 at 25°C to 2:1 at T>150°C [16]. While this would lower the impact of substrate thermal conductivity for heat extraction, it is important to remember that thermal generation in Si at these temperatures would decrease its resistivity by a large extent as compared to the wider band gap SiC substrates. It can be shown that for temperatures beyond 100°C, substrate resistivity drops steeply for the high-resistivity silicon which leads to capacitive coupling between the device regions and the conductive substrate at radio-frequencies thus leading to parasitic losses [103–106]. The increase in substrate loss with temperature lowers power amplifier efficiency, while the higher substrate conductivity is equivalent to moving the substrate ground plane upwards contributing to increased output capacitance and reducing bandwidth, all of which need to be factored into circuit design. It is also relevant to mention here that the presence of the parasitic channel at the substrate-epi interface, discussed previously, also contributes to RF loss and its conductivity could also be modulated with temperature. Therefore, it is quite challenging to experimentally separate out the effects of intrinsic substrate losses from those of the parasitic conduction path. We therefore discuss the use of device simulations validated against experimental data in order to unequivocally quantify the intrinsic temperature-dependent substrate losses for GaN-on-Si [106].

4.2. *Simulation Framework and Experimental Benchmarking of CPW Line Loss*

In order to evaluate temperature-dependent substrate loss for a range of operating frequencies and starting substrate resistivity, transmission lines in a CPW configuration were employed to estimate RF loss. The ensuing discussions would translate over to active devices as well. GaN-on-Si CPW lines with low losses of 0.8 dB/mm at 110 GHz have been demonstrated [54], but the temperature dependence of such losses had not been studied previously. A device simulation framework implemented in Silvaco ATLAS was used in order to fully capture the change in substrate resistance and capacitance contributions (see Fig. 8) and their effect on the small-signal S-parameters. We note that radiation and conductor losses are not accounted for in this simulation. The substrate loss in a transmission line was calculated by treating the substrate with parallel R and C contributions as a lossy dielectric. For a given loss tangent extracted from the small signal network parameters, the dielectric loss (dB/mm) for a CPW line with a center conductor width of W_C, gap spacing of S and substrate thickness t_{sub} is given by [107],

$$\alpha_d = \frac{8.686 q \varepsilon_r \tan\delta}{\varepsilon_{eff} \lambda_g}, \tag{2}$$

where ε_r is the relative permittivity of the substrate, λ_g is the guide wavelength given by $c/f\sqrt{\varepsilon_r}$, q is dielectric filling factor estimated as,

$$q = (\varepsilon_{eff} - 1) / (\varepsilon_{eff} - \varepsilon_{eff} / \varepsilon_r) \tag{3}$$

and the effective dielectric constant ε_{eff} is calculated using

$$\varepsilon_{eff} = 1 + \frac{1}{2}(\varepsilon_r - 1)\frac{K(k')}{K(k)}\frac{K(k_1)}{K(k_1')}, \tag{4}$$

where k and k' are calculated respectively as,

$$k = W_C/b,\ k' = \sqrt{(1-k2)},\ b = W_C + 2S, \tag{5}$$

and k_1 and k_1' as,

$$k_1 = \sinh(\pi W_t/4t_{sub}) / \sinh(\pi b_t/4t_{sub}),\ k_1' = \sqrt{(1-k_1 2)}, \tag{6}$$

with

$$W_t = W_C + 1.25 t_{metal}/\pi * [1 + \ln(4\pi W_C/t_{sub})], \tag{7}$$

$$b_t = b - 1.25 t_{metal}/\pi * [1 + \ln(4\pi W_C/t_{sub})], \tag{8}$$

and K(k) & K(k') are the complete elliptic integrals of the first kind whose ratios are computed using Hilberg's approximation as [108],

$$\frac{K(k)}{K(k')} = \frac{1}{2\pi}\ln(2\frac{\sqrt{1+k}+\sqrt[4]{4k}}{\sqrt{1+k}-\sqrt[4]{4k}}), \tag{9}$$

for $1 \leq K/(k)/K(k') \leq \infty$ and $1/\sqrt{2} \leq k \leq 1$, and

$$\frac{K(k)}{K(k')} = \frac{2\pi}{\ln(2\frac{\sqrt{1+k'}+\sqrt[4]{4k'}}{\sqrt{1+k'}-\sqrt[4]{4k'}})}, \tag{10}$$

for $0 \leq K/(k)/K(k') \leq 1$ and $0 \leq k \leq 1/\sqrt{2}$.

This model was then benchmarked against the published measurements and 3-D full-wave electromagnetic (EM) simulations of Eblabla *et al.* for CPWs placed directly on GaN on low-resistivity (LR) Si to verify its validity [109]. The substrate loss is high for LR substrates, since parasitic channel loss and auto-doping loss are insignificant, permitting a fair comparison to simulations once the same geometry and epitaxial stack thickness were used (in this case GaN of 1.4 μm, AlGaN of 750 nm, AlN and SiN_X of 200 nm each, and Si substrate of 40 Ω.cm and 675 μm thickness). In the 2-D device simulation, the attenuation constant was calculated at the central conductor accounting only for the resistive loss in the substrate. The results of such a comparison are shown in Fig. 8(b). It can be seen that the simulation captures all the major features of the measured data while consistently underestimating the total attenuation across the range of frequencies. This is because, unlike a complete EM simulation, conductor and radiation losses in these structures are not accounted for as mentioned earlier. Despite this, we observe that the substrate loss contribution constitutes the dominant component of the total attenuation constant (>90% at 67 GHz, or 0.76 dB/mm out of 0.83 dB/mm measured).

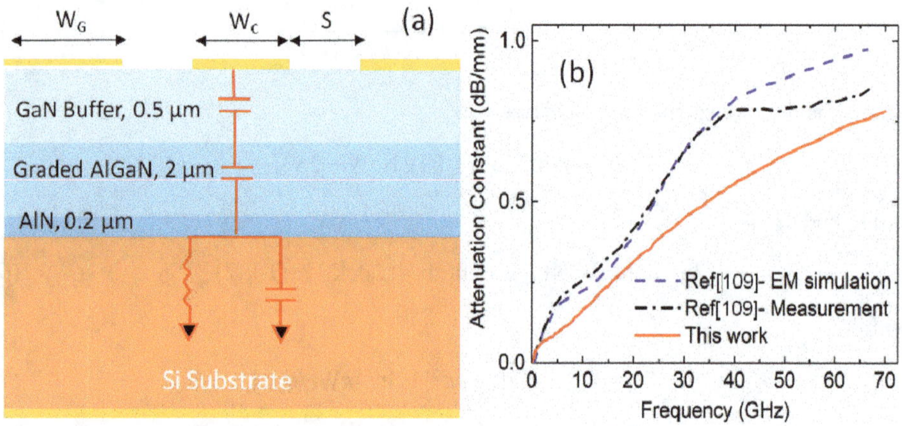

Fig. 8. (a) Cross-section of simulated CPW lines on GaN-on-Si with lumped R-C equivalent circuit. (b) Attenuation Constant (dB/mm) as a function of frequency for GaN-on-LR Si CPW lines – measured and EM simulations from Ref. 109 and substrate loss simulations of this work, for the same geometry and epitaxy. The current simulation of substrate loss reproduces all the trends of the measurement and is a significant fraction of measured line loss.

To further verify that the simulation fully captures the physics of the problem, CPW lines were fabricated on top of a silicon nitride layer on two silicon substrates of resistivity >5 kΩ.cm and 10-40 Ω.cm respectively. These lines had a centre conductor width (W_c) of 25 μm, spacing (S) of 15 μm and ground lines (W_g) of 100 μm and the same dimensions were used throughout this study. Since the interface between the SiN_x layer and silicon substrates is cleaner and does not suffer from parasitic channel effects, this configuration was preferred to CPW lines directly on GaN-on-Si for verifying the model. This makes it possible to compare the predictions of the model at different temperatures with measured values of insertion loss as shown in Fig. 9 below. The simulations correctly predict the trends of increasing substrate loss with temperature for high-resistivity Si substrates. We also see that the loss on low-resistivity substrates actually decreases with increasing temperature (experimental and simulated) due to the mild increase in Si resistivity with temperature (in turn caused by mobility degradation due to phonon scattering), a trend that is again captured by the simulation.

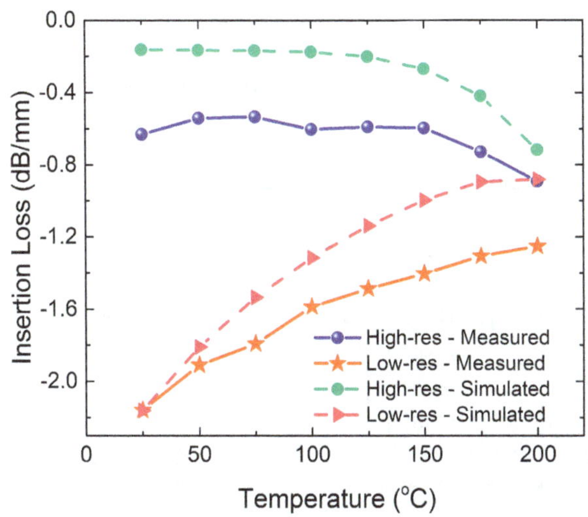

Fig. 9. Comparison of experimental and simulated line loss at 20 GHz on SiN_x-on-Si CPW structures for substrate resistivity of 10 kΩ.cm and 25 Ω.cm from 25-200°C. (© IEEE 2019, Reprinted with permission from *Quantifying Temperature-dependent Substrate Loss in GaN-on-Si RF Technology*, by *H. Chandrasekar et al.*, in *IEEE Transactions on Electron Devices 66(4), pg. 1681-1687, 2019*) [106].

4.3. *Substrate Loss – Temperature, Frequency and Starting Substrate Resistivity Dependencies*

Now that the simulation framework has been verified to accurately represent the substrate loss at various temperatures and frequencies, line losses for a representative GaN-on-Si stack of Fig. 8 were simulated for different temperature and frequencies and various starting substrate resistivity. Figure 10 shows the estimated substrate loss as a function of frequency for the three different Si substrates of HR, LR and very-low resistivity (ρ=10 kΩ.cm, 1 Ω.cm and 0.01 Ω.cm respectively).

It can be seen that not only is the substrate loss temperature-insensitive for low and very-low resistivity substrates in contrast to HR-Si which has a pronounced temperature dependence, the magnitude of such loss is also lower for these low-res substrates than HR-Si for frequencies <20 GHz, beyond which a sharp increase is observed. This is primarily due to the collective response of the substrate resistance and capacitance at the interface which causes a shift in the peak of the loss tangent and hence substrate loss to

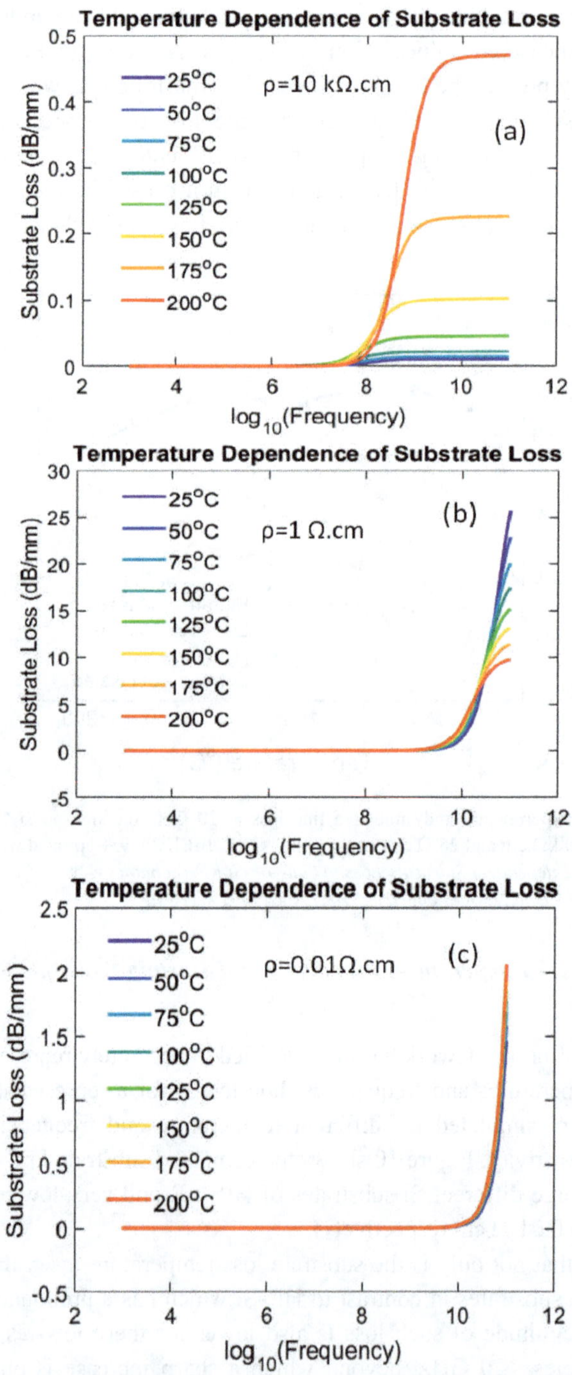

Fig. 10. Simulated substrate loss (dB/mm) contributions vs frequency for GaN CPW lines on (a) HR-Si, 10 kΩ.cm (b) LR-Si, 1 Ω.cm, and (c) very-low resistive Si, 0.01 Ω.cm for temperatures between 25-200°C.

higher frequencies for lower substrate resistivity. This data can be represented as shown in Fig. 11 below, which shows the simulated substrate loss with temperature for three different frequency bands and four different starting substrate resistivity.

It is easy to observe from Fig. 11 that apart from HR-Si all the low resistive Si substrates exhibit lesser variation in temperature-dependent loss irrespective of the frequencies considered. In fact, medium-res Si (100 Ω.cm) shows a decrease in loss with increasing temperature up to 150°C before reversing trend corresponding to the temperature dependent resistivity for this doping level. We see that HR-Si does not even exhibit lower absolute line losses when the whole temperature range is considered for S and X-bands. It can also be seen that LR-Si substrate of 1 Ω.cm resistivity has a higher loss only until 175°C at 2.5 GHz. Indeed such an observation has also been made experimentally by Takenaka *et al.* [110] for a comparison of 1-port pad loss between GaN-on-HR-Si and GaN-on-LR-Si with a loss crossover above 150°C at 2.2 GHz. But, at the same time HR-Si quite clearly shows lowest possible loss at 40 GHz and above for all temperatures and is hence the superior substrate for frequencies above the X-band, which represents an intermediate behavior between the S and K_a frequency bands.

We again stress that other factors need to be considered for efficient circuit design and our analysis of silicon substrate resistivity with frequency is based purely on substrate loss considerations. We have also not considered the temperature dependence of other loss mechanisms such as conductor/metal loss and radiation losses. While radiation loss at these frequencies are expected to be low, the same cannot be said of the conductor losses. However, it is relatively straightforward to separate out the dependencies of metal and substrate loss for different CPW designs as the metal loss increases for lower conductor spacing (dimension 'S' in Fig. 8) whereas the substrate loss reduces as expected because the extent of capacitive substrate coupling would be lower. The

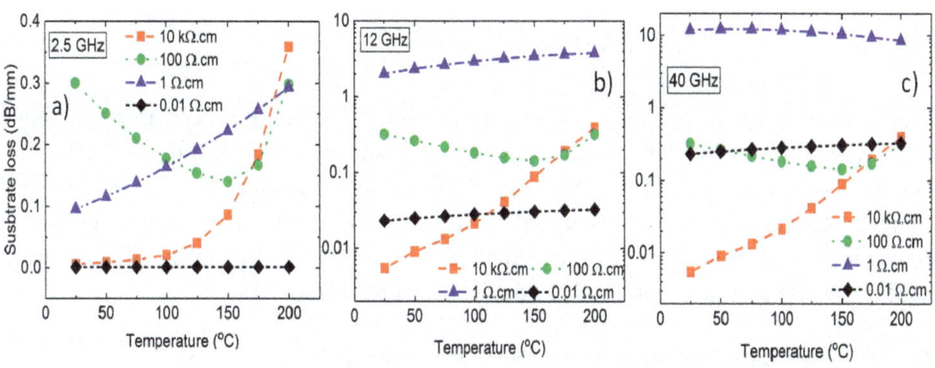

Fig. 11. Simulated substrate loss (dB/mm) as a function of temperature for Si substrate resistivity of 0.01 Ω.cm, 1 Ω.cm, 100 Ω.cm and 10 kΩ.cm at (a) 2.5 GHz (in S-band) (b) 12 GHz (in X-band) (c) 40 GHz (in K_a-band). (© IEEE 2019, Reprinted with permission from *Quantifying Temperature-dependent Substrate Loss in GaN-on-Si RF Technology*, by *H. Chandrasekar et al.*, in *IEEE Transactions on Electron Devices 66(4), pg. 1681-1687, 2019*) [106].

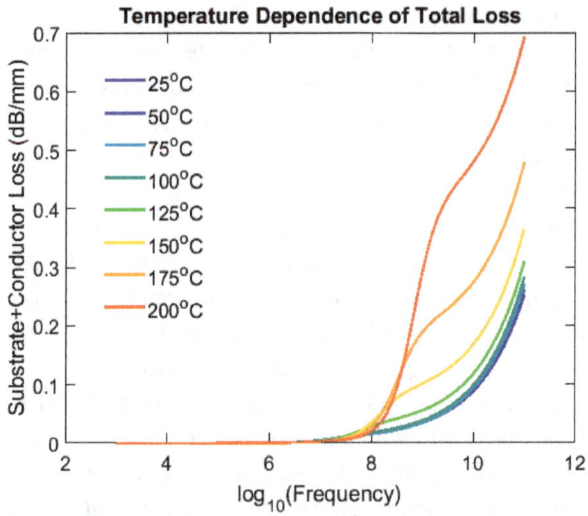

Fig. 12. Temperature dependence of total line loss consisting of substrate and conductor loss components as a function of frequency for CPW lines of 15 μm spacing and 25 μm conductor width on a typical GaN-on HR-Si stack.

conductor loss for a CPW line can be estimated as [107],

$$\alpha_c = \frac{8.686 R_s \sqrt{\varepsilon_{eff}}\,[\Phi(W_c) + \Phi(W_c + 2S)]}{480 K(k) K(k') k'} (dB/m), \tag{11}$$

where $R_s = \sqrt{(\pi f \mu_0 / \sigma)}$ and

$$\Phi(x) = \frac{\pi}{x} + \frac{\ln\left(\frac{8\pi x(1-x)}{t(1+k)}\right)}{x}, \tag{12}$$

and all other terms have their usual meanings. The total line loss accounting for both substrate and conductor loss for the typical line dimensions used in this study (W_c=25 μm, S=15 μm) are shown in Fig. 12 for the representative GaN-on-HR-Si CPW lines. As compared to substrate loss component alone shown in Fig. 10(a), the conductor loss is the dominant loss mechanism at high temperatures for frequencies >10 GHz and also needs to be taken into consideration for these dimensions and frequencies.

4.4. *Comparison of Strategies to Minimize RF Substrate Loss in GaN-on-Si*

Two promising approaches documented for reducing the effect of substrate-induced RF loss in GaN-on-Si involve the use of low-resistivity silicon substrates and back-side etch of the silicon substrate. We have already discussed the use of LR-Si above and the use of such conductive substrate gives rise to relatively temperature insensitive resistivity variations in Si, but at the cost of increasing the much-desired low output capacitances.

At this juncture it is important to mention that Takenaka *et al.* [110], have demonstrated superlative performances for a GaN-on-LR Si inverse Doherty power amplifier at 2.14 GHz with superior drain efficiencies as a function of temperature as compared to GaN-on-HR-Si HEMT device. GaN-on-LR Si is hence an approach that promises to be of value, especially for frequencies <12 GHz as discussed in the previous section and needs further investigation.

Another approach that has been mooted for parasitic channel suppression primarily, is Si substrate removal. As discussed in an earlier section, this approach has been shown to be effective in eliminating the effect of the substrate parasitic channel by removing Si underneath portions of the active device area, retaining them as suspended HEMTs or backfilling it with epoxy [12, 61]. We further evaluate the efficacy of such an approach here to reduce temperature dependent substrate loss using our simulation framework for a GaN HEMT active device. Three scenarios were compared – 1) removal of the substrate under the channel, 2) removal of the substrate underneath the channel and drain pads and 3) complete substrate removal under the active device, and in each case replacing it with an SiO_2 layer which has a similar dielectric constant to most epoxy resins and hence similar capacitances. To draw parallels with our CPW transmission line structures discussed above, these scenarios are analogous to removing the substrate from underneath the spacing between the central conductor and ground lines (region denoted by S in Fig. 8), central conductor and spacings between central and ground lines (regions marked W_c and S in Fig. 8 and complete removal from beneath the centre conductor, spacing and beneath the ground pads respectively. The substrate loss for these three different configurations is shown in Fig. 13.

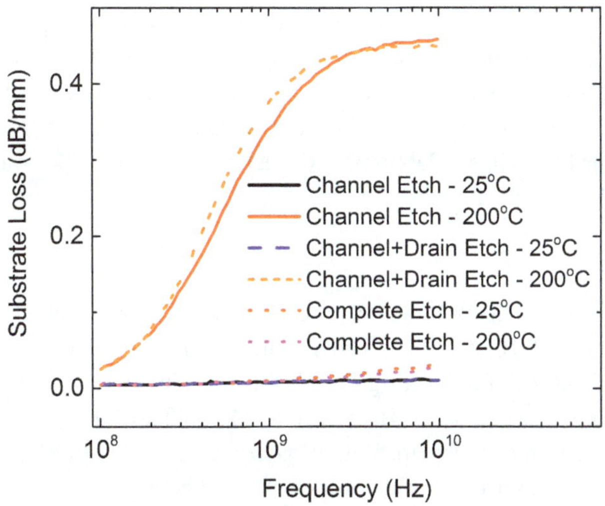

Fig. 13. Substrate Loss (dB/mm) vs frequency (0.1-10 GHz) on HR-Si with back-side substrate removal schemes – under the channel region, channel+drain pads and complete removal from underneath channel+ drain+source pads.

We see that substrate removal and back-fill under the channel and channel & drain pads confer minimal benefits towards the reduction of temperature-dependent substrate loss. Only a complete substrate removal scheme minimizes substrate loss in the range of operating frequencies investigated here (0.1-10 GHz). Clearly such an approach will also result in thermal challenges due to the thermal conductivity and interfacial resistances associated with such a scheme and the use of higher thermal conductivity materials for such back-filling might be a better solution both from a thermal management and hence substrate loss standpoint, but with added process complexity and costs. It is also important to note that for the use of HR-Si substrates, the substrate loss component only increases by a very large extent for temperatures beyond 125°C. Hence, thermal management strategies at the device and package level to keep the substrate temperatures below this value would be greatly beneficial in managing the temperature-dependent characteristics of GaN-on-Si PAs and this remains an area of active research.

5. Conclusions

Three device aspects of using a narrow bandgap, semiconducting Si substrate on the performance of GaN-on-Si RF technology have been reviewed. The presence of a parasitic channel at the epitaxy-substrate interface giving rise to RF losses has been discussed along with a detailed look at the physical mechanisms for its formation and its effect on RF loss. Several solutions proposed to address this effect have also been discussed. The use of Si substrates together with the presence of the parasitic channel has been shown to make GaN-on-Si epitaxy more susceptible to current collapse as compared to GaN-on-insulating SiC and potential solutions to mitigate this problem have also been presented. Finally, temperature-dependent change in substrate resistivity, also giving rise to RF losses, due to heat flow from active devices as well as higher rate operating temperatures have been quantified. Methods proposed to reduce this effect have also been elaborated upon. It is anticipated that the discussions presented here offer an overview of an important subset of potential issues that need to be considered, along with their likely solutions, in order to establish GaN-on-Si RF technology as a viable solution for power amplifier and MMIC applications.

Acknowledgments

The author would like to thank Profs. Michael J Uren and Martin Kuball at the School of Physics, University of Bristol, UK; Profs. Paul J Tasker and Khaled Elgaid at the Department of Electrical Engineering, Cardiff University, UK; and Profs. Srinivasan Raghavan, Navakanta Bhat, K N Bhat and R Muralidharan at the Centre for Nano Science and Engineering, Indian Institute of Science, India for fruitful discussions and collaborations on various aspects of the work presented here.

References

1. R. S. Pengelly, S. M. Wood, J. W. Milligan, S. T. Sheppard and W. L. Pribble, IEEE Transactions on Microwave Theory and Techniques **60** (6), 1764-1783 (2012).
2. U. K. Mishra, L. Shen, T. E. Kazior and Y.-F. Wu, Proceedings of the IEEE **96** (2), 287-305 (2008).
3. J. Shealy, J. Smart, M. Poulton, R. Sadler, D. Grider, S. Gibb, B. Hosse, B. Sousa, D. Halchin and V. Steel, presented at the 24th Annual Technical Digest Gallium Arsenide Integrated Circuit (GaAs IC) Symposium, 2002 (unpublished).
4. E. Feltin, B. Beaumont, M. Laügt, P. De Mierry, P. Vennégues, H. Lahreche, M. Leroux and P. Gibart, Applied Physics Letters **79** (20), 3230-3232 (2001).
5. A. Dadgar, M. Poschenrieder, J. Bläsing, O. Contreras, F. Bertram, T. Riemann, A. Reiher, M. Kunze, I. Daumiller and A. Krtschil, Journal of Crystal Growth **248**, 556-562 (2003).
6. S. Raghavan and J. M. Redwing, (Boca Raton, FL: CRC Press, 2011), pp. 209Ã256.
7. J. Brown, R. Borges, E. Piner, A. Vescan, S. Singhal and R. Therrien, Solid-State Electronics **46** (10), 1535-1539 (2002).
8. A. Vescan, J. Brown, J. Johnson, R. Therrien, T. Gehrke, P. Rajagopal, J. Roberts, S. Singhal, W. Nagy and R. Borges, physica status solidi (c) (1), 52-56 (2003).
9. H. Chandrasekar, N. Mohan, A. Bardhan, K. Bhat, N. Bhat, N. Ravishankar and S. Raghavan, Applied Physics Letters **103** (21), 211902 (2013).
10. B. J. Baliga, Semiconductor Science and Technology **28** (7), 074011 (2013).
11. K. J. Chen, O. Häberlen, A. Lidow, C. lin Tsai, T. Ueda, Y. Uemoto and Y. Wu, IEEE Transactions on Electron Devices **64** (3), 779-795 (2017).
12. E. M. Chumbes, A. Schremer, J. A. Smart, Y. Wang, N. C. MacDonald, D. Hogue, J. J. Komiak, S. J. Lichwalla, R. E. Leoni and J. R. Shealy, IEEE Transactions on electron devices **48** (3), 420-426 (2001).
13. A. Minko, V. Hoel, S. Lepilliet, G. Dambrine, J. De Jaeger, Y. Cordier, F. Semond, F. Natali and J. Massies, IEEE Electron Device Letters **25** (4), 167-169 (2004).
14. J. Johnson, E. Piner, A. Vescan, R. Therrien, P. Rajagopal, J. Roberts, J. Brown, S. Singhal and K. Linthicum, IEEE Electron Device Letters **25** (7), 459-461 (2004).
15. J. Brown, W. Nagy, S. Singhal, S. Peters, A. Chaudhari, T. Li, R. Nichols, R. Borges, P. Rajagopal and J. Johnson, presented at the 2004 IEEE MTT-S International Microwave Symposium Digest (IEEE Cat. No. 04CH37535), 2004 (unpublished).
16. F. Iucolano and T. Boles, Materials Science in Semiconductor Processing **98**, 100-105 (2019).
17. S. Tirelli, D. Marti, H. Sun, A. R. Alt, H. Benedickter, E. L. Piner and C. Bolognesi, IEEE Electron Device Letters **31** (4), 296-298 (2010).
18. F. Medjdoub, M. Zegaoui, B. Grimbert, D. Ducatteau, N. Rolland and P. Rolland, IEEE Electron Device Letters **33** (8), 1168-1170 (2012).
19. S. Singhal, T. Li, A. Chaudhari, A. Hanson, R. Therrien, J. Johnson, W. Nagy, J. Marquart, P. Rajagopal and J. Roberts, Microelectronics Reliability **46** (8), 1247-1253 (2006).
20. G. Meneghesso, M. Meneghini, F. Medjdoub, Y. Tagro, B. Grimbert, D. Ducatteau, N. Rolland, R. Silvestri and E. Zanoni, IEEE Transactions on Device and Materials Reliability **13** (4), 480-488 (2013).
21. H. Sun, A. R. Alt, H. Benedickter, C. Bolognesi, E. Feltin, J.-F. Carlin, M. Gonschorek, N. Grandjean, T. Maier and R. Quay, IEEE Electron Device Letters **30** (8), 796-798 (2009).

22. S. Hoshi, M. Itoh, T. Marui, H. Okita, Y. Morino, I. Tamai, F. Toda, S. Seki and T. Egawa, Applied Physics Express **2** (6), 061001 (2009).
23. A. W. Hanson, J. C. Roberts, E. L. Piner and P. Rajagopal, (Google Patents, 2007).
24. D. Xiao, D. Schreurs, W. De Raedt, J. Derluyn, M. Germain, B. Nauwelaers and G. Borghs, Solid-State Electronics **53** (2), 185-189 (2009).
25. H. Umeda, A. Suzuki, Y. Anda, M. Ishida, T. Ueda, T. Tanaka and D. Ueda, presented at the 2010 International Electron Devices Meeting, 2010 (unpublished).
26. H. Yacoub, D. Fahle, M. Finken, H. Hahn, C. Blumberg, W. Prost, H. Kalisch, M. Heuken and A. Vescan, Semiconductor Science and Technology **29** (11), 115012 (2014).
27. T. T. Luong, F. Lumbantoruan, Y. Y. Chen, Y. T. Ho, Y. C. Weng, Y. C. Lin, S. Chang and E. Y. Chang, Physica status solidi (a) **214** (7), 1600944 (2017).
28. E. Calleja, M. Sanchez-Garcia, D. Basak, F. Sánchez, F. Calle, P. Youinou, E. Munoz, J. Serrano, J. Blanco and C. Villar, Physical Review B **58** (3), 1550 (1998).
29. H. Chandrasekar, K. Bhat, M. Rangarajan, S. Raghavan and N. Bhat, Scientific reports **7** (1), 15749 (2017).
30. X. Zhang, D. Walker, A. Saxler, P. Kung, J. Xu and M. Razeghi, Electronics Letters **32** (17), 1622-1623 (1996).
31. S. Raghavan and J. M. Redwing, Journal of applied physics **98** (2), 023514 (2005).
32. S. Raghavan and J. Redwing, Journal of applied physics **98** (2), 023515 (2005).
33. R. Liu, F. Ponce, A. Dadgar and A. Krost, Applied Physics Letters **83** (5), 860-862 (2003).
34. K. S. Stevens, M. Kinniburgh, A. F. Schwartzman, A. Ohtani and R. Beresford, Applied Physics Letters **66** (23), 3179-3181 (1995).
35. D. Xi, Y. Zheng, P. Chen, R. Chu, S. Gu, B. Shen and R. Zhang, Optical Materials **23** (1-2), 143-146 (2003).
36. F. Engelmark, J. Westlinder, G. F. Iriarte, I. V. Katardjiev and J. Olsson, IEEE Transactions on Electron Devices **50** (5), 1214-1219 (2003).
37. M. Morita, S. Isogai, K. Tsubouchi and N. Mikoshiba, Applied Physics Letters **38** (1), 50-52 (1981).
38. E. Simoen, D. Visalli, M. Van Hove, M. Leys and G. Borghs, Journal of Physics D: Applied Physics **44** (47), 475104 (2011).
39. E. Simoen, D. Visalli, M. Van Hove, M. Leys, P. Favia, H. Bender, G. Borghs, A. P. D. Nguyen and A. Stesmans, physica status solidi (a) **209** (10), 1851-1856 (2012).
40. A. Krost and A. Dadgar, physica status solidi (a) **194** (2), 361-375 (2002).
41. A. Dadgar, M. Poschenrieder, A. Reiher, J. Bläsing, J. Christen, A. Krtschil, T. Finger, T. Hempel, A. Diez and A. Krost, Applied physics letters **82** (1), 28-30 (2003).
42. A. Le Louarn, S. Vézian, F. Semond and J. Massies, Journal of Crystal Growth **311** (12), 3278-3284 (2009).
43. A. Lange, X. Tan, C. Fadley and S. Mahajan, Acta Materialia **115**, 94-103 (2016).
44. H. Lahreche, P. Vennéguès, O. Tottereau, M. Laügt, P. Lorenzini, M. Leroux, B. Beaumont and P. Gibart, Journal of Crystal Growth **217** (1-2), 13-25 (2000).
45. D. Xi, Y. Zheng, P. Chen, Z. Zhao, P. Chen, S. Xie, B. Shen, S. Gu and R. Zhang, physica status solidi (a) **191** (1), 137-142 (2002).
46. J. Freedsman, A. Watanabe, Y. Yamaoka, T. Kubo and T. Egawa, physica status solidi (a) **213** (2), 424-428 (2016).
47. F. Shimura, H. Tsuya and T. Kawamura, Applied Physics Letters **37** (5), 483-486 (1980).

48. J. M. Zavada, R. G. Wilson, C. R. Abernathy and S. J. Pearton, Applied Physics Letters **64** (20), 2724-2726 (1994).
49. D. Green, U. Mishra and J. Speck, Journal of applied physics **95** (12), 8456-8462 (2004).
50. D. Koleske, A. Wickenden, R. Henry and M. Twigg, Journal of crystal growth **242** (1-2), 55-69 (2002).
51. H.-H. Vuong, H.-J. Gossmann, L. Pelaz, G. Celler, D. Jacobson, D. Barr, J. Hergenrother, D. Monroe, V. Venezia and C. Rafferty, Applied physics letters **75** (8), 1083-1085 (1999).
52. H. Park, K. S. Jones, J. A. Slinkman and M. E. Law, Journal of Applied Physics **78** (6), 3664-3670 (1995).
53. F. Berber, D. W. Johnson, K. M. Sundqvist, E. L. Piner, G. H. Huff and H. R. Harris, IEEE Transactions on Microwave Theory and Techniques **65** (5), 1465-1470 (2017).
54. D. Marti, M. Vetter, A. R. Alt, H. Benedickter and C. Bolognesi, Applied physics express **3** (12), 124101 (2010).
55. D. Marcon, J. Viaene, F. Vanaverbeke, X. Kang, S. Lenci, S. Stoffels, R. Venegas, P. Srivastava and S. Decoutere, presented at the 2012 7th European Microwave Integrated Circuit Conference, 2012 (unpublished).
56. J. C. Roberts, K. J. Linthicum, A. W. Hanson and J. W. Cook Jr, (Google Patents, 2017).
57. L. Cao, C. Lo, H. Marchand, W. Johnson and P. Fay, IEEE Microwave and Wireless Components Letters **28** (10), 861-863 (2018).
58. A. K. Sahoo, N. K. Subramani, J. C. Nallatamby, L. Sylvain, C. Loyez, R. Quere and F. Medjdoub, Solid-State Electronics **115**, 12-16 (2016).
59. A. Jarndal, International Journal of RF and Microwave Computer-Aided Engineering **24** (3), 389-400 (2014).
60. A. Jarndal, A. Z. Markos and G. Kompa, IEEE transactions on microwave theory and techniques **59** (3), 644-651 (2011).
61. L. Pattison, T. Boles, N. Tuffy and G. Lopes, presented at the 2014 9th European Microwave Integrated Circuit Conference, 2014 (unpublished).
62. R. Vetury, N. Q. Zhang, S. Keller and U. K. Mishra, IEEE Transactions on Electron Devices **48** (3), 560-566 (2001).
63. M. Faqir, G. Verzellesi, A. Chini, F. Fantini, F. Danesin, G. Meneghesso, E. Zanoni and C. Dua, IEEE Transactions on Device and Materials Reliability **8** (2), 240-247 (2008).
64. W. Lu, V. Kumar, R. Schwindt, E. Piner and I. Adesida, Solid-State Electronics **46** (9), 1441-1444 (2002).
65. W. Saito, Y. Kakiuchi, T. Nitta, Y. Saito, T. Noda, H. Fujimoto, A. Yoshioka, T. Ohno and M. Yamaguchi, IEEE Electron Device Letters **31** (7), 659-661 (2010).
66. M. T. Hasan, T. Asano, H. Tokuda and M. Kuzuhara, IEEE Electron Device Letters **34** (11), 1379-1381 (2013).
67. W. M. Waller, M. Gajda, S. Pandey, J. J. Donkers, D. Calton, J. Croon, J. Šonský, M. J. Uren and M. Kuball, IEEE Transactions on Electron Devices **64** (10), 4044-4049 (2017).
68. M. J. Uren, J. Moreke and M. Kuball, IEEE Transactions on Electron Devices **59** (12), 3327-3333 (2012).
69. S. Gustafsson, J.-T. Chen, J. Bergsten, U. Forsberg, M. Thorsell, E. Janzén and N. Rorsman, IEEE Transactions on Electron Devices **62** (7), 2162-2169 (2015).
70. D. Bisi, A. Stocco, I. Rossetto, M. Meneghini, F. Rampazzo, A. Chini, F. Soci, A. Pantellini, C. Lanzieri, P. Gamarra, C. Lacam, M. Tordjman, M. A. di Forte-Poisson, D. De Salvador, M.

Bazzan, G. Meneghesso and E. Zanoni, Microelectronics Reliability **55** (9), 1662-1666 (2015).

71. O. Axelsson, S. Gustafsson, H. Hjelmgren, N. Rorsman, H. Blanck, J. Splettstoesser, J. Thorpe, T. Roedle and M. Thorsell, IEEE Transactions on Electron Devices **63** (1), 326-332 (2016).
72. H. Chandrasekar, M. Singh, S. Raghavan and N. Bhat, Semiconductor Science and Technology **30** (11), 115018 (2015).
73. M. Silvestri, M. J. Uren and M. Kuball, Applied Physics Letters **102** (7), 073501 (2013).
74. J. L. Lyons, A. Janotti and C. G. Van de Walle, Physical Review B **89** (3), 035204 (2014).
75. A. Fariza, A. Lesnik, S. Neugebauer, M. Wieneke, J. Hennig, J. Bläsing, H. Witte, A. Dadgar and A. Strittmatter, Journal of Applied Physics **122** (2), 025704 (2017).
76. B. Rackauskas, M. J. Uren, S. Stoffels, M. Zhao, S. Decoutere and M. Kuball, IEEE Transactions on Electron Devices **65** (5), 1838-1842 (2018).
77. V. Joshi, S. P. Tiwari and M. Shrivastava, IEEE Transactions on Electron Devices **66** (1), 570-577 (2019).
78. M. J. Uren, S. Karboyan, I. Chatterjee, A. Pooth, P. Moens, A. Banerjee and M. Kuball, IEEE Transactions on Electron Devices **64** (7), 2826-2834 (2017).
79. M. Kuball, M. J. Uren, A. Pooth, S. Karboyan, W. M. Waller and I. Chatterjee, presented at the 2015 IEEE 3rd Workshop on Wide Bandgap Power Devices and Applications (WiPDA), 2015 (unpublished).
80. M. Singh, M. J. Uren, T. Martin, S. Karboyan, H. Chandrasekar and M. Kuball, IEEE Transactions on Electron Devices **65** (9), 3746-3753 (2018).
81. M. J. Uren, M. Silvestri, M. Cäsar, G. A. M. Hurkx, J. A. Croon, J. Šonský and M. Kuball, IEEE Electron Device Letters **35** (3), 327-329 (2014).
82. M. Borga, M. Meneghini, S. Stoffels, X. Li, N. Posthuma, M. Van Hove, S. Decoutere, G. Meneghesso and E. Zanoni, IEEE Transactions on Electron Devices **65** (7), 2765-2770 (2018).
83. H. Chandrasekar, M. J. Uren, A. Eblabla, H. Hirshy, M. A. Casbon, P. J. Tasker, K. Elgaid and M. Kuball, IEEE Electron Device Letters **39** (10), 1556-1559 (2018).
84. S. Glunz, S. Rein, J. Lee and W. Warta, Journal of Applied Physics **90** (5), 2397-2404 (2001).
85. B. Van Zeghbroeck, Colarado University **34** (2004).
86. S. Kaneko, M. Kuroda, M. Yanagihara, A. Ikoshi, H. Okita, T. Morita, K. Tanaka, M. Hikita, Y. Uemoto and S. Takahashi, presented at the 2015 IEEE 27th International Symposium on Power Semiconductor Devices & IC's (ISPSD), 2015 (unpublished).
87. X. Tang, B. Li, Y. Lu, H. Wang, C. Liu, J. Wei and K. J. Chen, presented at the 2015 IEEE International Electron Devices Meeting (IEDM), 2015 (unpublished).
88. Y.-F. Wu, B. Keller, S. Keller, D. Kapolnek, P. Kozodoy, S. Denbaars and U. Mishra, Solid-State Electronics **41** (10), 1569-1574 (1997).
89. J. Kuzmik, R. Javorka, A. Alam, M. Marso, M. Heuken and P. Kordos, IEEE transactions on electron devices **49** (8), 1496-1498 (2002).
90. J. Kuzmík, S. Bychikhin, M. Neuburger, A. Dadgar, A. Krost, E. Kohn and D. Pogany, IEEE Transactions on Electron Devices **52** (8), 1698-1705 (2005).
91. R. Aubry, J.-C. Jacquet, J. Weaver, O. Durand, P. Dobson, G. Mills, M.-A. di Forte-Poisson, S. Cassette and S.-L. Delage, IEEE transactions on electron devices **54** (3), 385-390 (2007).
92. R. J. Simms, J. W. Pomeroy, M. J. Uren, T. Martin and M. Kuball, IEEE Transactions on Electron Devices **55** (2), 478-482 (2008).

93. A. Sarua, H. Ji, M. Kuball, M. J. Uren, T. Martin, K. P. Hilton and R. S. Balmer, IEEE Transactions on Electron Devices **53** (10), 2438-2447 (2006).
94. N. Killat, M. Kuball, T.-M. Chou, U. Chowdhury and J. Jimenez, presented at the 2010 IEEE International Reliability Physics Symposium, 2010 (unpublished).
95. M. Kuball, S. Rajasingam, A. Sarua, M. Uren, T. Martin, B. Hughes, K. Hilton and R. Balmer, Applied physics letters **82** (1), 124-126 (2003).
96. M. Faqir, T. Batten, T. Mrotzek, S. Knippscheer, M. Massiot, M. Buchta, H. Blanck, S. Rochette, O. Vendier and M. Kuball, Microelectronics Reliability **52** (12), 3022-3025 (2012).
97. J. W. Pomeroy, M. J. Uren, B. Lambert and M. Kuball, Microelectronics Reliability **55** (12), 2505-2510 (2015).
98. D. Altman, M. Tyhach, J. McClymonds, S. Kim, S. Graham, J. Cho, K. Goodson, D. Francis, F. Faili and F. Ejeckam, presented at the Fourteenth Intersociety Conference on Thermal and Thermomechanical Phenomena in Electronic Systems (ITherm), 2014 (unpublished).
99. J. Pomeroy, M. Bernardoni, A. Sarua, A. Manoi, D. Dumka, D. Fanning and M. Kuball, presented at the 2013 IEEE Compound Semiconductor Integrated Circuit Symposium (CSICS), 2013 (unpublished).
100. D. Dumka, T. Chou, J. Jimenez, D. Fanning, D. Francis, F. Faili, F. Ejeckam, M. Bernardoni, J. Pomeroy and M. Kuball, presented at the 2013 IEEE Compound Semiconductor Integrated Circuit Symposium (CSICS), 2013 (unpublished).
101. C. Middleton, H. Chandrasekar, M. Singh, J. W. Pomeroy, M. J. Uren, D. Francis and M. Kuball, Applied Physics Express **12** (2), 024003 (2019).
102. S. Singhal, J. Brown, R. Borges, E. Piner, W. Nagy and A. Vescan, (2002).
103. C. R. Neve, K. B. Ali, P. Sarafis, E. Hourdakis, A. Nassiopoulou and J.-P. Raskin, Microelectronic Engineering **120**, 205-209 (2014).
104. D. Lederer and J.-P. Raskin, in Science and Technology of Semiconductor-On-Insulator Structures and Devices Operating in a Harsh Environment (Springer, 2005), pp. 191-196.
105. M. Rack, Y. Belaroussi, K. B. Ali, G. Scheen, B. K. Esfeh and J.-P. Raskin, IEEE Transactions on Electron Devices **65** (5), 1887-1895 (2018).
106. H. Chandrasekar, M. J. Uren, M. A. Casbon, H. Hirshy, A. Eblabla, K. Elgaid, J. W. Pomeroy, P. J. Tasker and M. Kuball, IEEE Transactions on Electron Devices **66** (4), 1681-1687 (2019).
107. B. C. Wadell, *Transmission line design handbook.* (Artech House, 1991).
108. W. Hilberg, IEEE Transactions on Microwave Theory and techniques **17** (5), 259-265 (1969).
109. A. Eblabla, D. J. Wallis, I. Guiney and K. Elgaid, IEEE microwave and wireless components letters **25** (7), 427-429 (2015).
110. I. Takenaka, K. Ishikura, K. Asano, S. Takahashi, Y. Murase, Y. Ando, H. Takahashi and C. Sasaoka, IEEE Transactions on Microwave Theory and Techniques **62** (3), 502-512 (2014).

Gallium Oxide Field Effect Transistors — Establishing New Frontiers of Power Switching and Radiation-Hard Electronics

Man Hoi Wong* and Masataka Higashiwaki

National Institute of Information and Communications Technology,
4-2-1 Nukui-Kitamachi, Koganei, Tokyo 184-8795, Japan
**mhwong@nict.go.jp*

Ga_2O_3 has gained worldwide interests for next-generation power electronics because its large critical field strength stemming from an ultra-wide bandgap promises miniaturized circuits and systems with high conversion efficiency. Intense pursuit of Ga_2O_3 power devices stimulated by early demonstrations of high-voltage Ga_2O_3 field effect transistors (FETs) has brought about tremendous advancements in this new technology, whose strong radiation tolerance and high thermal stability also befit harsh-environment applications. In this paper, we review the various types of depletion- and enhancement-mode Ga_2O_3 FETs — which have predominantly been lateral devices — for power switching and radiation-hard electronics. The development of vertical Ga_2O_3 transistors based on a low-cost, highly-manufacturable ion implantation doping process will also be discussed.

Keywords: Ga_2O_3; gallium oxide; MESFET; MOSFET; molecular beam epitaxy (MBE); power device; ion implantation; field plate; breakdown; channel temperature; thermal resistance; self-heating; radiation-hard; gamma ray; enhancement mode; normally off; halide vapor phase epitaxy (HVPE); current aperture; current blocking layer; vertical MOSFET.

1. Introduction

The advancement of modern societies depends critically on the effective use of electrical energy. At the heart of electrical systems are power rectifiers and switches for the control and conversion of electricity to provide optimal conditions for transmission, distribution, and loadside consumption. Today, Si power devices are the mainstream but they are rapidly approaching fundamental performance limitations, rendering commercial power systems bulky and inefficient. These challenges are converging with technology opportunities created by wide-bandgap semiconductors to define a new landscape for power conversion.

Recent years have seen an upsurge in research activities on beta-phase gallium oxide (β-Ga_2O_3, referred to hereinafter as Ga_2O_3) for power electronics [1]. Boasting a large bandgap of 4.5–4.9 eV [2–4] and resultant experimental critical fields of 5.1–5.9 MV/cm that surpass the theoretical bulk values for GaN and SiC [5–9], Ga_2O_3 promises dramatic reductions in the size, weight, and energy consumption of power systems by simultaneously increasing power density and power conversion efficiency at the device level.

Ga_2O_3 is also thermally stable and radiation hard [10], thus rendering it particularly suitable for constructing devices and systems for aerospace, detector, survey, and nuclear technologies that demand stable operations in harsh environments [11, 12]. The impact of these emerging Ga_2O_3 technologies is bolstered by the availability of melt-grown native substrates (floating zone [13–15], Czochralski [16–19], edge-defined film-fed growth [20, 21], vertical Bridgman [22]), whose high quality and scalability offer Ga_2O_3 a unique platform that carries a compelling cost advantage over the incumbent wide-bandgap technologies. This paper summarizes the progress made and lessons learnt on both lateral and vertical Ga_2O_3 field effect transistors (FETs) for power switching and radiation-hard applications, with a particular focus on device technologies developed in our group.

2. Lateral Depletion-Mode Ga_2O_3 FETs

2.1. *MESFET*

The first single-crystal, high-voltage Ga_2O_3 transistor is a depletion-mode (D-mode) Ga_2O_3 metal-semiconductor FET (MESFET) consisting of a 300-nm-thick Sn-doped *n*-type Ga_2O_3 channel layer (Fig. 1) [23]. The device, which is grown by ozone molecular beam epitaxy (MBE) on a Mg-doped semi-insulating Ga_2O_3 (010) substrate [24, 25], demonstrates successful transistor action with sharp drain current (I_D) pinch-off [Fig. 2(a)]. A three-terminal off-state breakdown voltage (V_{br}) of 257 V is obtained for a gate–drain spacing (L_{GD}) of 8 μm [Fig. 2(a)], which is notable considering that this is the first attempt on a high-voltage Ga_2O_3 transistor. However, the I_D on/off ratio (I_{on}/I_{off}) is limited to about 10^4 by a small leakage current through the unpassivated Ga_2O_3 surface [Fig. 2(b)]. The ohmic contact resistance also leaves much room for improvement.

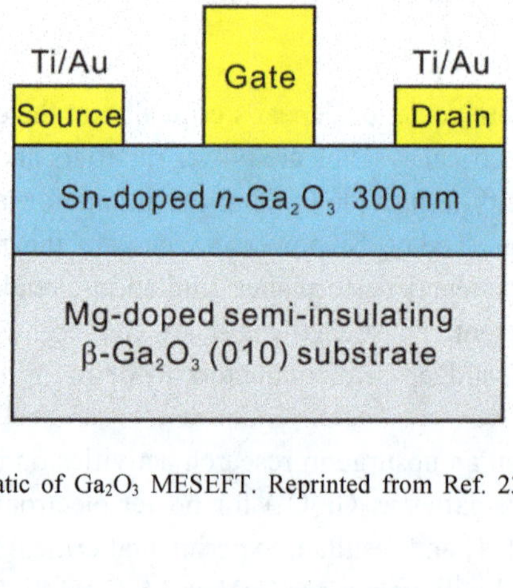

Fig. 1. Cross-sectional schematic of Ga_2O_3 MESEFT. Reprinted from Ref. 23, with the permission of AIP Publishing.

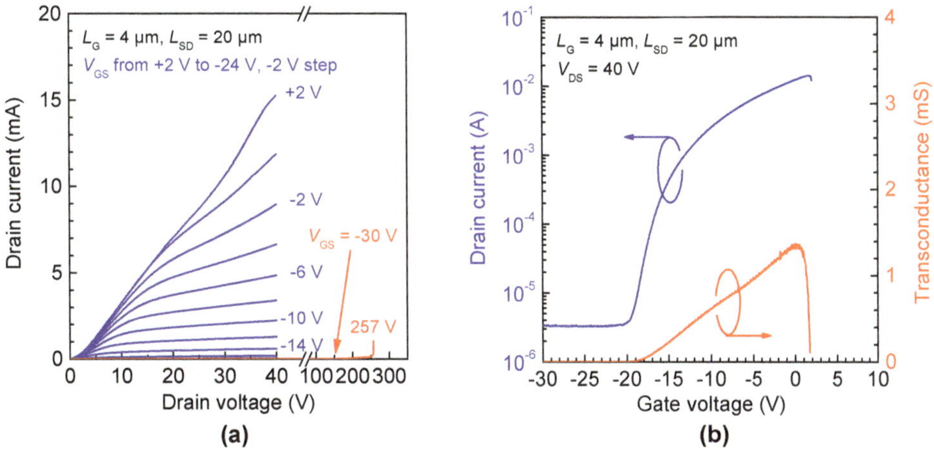

Fig. 2. DC characteristics of Ga_2O_3 MESFET with gate length (L_G) of 4 µm and source-drain spacing (L_{SD}) of 20 µm: (a) I_D–V_D and (b) I_D–V_G. Reprinted from Ref. 23, with the permission of AIP Publishing.

2.2. *MOSFET with Si-ion-implanted source/drain ohmic contacts*

To address the two major shortcomings of the MESFET, a metal-oxide-semiconductor FET (MOSFET) structure engineered with an Al_2O_3 dielectric grown by plasma-assisted atomic layer deposition (ALD) and n^+ source/drain contact layers formed by Si-ion (Si^+) implantation is fabricated on an Fe-doped semi-insulating Ga_2O_3 (010) substrate (Fig. 3) [26, 27]. The Al_2O_3 dielectric suppresses I_{off} by functioning as both a surface passivation layer and a gate insulator, leading to a dramatic increase in I_{on}/I_{off} to over 10^{10}. Degenerate doping of the ohmic contacts results in a small specific contact resistivity of 8.1×10^{-6} $\Omega\cdot cm^2$, which is comparable to the typical resistivity values of Ti/Al-based contacts made to *n*-GaN and AlGaN/GaN heterostructures. The MOSFET, which has a Sn-doped channel layer nominally the same as that of the MESFET introduced in Sec. 2.1, delivers a maximum I_D of 39 mA/mm and an off-state V_{br} of 370 V [Fig. 4(a)].

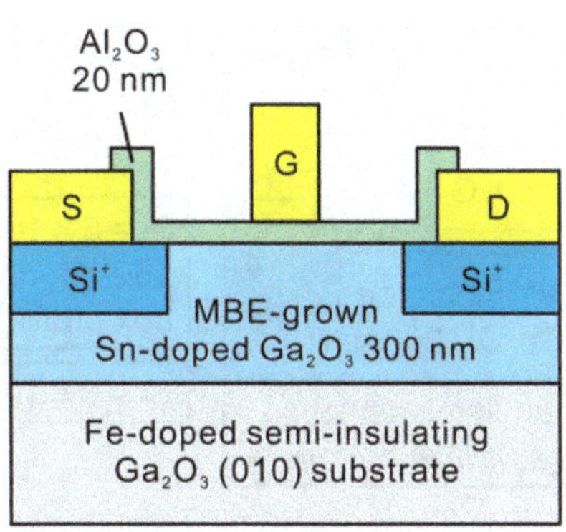

Fig. 3. Cross-sectional schematic of Ga_2O_3 MOSFET with Si^+-implanted source/drain ohmic contacts. Reprinted from Ref. 26, with the permission of AIP Publishing.

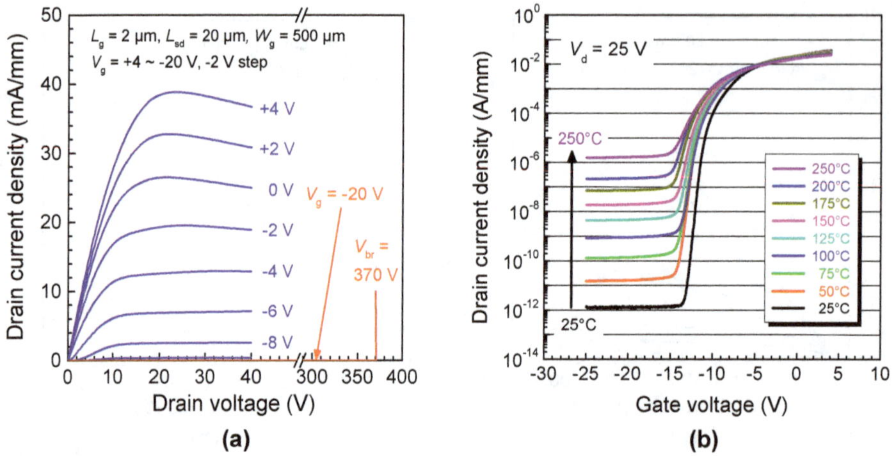

Fig. 4. DC characteristics of Ga_2O_3 MOSFET with Si^+-implanted source/drain ohmic contacts: (a) I_D–V_D and (b) temperature-dependent I_D–V_G. Reprinted from Ref. 26, with the permission of AIP Publishing.

High-temperature operation at 250°C is demonstrated without permanent degradation in electrical characteristics, albeit with a lower I_{on}/I_{off} of 10^4 [Fig. 4(b)].

2.3. *MOSFET with Si-ion-implanted channel and source/drain ohmic contacts*

The Ga_2O_3 MESFET and MOSFET with a Sn-doped channel have shown great promise as an emerging wide-bandgap device technology, yet further advancement is hampered by a narrow MBE growth temperature window required to promote Sn incorporation into the Ga_2O_3 epitaxial layer while maintaining high crystal quality [25]. Oftentimes, delayed Sn incorporation due to surface segregation of the dopant leads to poor control of channel thickness and non-uniform distribution of in-plane carrier concentration. These challenges are circumvented by applying Si^+ implantation doping to form the channel in an unintentionally-doped (UID) Ga_2O_3 epilayer (Fig. 5) [28], which allows higher growth

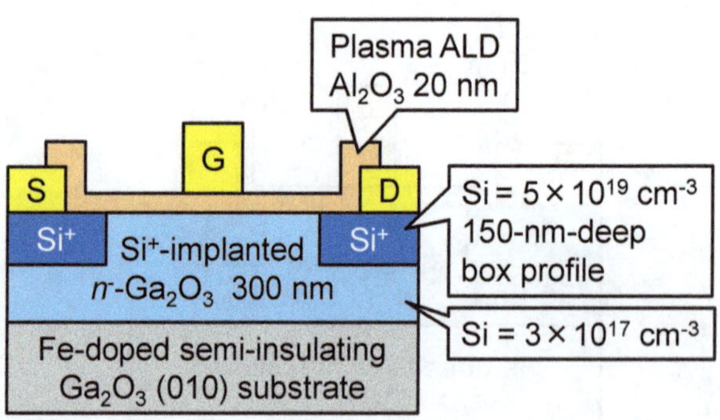

Fig. 5. Cross-sectional schematic of Ga_2O_3 MOSFET with Si^+-implanted channel and source/drain ohmic contacts. © 2013 IEEE. Reprinted, with permission, from Ref. 28.

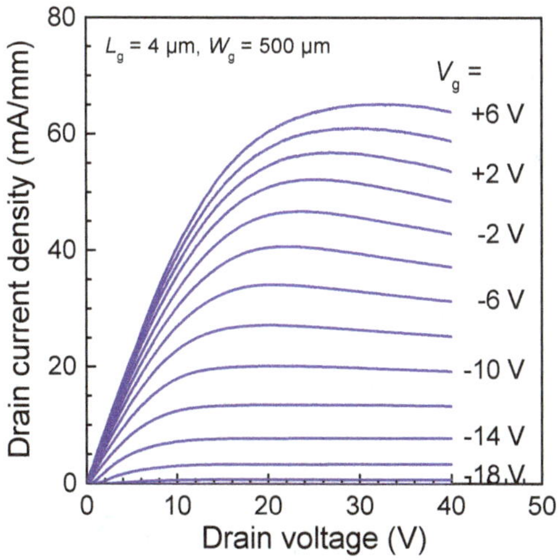

Fig. 6. DC I_D–V_D characteristics of Ga_2O_3 MOSFET with Si^+-implanted channel and source/drain ohmic contacts. © 2013 IEEE. Reprinted, with permission, from Ref. 28.

temperatures to ensure optimal crystal quality without concern for irreproducible dopant profiles. The MOSFET with a Si^+-implanted channel delivers comparable performance to its Sn-doped counterpart, including a maximum I_D of 65 mA/mm (Fig. 6), an off-state V_{br} of 415 V, and an I_{on}/I_{off} of over 10^{10}.

2.4. *Field-plated MOSFETs*

Field-plating is an edge termination technique commonly employed in FETs to enhance V_{br}. Accordingly, a gate-connected field plate (FP) is introduced into the Si^+-implanted MOSFET discussed in Sec. 2.3 for a target V_{br} of over 1 kV [29]. The field-plated MOSFET (FP-MOSFET), illustrated schematically in Fig. 7, incorporates a 0.9-μm-thick UID Ga_2O_3 buffer layer under its Si^+-implanted channel to protect the substrate from implantation damage, thereby preventing defect-assisted outdiffusion of Fe toward the channel upon implant activation annealing and the consequent charge compensation [30, 31]. Proximity of Fe doping to the electron channel has also been ascribed to mobility degradation, DC–RF dispersion, and threshold voltage (V_T) instability in Ga_2O_3 MOSFETs with non-ion-implanted channels [32, 33]. At room temperature, the FP-MOSFET delivers a maximum I_D of 78 mA/mm and an I_{on}/I_{off} of over 10^9 (Fig. 8). Stability against thermal stress is sustained up to at least 300°C, where an I_{on}/I_{off} higher than 10^3 is maintained [Fig. 8(b)]. Gate- and drain-lag measurements reveal negligible bulk trapping and a well-passivated surface (Fig. 9) [29, 34]. Compared with the non-field-plated baseline device, a 2.5-μm-long FP contributes to an 80% increase in off-state V_{br} from 415 V to 755 V [Fig. 8(a)]; however, hard breakdown of this device is limited by parasitic source–drain punchthrough instead of peak electric fields in the gate–drain depletion region, which is consistent with the off-state drain leakage current being one to two orders of magnitude higher than the gate leakage current (I_G) and suggests that the intrinsic V_{br} should have been much higher, as to be expected from simulations. By developing a high-density composite oxide stack for the field plate as well as adopting a SiO_2 gate insulator to provide a larger conduction band offset than with Al_2O_3 [35, 36], Zeng *et al.* have obtained a considerably improved V_{br} of 1.85 kV for a FP-MOSFET that utilizes an MBE-grown Sn-doped channel [37] and spin-on-glass source/drain doping [38], yet the accompanying specific on-resistance ($R_{on,sp}$) is almost two orders of magnitude larger than that of the 755-V device.

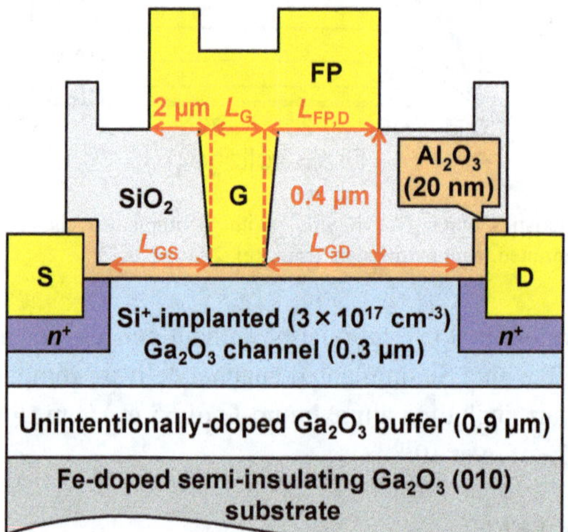

Fig. 7. Cross-sectional schematic of Ga_2O_3 FP-MOSFET with gate-connected FP of length $L_{FP,D}$ extended toward the drain.

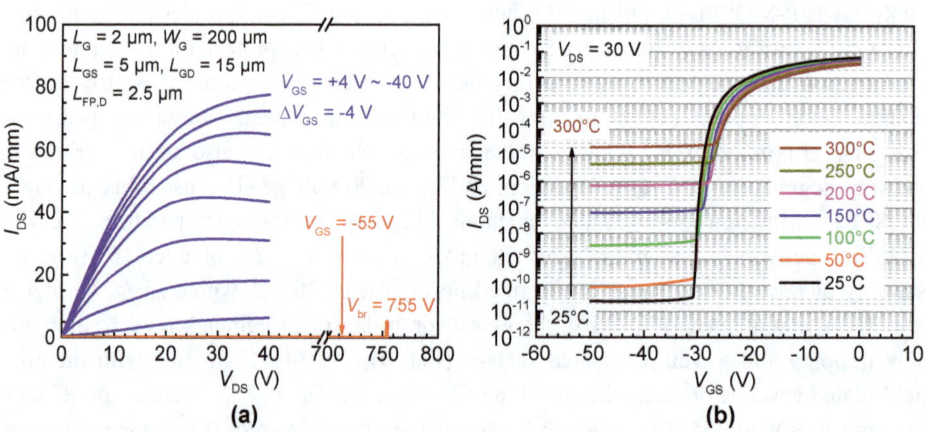

Fig. 8. DC characteristics of Ga_2O_3 FP-MOSFET with 2.5-µm-long FP: (a) I_D–V_D and (b) temperature-dependent I_D–V_G. © 2016 IEEE. Reprinted, with permission, from Ref. 29.

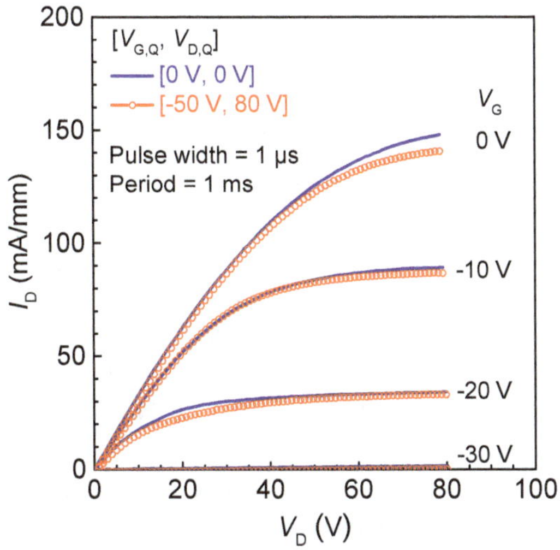

Fig. 9. Pulsed I_D–V_D characteristics of Ga_2O_3 FP-MOSFET measured at quiescent gate and drain biases ([$V_{G,Q}$, $V_{D,Q}$]) of [0 V, 0 V] (line) and [-50 V, 80 V] (symbol) that correspond to dispersion-free and drain-lag conditions, respectively [34].

Ga_2O_3 FP-MOSFETs have also been fabricated with a source-connected FP [39, 40], which is often preferred to a gate-connected FP from the perspective of high-speed performance since comparable electric field profiles can be engineered without introducing a parasitic gate–drain feedback capacitance. Devices reported by Lv *et al.* [39] feature Si^+-implanted source/drain contacts and a highly-doped channel containing at least twice as much charge as in the aforementioned FP-MOSFETs, thereby giving rise to a high I_D of >200 mA/mm. A V_{br} of 480 V and a corresponding $R_{on,sp}$ of 4.57 mΩ·cm^2 translate to a large power figure of merit ($V_{br}^2/R_{on,sp}$) of 50.4 MW/cm^2. In a separate work by Mun *et al.* [40], field-plated devices show enhancement in V_{br} up to 2.32 kV (corresponding $R_{on,sp}$ = 959 mΩ·cm^2), which is the highest breakdown reported to date for a lateral Ga_2O_3 MOSFET.

2.5. *Critical field strength in Ga_2O_3 MOSFETs*

The experimental critical electric fields of the Ga_2O_3 MOSFETs introduced so far in this section, calculated by approximating a linear potential gradient between the gate and drain terminals, are smaller than 1 MV/cm and thus fall far short of the projected theoretical value of 8 MV/cm [23]. This is because those devices have an L_{GD} larger than the minimum required to support the V_{br} achievable at their respective channel doping concentrations. In the limit of a very large L_{GD}, the average breakdown field approaches zero while the peak field is determined by the extent of the gate–drain depletion region; on the other hand, the lateral electric field in a scaled gate–drain drift region becomes more uniform in magnitude such that the averaged field more accurately reflects the peak field. Indeed, improved average field strengths of 3.8 and 2.2 MV/cm – the former of which being the highest ever measured in a lateral transistor – have been reported in MOSFETs possessing smaller L_{GD} of 0.6 and 1.8 μm, respectively [5, 37].

3. Thermal Characteristics of Ga_2O_3 MOSFETs

MOSFETs in the on-state experience highly localized power dissipation and heat generation within a submicron region of the channel at the drain side of the gate where the maximum electric field occurs. Self-heating in Ga_2O_3 devices is exacerbated by the low thermal conductivity of Ga_2O_3 that ranges from 11 to 27 W/m·K at room temperature along the three principal crystallographic directions [18, 41–44]. Inefficient heat transport leads to higher temperatures in the active region of Ga_2O_3 transistors than for other technologies at comparable power densities (P_D). A large thermal resistance (R_{th}) of 48 mm·K/W (i.e. a temperature rise of 48 K per 1 W/mm of P_D) is extracted for a Ga_2O_3 FP-MOSFET (Fig. 7) under room temperature operation by an electrical method that determines a spatially-averaged channel temperature (T_{ch}) via isothermal pulsed current–voltage measurements with quiescent biases and sweep parameters chosen to minimize dispersion effects related to trapping (Fig. 10) [45]. Raman nanoparticle thermography, which enables localized measurement of the peak T_{ch}, yields a larger R_{th} of 88 mm·K/W for a similar device structure [46]. These measured R_{th} values exceed those of AlGaN/GaN high electron mobility transistors fabricated on sapphire substrates [47], and are a factor of three to five larger than those of GaN devices that benefit from efficient heat sinking through thermally-conductive substrates like SiC [47] or diamond [48, 49]. Significantly higher junction temperatures have also been revealed in a Ga_2O_3 Schottky barrier diode than in a SiC diode at comparable P_D [50].

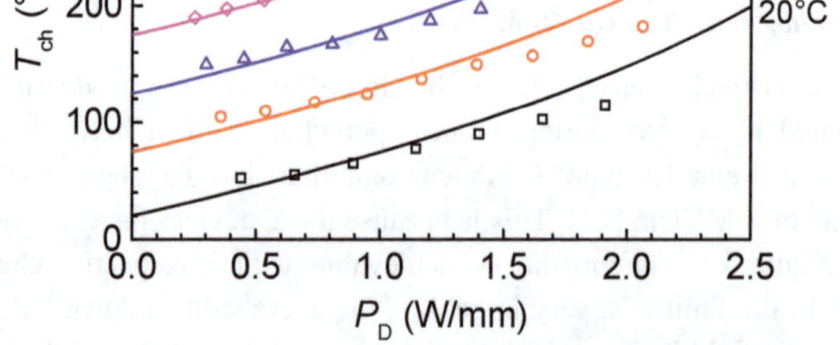

Fig. 10. Dependence of T_{ch} (symbol: measurement; line: simulation) in Ga_2O_3 FP-MOSFET on P_D at ambient temperatures (T_{amb}) of 20, 75, 125, and 175°C. Reprinted from Ref. 45, with the permission of AIP Publishing.

Performance boost in Ga_2O_3 FETs is evident when self-heating effects due to the material's low thermal conductivity are mitigated by means of pulsed-mode operation [34, 51], which indicates that proper thermal mounting and heat spreading techniques will dramatically improve the electrical performance of Ga_2O_3 devices. It has recently been demonstrated that direct attachment of an n^+-Ga_2O_3 wafer to a thermally-conductive polycrystalline n^+-SiC (poly-SiC) substrate via surface activated bonding forms a void-free interface with a negligible R_{th} (Fig. 11) [52]. The effective out-of-plane thermal conductivity of the composite Ga_2O_3/poly-SiC structure matches that of bulk Si (156 W/m·K) for a Ga_2O_3 thickness of 21 μm, and is expected to be comparable to that of GaN (220 W/m·K) if the Ga_2O_3 is further thinned down to about 10 μm. Furthermore, the heterointerface exhibits only a small specific electrical resistivity of 2×10^{-4} $\Omega\cdot cm^2$, which can be further reduced with higher doping. This wafer bonding technique, in combination with top-side heat extraction strategies, thus provides a viable thermal solution for vertical Ga_2O_3 power devices.

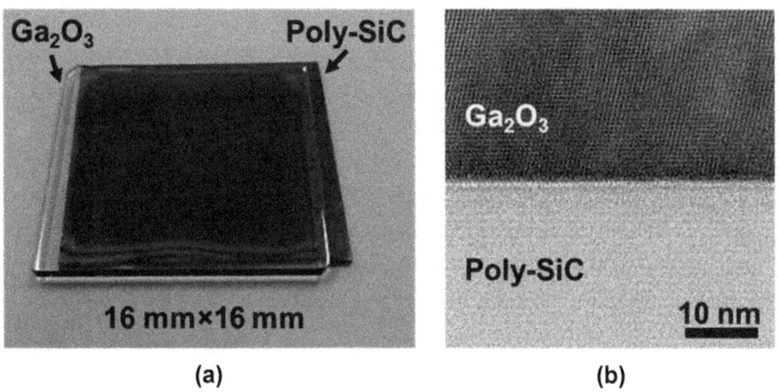

Fig. 11. Surface activated bonding of Ga_2O_3 and poly-SiC: (a) Photograph of bonded substrate and (b) cross-sectional transmission electron micrograph of bonded heterointerface between Ga_2O_3 and poly-SiC. Reprinted from Ref. 52, with the permission of AIP Publishing.

4. Ga_2O_3 MOSFETs for Radiation-Hard Electronics

Intense particle and electromagnetic radiations present severe challenges to the performance and long-term reliability of semiconductor electronic devices by causing physical damage in the constituent materials and/or transient effects related to ionization. It has been established that wide-bandgap semiconductors (notably SiC [53], GaN [54, 55], and diamond [56, 57]) are much less susceptible to displacement damage by particle irradiation than elemental and compound semiconductors with narrower bandgaps. A higher displacement threshold reflects stronger bond strength and serves as an indication of enhanced radiation hardness. Demonstrations of operational Ga_2O_3 solar-blind ultraviolet photodetectors [58], Schottky rectifiers [59–61], and exfoliated quasi-two-dimensional nanobelt FETs [62] that have been exposed to energetic protons or electrons at high fluences ($\geq 10^{15}$ cm^{-2}) reveal the strong prospects of Ga_2O_3 for radiation-hard applications.

Effects of ionizing radiation on epitaxial Ga_2O_3 FETs have been studied by exposing MOSFETs to gamma-rays (γ-rays) emitted from a ^{60}Co source with an average photon energy of 1.25 MeV [63]. The devices show no discernible increase in on-resistance (R_{on}), reduction in I_D, or shift in V_T up to a cumulative γ-ray dose of 500 kGy(SiO_2), where a dose of 1 Gray (Gy) corresponds to an absorbed radiation energy of 1 J per kg of mass (SiO_2 being a legacy reference material). The unchanged R_{on} and I_D reflect insignificant generation of radiation-induced bulk traps or scattering centers in the Ga_2O_3 channel, while a stable V_T indicates insignificant carrier trapping or fixed charge accumulation under the gate. All tested devices remain functional at cumulative doses of 1 MGy(SiO_2) and beyond, which are more than ten times the target γ-ray tolerance for space and radiation detector applications [64]. At the highest cumulative dose of 1.6 MGy(SiO_2), the devices preserve the small hysteresis of the transfer characteristics from their virgin state, albeit showing slight reductions of less than 5% in the maximum I_D and peak transconductance (g_m) with an associated positive V_T shift that are ascribable to bulk carrier trapping (Fig. 12). Such heavy irradiation also causes an increase in off-state I_G and degradation in dynamic R_{on} due to gate dielectric damage and radiation-induced electron trapping in the drain access region, respectively. On the basis of these electrical characteristics, it can be deduced that the total-dose γ-ray response of Ga_2O_3 MOSFETs is dominated by the deleterious effects of dielectric damage and interface charge trapping instead of the intrinsic radiation hardness of Ga_2O_3. Robust processes for the gate and passivating dielectrics are therefore paramount to enhancing the overall radiation hardness of Ga_2O_3 MOSFETs.

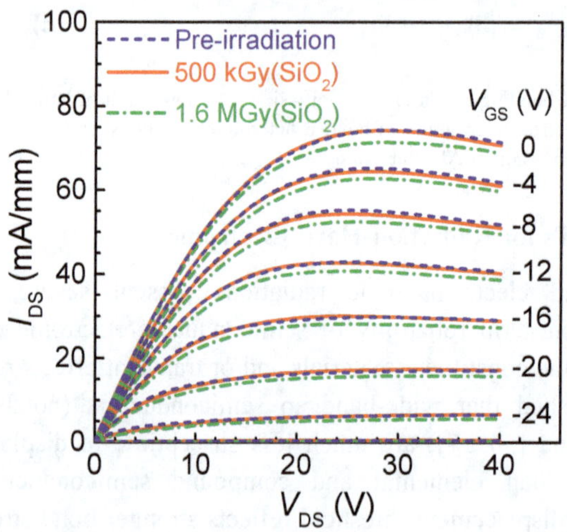

Fig. 12. DC I_D–V_D characteristics of Ga_2O_3 MOSFET before γ-ray irradiation (blue dashed line) and after exposure to cumulative γ-ray doses of 500 kGy(SiO_2) (red solid line) and 1.6 MGy(SiO_2) (green dashed-dotted line). Reprinted from Ref. 63, with the permission of AIP Publishing.

5. Lateral Enhancement-Mode Ga_2O_3 FETs

5.1. *Overview*

While the rapid progress of D-mode Ga_2O_3 transistors has established the viability of Ga_2O_3 for power electronics, normally-off power switches are preferred for practical applications to ensure fail-safe operation and eliminate negative-polarity power supplies for simplified circuit designs and system architecture. The majority of enhancement-mode (E-mode) Ga_2O_3 MOSFETs have been implemented as *n*-channel accumulation-mode devices, until very recently when inversion-mode operation via acceptor doping of the Ga_2O_3 channel layer is also proposed. Both planar and non-planar topologies have been pursued.

5.2. *Planar MOSFETs*

Planar E-mode Ga_2O_3 MOSFETs can be realized by decreasing the thickness and/or doping concentration of the channel layer to enable full depletion under the gate at a gate voltage (V_G) of 0 V. The earliest device fabricated following this simple consideration shows a V_T of +3 V and a respectable V_{br} of 390 V, but large parasitic resistances in the ungated source and drain access regions render its I_{on} extremely low (~0.1 mA/mm) [65]. One potential approach to high-current operation involves overlapping the gate with n^+ source and drain contacts formed by selective-area Si^+ implantation doping of a UID Ga_2O_3 channel grown by ozone MBE with a low background carrier concentration ($\leq 4\times10^{14}$ cm^{-3}) (Fig. 13) [66]. Despite a significant reduction in the total parasitic resistance to a mere 2.2 Ω·mm, severe charge trapping effects at the dielectric interface limit the maximum I_D to 1.4 mA/mm (Fig. 14) and the corresponding I_{on}/I_{off} to 10^6 while causing large positive V_T shifts. Based on the device structure in Ref. 66, normally-off Ga_2O_3 MOSFETs aiming at inversion-mode operation to stabilize the V_T at a larger value are fabricated by employing plasma-assisted MBE to grow a Ga_2O_3 channel layer doped unintentionally with deep nitrogen (N) acceptors at a concentration of 1×10^{18} cm^{-3} (Fig. 15) [67]. A large V_T of over +8 V is achieved. Even though on-state device operation is restricted to within the subthreshold regime owing to limited V_G swing, an I_{on}/I_{off} ratio exceeding five orders of magnitude is obtained (Fig. 16).

E-mode Ga_2O_3 MOSFETs that feature gate–source and gate–drain overlaps resemble prototypical Si transistors for complementary logic and are therefore unamenable to high-voltage operation. Devices with both high V_{br} and high I_D are demonstrated by using a gate-recess process to partially remove a highly-doped channel under the gate while the ungated access regions remain at full thickness to minimize R_{on} [68]. This process produces T-shaped gates with a built-in FP. Laterally-scaled recessed-gate MOSFETs (V_T = +2 V) show an R_{on} of 215 Ω·mm, I_D of 40 mA/mm, and I_{on}/I_{off} of 10^9, which are the best reported for homoepitaxial E-mode Ga_2O_3 devices and have only been surpassed by non-epitaxial nanomembrane FETs [69, 70]. An off-state V_{br} of 505 V is measured at V_G = 0 V.

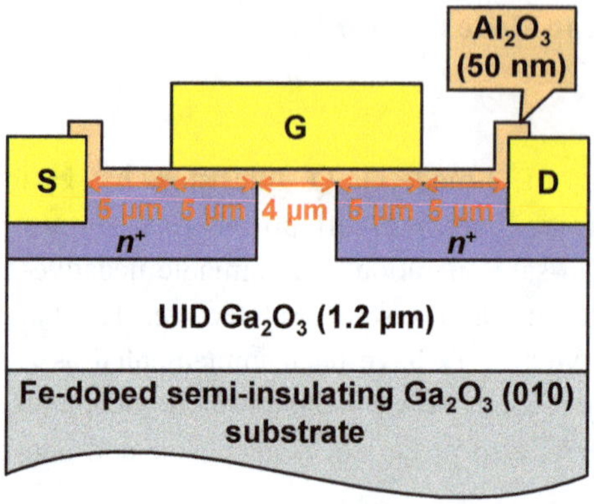

Fig. 13. Cross-sectional schematic of E-mode Ga_2O_3 MOSFET with UID channel and Si^+-implanted source/drain contacts [66]. Copyright 2017, the Japan Society of Applied Physics.

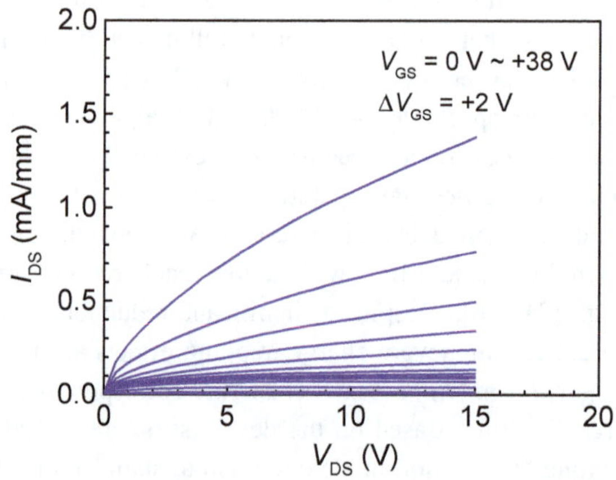

Fig. 14. DC I_D–V_D characteristics of E-mode Ga_2O_3 MOSFET with UID channel and Si^+-implanted source/drain contacts [66]. Copyright 2017, the Japan Society of Applied Physics.

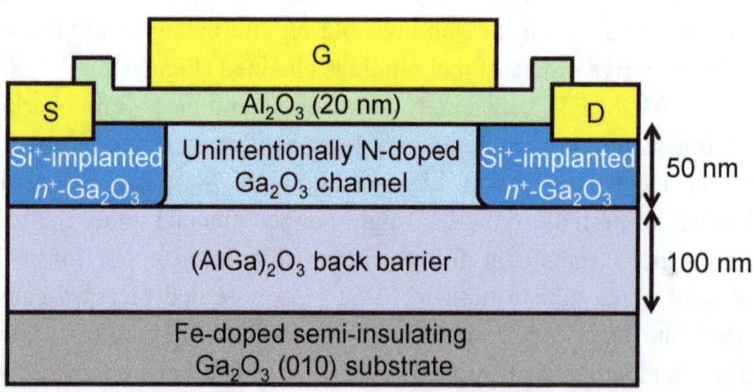

Fig. 15. Cross-sectional schematic of E-mode Ga_2O_3 MOSFET with N-doped channel and Si^+-implanted source/drain contacts. © 2019 IEEE. Reprinted, with permission, from Ref. 67.

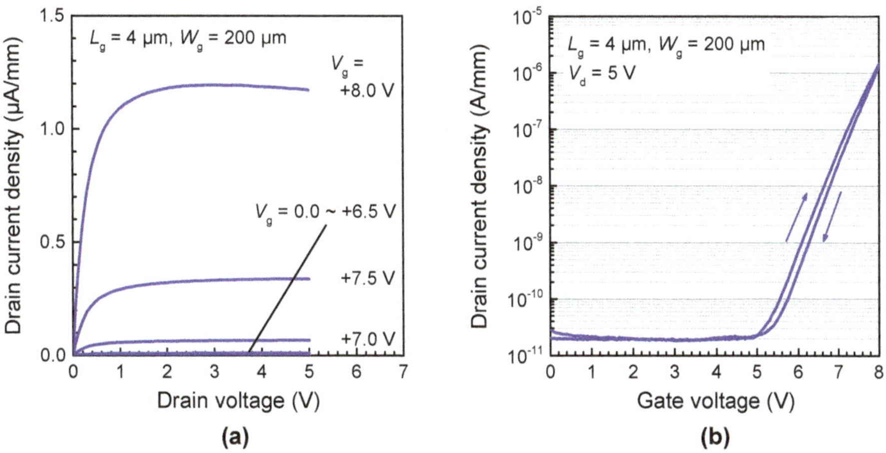

Fig. 16. DC characteristics of E-mode Ga_2O_3 MOSFET with N-doped channel and Si^+-implanted source/drain contacts: (a) I_D–V_D and (b) I_D–V_G. © 2019 IEEE. Reprinted, with permission, from Ref. 67.

5.3. *Fin-array MOSFETs*

A non-planar conception of normally-off Ga_2O_3 FETs capitalizes on the enhanced electrostatic control offered by wrap-gate architecture to deplete non-planar fin-shaped channels from the sidewalls, thereby shifting the V_T to a positive value without sacrificing doping concentration or volume current density [71]. FinFETs with a 20-nm-thick Al_2O_3 dielectric and a 2-μm-long wrap-gate demonstrate normally-off operation with a V_T between 0 and +1 V. The I_{on}/I_{off} is larger than 10^5 and is mainly limited by a low I_{on} of less than 1 mA/mm, but substantial improvements can be expected with R_{on} optimization. Due to a lack of proper sidewall surface treatment, plasma etch damage of the fin sidewalls results in trapping effects and a relatively low electron mobility of 24 cm^2/V·s, in spite of which an acceptable subthreshold slope of 158 mV/dec is obtained. At V_G = 0 V, a V_{br} of 612 V is achieved without a field plate. This performance represents the highest breakdown for any transistor technology utilizing non-planar device channels.

6. Current Aperture Vertical Ga_2O_3 MOSFET

The Ga_2O_3 transistors reported to date have predominantly taken on a lateral geometry. However, lateral devices are not amenable to the large currents and voltages required for many applications owing to their sizeable layout areas and potential reliability concerns arising from self-heating and surface instabilities. For applications demanding high voltage ratings and high power levels, vertical switching devices are desirable since they allow for higher current drives, simplified thermal management, and far superior field termination.

Two types of vertical Ga_2O_3 MOSFETs have been proposed. Owing to a lack of shallow acceptor dopants for Ga_2O_3 [72–74], a number of early devices have pursued a non-planar fin architecture featuring side-gate modulation to circumvent the need for *pn* junction isolation between source and drain [75–78]. Normally-off operation with an I_D exceeding 1 kA/cm^2, an I_{on}/I_{off} of 10^9, and a V_{br} of 1.6 kV has been demonstrated in these transistors [76–78]. The other implementation — known as the current aperture vertical MOSFET [79, 80] — is inspired by the commercially successful SiC double-implanted MOSFET [81, 82] and modeled on the evolving GaN current aperture vertical electron transistor [83–85], both of which have been conceived from the Si vertical double-diffused MOSFET [86]. Among the family of ultra-wide-bandgap semiconductors, Ga_2O_3 is distinctively amenable to doping with both shallow donors (Si) and deep acceptors (Mg, N) by ion implantation with efficient dopant activation at a low thermal budget [27, 87], thereby enabling the fabrication of current aperture vertical Ga_2O_3 MOSFETs based on a manufacturable all-ion-implanted process resembling that for Si and SiC MOSFETs.

Figure 17 illustrates the generic structure and operating principles of a D-mode current aperture vertical Ga_2O_3 MOSFET. In this implementation, the top source contact is electrically isolated from the bottom drain contact by a buried electron barrier called a current blocking layer (CBL), except at an aperture opening (L_{ap}) bounded by CBLs through which I_D is conducted. The CBL also serves as a back-barrier for a top-gated lateral channel, where the gate–CBL overlap dimension (L_{go}) defines the gate length. Initial devices have adopted a Mg-ion-implanted CBL, but they suffer a large source–drain leakage due to rapid thermal diffusion and a concomitant loss of Mg from the CBL upon implant activation annealing [79]. In contrast to Mg, N exhibits much lower thermal diffusivity in Ga_2O_3, thus enabling the use of higher annealing temperatures to maximize N activation efficiency without drastically affecting the dopant profile [87, 88]. A D-mode current aperture vertical MOSFET is subsequently accomplished through the integration of N-ion (N^{++}) and Si^+ implantation doping [80].

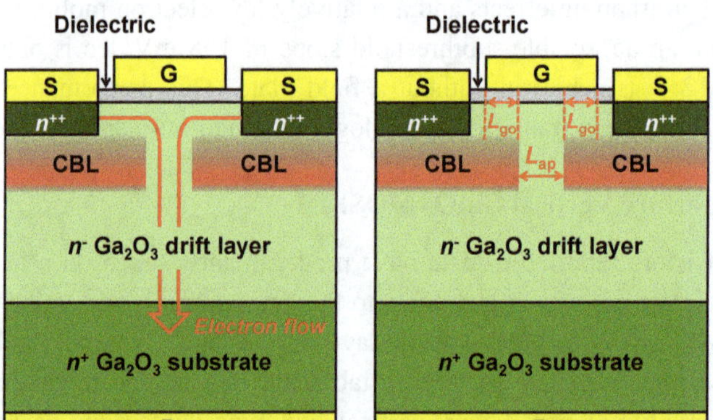

Fig. 17. Structure and operation of a generic D-mode current aperture vertical Ga_2O_3 MOSFET [79]. Copyright 2018, the Japan Society of Applied Physics.

The current aperture vertical Ga_2O_3 MOSFET with N^{++}-implanted CBLs consists of a 5-μm-thick Si-doped (2.5×10^{16} cm^{-3}) n^--Ga_2O_3 drift layer grown by halide vapor phase epitaxy (HVPE) on a Sn-doped n^+-Ga_2O_3 (001) substrate. HVPE is the epitaxial technique of choice since its capabilities of growing single-crystal Ga_2O_3 films with a low background electron density ($<10^{13}$ cm^{-3}) and at high speeds (>10 μm/h) are desirable for obtaining thick, lightly-doped drift layers required for high V_{br} [89–91]. Three ion implantation steps are employed to form the Si-doped n^{++} source contacts, Si-doped n-type channel, and nitrogen-doped CBLs. Figure 18 shows a cross-sectional schematic of a device with a 150-nm-thick channel, a 50-nm-thick Al_2O_3 gate dielectric, an L_{ap} of 20 μm, and an L_{go} of 2.5 μm. This MOSFET delivers a maximum I_D of 0.42 kA/cm² (normalized to the Si channel implant area) and an $R_{on,sp}$ of 31.5 mΩ·cm² [Fig. 19(a)]; the corresponding values are 1.37 kA/cm² and 9.7 mΩ·cm² when normalized

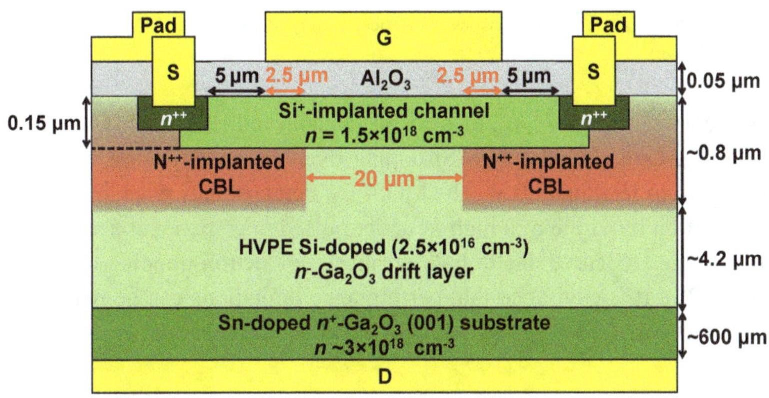

Fig. 18. Cross-sectional schematic of D-mode current aperture vertical Ga_2O_3 MOSFET with L_{go} = 2.5 μm and L_{ap} = 20 μm.

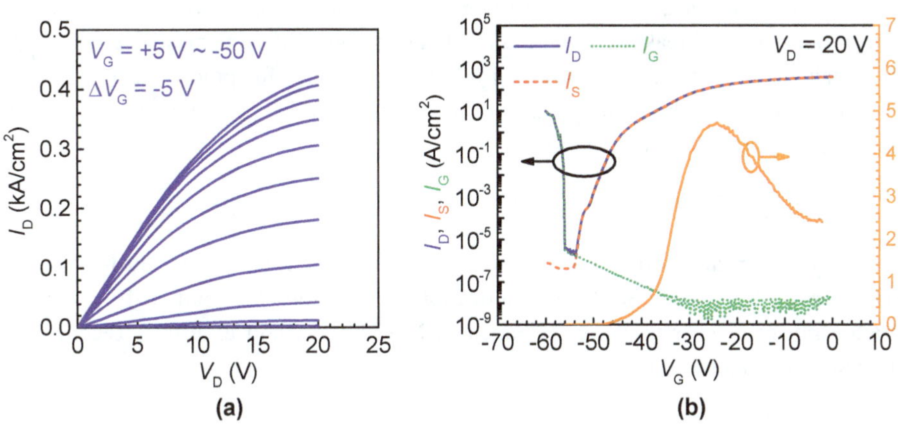

Fig. 19. DC characteristics of vertical Ga_2O_3 MOSFET: (a) I_D–V_D and (b) I_D–V_G. © 2019 IEEE. Reprinted, with permission, from Ref. 80.

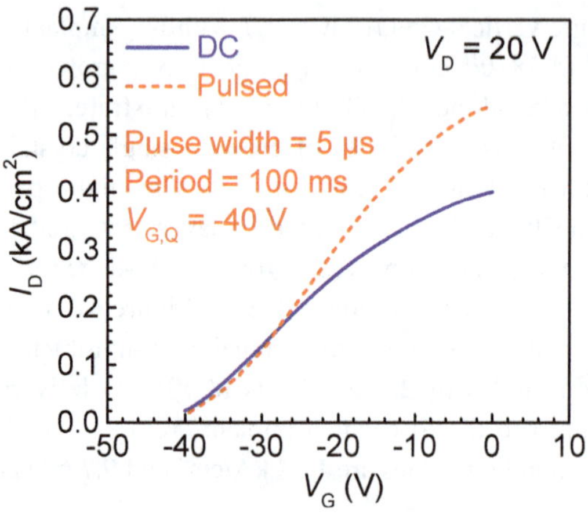

Fig. 20. DC and pulsed transfer characteristics of vertical Ga_2O_3 MOSFET. © 2019 IEEE. Reprinted, with permission, from Ref. 80.

to the aperture area. A high I_{on}/I_{off} of over 10^8 is achieved [Fig. 19(b)], where the I_{off} is dominated by I_G. Owing to the high Si doping over the aperture, the V_{br} is limited by Al_2O_3 breakdown to less than 30 V. The MOSFET demonstrates potential for high-speed switching operation by virtue of a high I_D under pulsed operation, yet a small positive V_T shift relative to the DC curve due to bulk or interface electron trapping under the gate is observed (Fig. 20). It is envisaged that performance limitations can be readily overcome with improved gate dielectric quality and optimized doping schemes.

7. Summary

Ga_2O_3 is poised to engender paradigm shifts in power conversion and harsh-environment applications by offering higher efficiency and enhanced functionality at an affordable cost. In a matter of years, Ga_2O_3 transistors have taken enormous strides including demonstrations of kV breakdown, normally-off operation, and vertical device concepts, all of which serve to position Ga_2O_3 as a relevant contender for practical applications. Continued efforts to establish the physics of materials and devices will pave the way for widespread adoption of Ga_2O_3 electronics.

Acknowledgment

This work was partially supported by Council for Science, Technology and Innovation (CSTI), Cross-ministerial Strategic Innovation Promotion Program (SIP), "Next-generation power electronics" (funding agency: New Energy and Industrial Technology Development Organization).

References

1. M. Higashiwaki and G. H. Jessen, *Appl. Phys. Lett.* **112**, 060401 (2018).
2. H. H. Tippins, *Phys. Rev.* **140**, A316 (1965).
3. M. Orita, H. Ohta, M. Hirano, and H. Hosono, *Appl. Phys. Lett.* **77**, 4166 (2000).
4. T. Onuma, S. Saito, K. Sasaki, T. Masui, T. Yamaguchi, T. Honda, and M. Higashiwaki, *Jpn. J. Appl. Phys.* **54**, 112601 (2015).
5. A. J. Green, K. D. Chabak, E. R. Heller, R. C. Fitch, Jr., M. Baldini, A. Fiedler, K. Irmscher, G. Wagner, Z. Galazka, S. E. Tetlak, A. Crespo, K. Leedy, and G. H. Jessen, *IEEE Electron Device Lett.* **37**, 902 (2016).
6. K. Konishi, K. Goto, H. Murakami, Y. Kumagai, A. Kuramata, S. Yamakoshi, and M. Higashiwaki, *Appl. Phys. Lett.* **110**, 103506 (2017).
7. C. Joishi, S. Rafique, Z. Xia, L. Han, S. Krishnamoorthy, Y. Zhang, S. Lodha, H. Zhao, and S. Rajan, *Appl. Phys. Express* **11**, 031101 (2018).
8. X. Yan, I. S. Esqueda, J. Ma, J. Tice, and H. Wang, *Appl. Phys. Lett.* **112**, 032101 (2018).
9. W. Li, Z. Hu, K. Nomoto, R. Jinno, Z. Zhang, Q. T. Thieu, K. Sasaki, A. Kuramata, D. Jena, and H. G. Xing, in *Proc. IEEE International Electron Devices Meeting*, San Francisco, CA, USA, 1–5 Dec. 2018, pp. 8.5.1–8.5.4.
10. J. Kim, S. J. Pearton, C. Fares, J. Yang, F. Ren, S. Kim, and A. Y. Polyakov, *J. Mater. Chem. C* **7**, 10 (2019).
11. D. Szalkai, Z. Galazka, K. Irmscher, P. Tüttő, A. Klix, and D. Gehre, *IEEE Trans. Nucl. Sci.* **64**, 1574 (2017).
12. X. Lu, L. Zhou, L. Chen, X. Ouyang, B. Liu, J. Xu, and H. Tang, *Appl. Phys. Lett.* **112**, 103502 (2018).
13. N. Ueda, H. Hosono, R. Waseda, and H. Kawazoe, *Appl. Phys. Lett.* **70**, 3561 (1997).
14. E. G. Víllora, K. Shimamura, Y. Yoshikawa, K. Aoki, and N. Ichinose, *J. Cryst. Growth* **270**, 420 (2004).
15. N. Suzuki, S. Ohira, M. Tanaka, T. Sugawara, K. Nakajima, and T. Shishido, *Phys. Status Solidi C* **4**, 2310 (2007).
16. Y. Tomm, P. Reiche, D. Klimm, and T. Fukuda, *J. Cryst. Growth* **220**, 510 (2000).
17. Z. Galazka, R. Uecker, K. Irmscher, M. Albrecht, D. Klimm, M. Pietsch, M. Brützam, R. Bertram, S. Ganschow, and R. Fornari, *Cryst. Res. Technol.* **45**, 1229 (2010).
18. Z. Galazka, K. Irmscher, R. Uecker, R. Bertram, M. Pietsch, A. Kwasniewski, M. Naumann, T. Schulz, R. Schewski, D. Klimm, and M. Bickermann, *J. Cryst. Growth* **404**, 184 (2014).
19. Z. Galazka, R. Uecker, D. Klimm, K. Irmscher, M. Naumann, M. Pietsch, A. Kwasniewski, R. Bertram, S. Ganschow, and M. Bickermann, *ECS J. Solid State Sci. Technol.* **6**, Q3007 (2017).
20. H. Aida, K. Nishiguchi, H. Takeda, N. Aota, K. Sunakawa, and Y. Yaguchi, *Jpn. J. Appl. Phys.* **47**, 8506 (2008).
21. A. Kuramata, K. Koshi, S. Watanabe, Y. Yamaoka, T. Masui, and S. Yamakoshi, *Jpn. J. Appl. Phys.* **55**, 1202A2 (2016).
22. K. Hoshikawa, E. Ohba, T. Kobayashi, J. Yanagisawa, C. Miyagawa, and Y. Nakamura, *J. Cryst. Growth* **447**, 36 (2016).
23. M. Higashiwaki, K. Sasaki, A. Kuramata, T. Masui, and S. Yamakoshi, *Appl. Phys. Lett.* **100**, 013504 (2012).
24. K. Sasaki, A. Kuramata, T. Masui, E. G. Víllora, K. Shimamura, and S. Yamakoshi, *Appl. Phys. Express* **5**, 035502 (2012).

25. K. Sasaki, M. Higashiwaki, A. Kuramata, T. Masui, and S. Yamakoshi, *J. Cryst. Growth* **392**, 30 (2014).
26. M. Higashiwaki, K. Sasaki, T. Kamimura, M. H. Wong, D. Krishnamurthy, A. Kuramata, T. Masui, and S. Yamakoshi, *Appl. Phys. Lett.* **103**, 123511 (2013).
27. K. Sasaki, M. Higashiwaki, A. Kuramata, T. Masui, and S. Yamakoshi, *Appl. Phys. Express* **6**, 086502 (2013).
28. M. Higashiwaki, K. Sasaki, M. H. Wong, T. Kamimura, D. Krishnamurthy, A. Kuramata, T. Masui, and S. Yamakoshi, in *Proc. IEEE International Electron Devices Meeting*, Washington, DC, USA, 9–11 Dec. 2013, pp. 28.7.1–28.7.4.
29. M. H. Wong, K. Sasaki, A. Kuramata, S. Yamakoshi, and M. Higashiwaki, *IEEE Electron Device Lett.* **37**, 212 (2016).
30. M. H. Wong, K. Sasaki, A. Kuramata, S. Yamakoshi, and M. Higashiwaki, *Appl. Phys. Lett.* **106**, 032105 (2015).
31. M. H. Wong, K. Sasaki, A. Kuramata, S. Yamakoshi, and M. Higashiwaki, *Jpn. J. Appl. Phys.* **55**, 1202B9 (2016).
32. J. F. McGlone, Z. Xia, Y. Zhang, C. Joishi, S. Lodha, S. Rajan, S. A. Ringel, and A. R. Arehart, *IEEE Electron Device Lett.* **39**, 1042 (2018).
33. C. Joishi, Z. Xia, J. McGlone, Y. Zhang, A. R. Arehart, S. Ringel, S. Lodha, and S. Rajan, *Appl. Phys. Lett.* **113**, 123501 (2018).
34. M. Singh, M. A. Casbon, M. J. Uren, J. W. Pomeroy, S. Dalcanale, S. Karboyan, P. J. Tasker, M. H. Wong, K. Sasaki, A. Kuramata, S. Yamakoshi, M. Higashiwaki, and M. Kuball, *IEEE Electron Device Lett.* **39**, 1572 (2018).
35. T. Kamimura, K. Sasaki, M. H. Wong, D. Krishnamurthy, A. Kuramata, T. Masui, S. Yamakoshi, and M. Higashiwaki, *Appl. Phys. Lett.* **104**, 192104 (2014).
36. K. Konishi, T. Kamimura, M. H. Wong, K. Sasaki, A. Kuramata, S. Yamakoshi, and M. Higashiwaki, *Phys. Status Solidi B* **253**, 623 (2016).
37. K. Zeng, A. Vaidya, and U. Singisetti, *IEEE Electron Device Lett.* **39**, 1385 (2018).
38. K. Zeng, J. S. Wallace, C. Heimburger, K. Sasaki, A. Kuramata, T. Masui, J. A. Gardella, Jr., and U. Singisetti, *IEEE Electron Device Lett.* **38**, 513 (2017).
39. Y. Lv, X. Zhou, S. Long, X. Song, Y. Wang, S. Liang, Z. He, T. Han, X. Tan, Z. Feng, H. Dong, X. Zhou, Y. Yu, S. Cai, and M. Liu, *IEEE Electron Device Lett.* **40**, 83 (2019).
40. J. K. Mun, K. Cho, W. Chang, H.-W. Jung, and J. Do, *ECS J. Solid State Sci. Technol.* **8**, Q3079 (2019).
41. E. G. Víllora, K. Shimamura, T. Ujiie, and K. Aoki, *Appl. Phys. Lett.* **92**, 202118 (2008).
42. M. Handwerg, R. Mitdank, Z. Galazka, and S. F. Fischer, *Semicond. Sci. Technol.* **30**, 024006 (2015).
43. Z. Guo, A. Verma, X. Wu, F. Sun, A. Hickman, T. Masui, A. Kuramata, M. Higashiwaki, D. Jena, and T. Luo, *Appl. Phys. Lett.* **106**, 111909 (2015).
44. M. D. Santia, N. Tandon, and J. D. Albrecht, *Appl. Phys. Lett.* **107**, 041907 (2015).
45. M. H. Wong, Y. Morikawa, K. Sasaki, A. Kuramata, S. Yamakoshi, and M. Higashiwaki, *Appl. Phys. Lett.* **109**, 193503 (2016).
46. J. W. Pomeroy, C. Middleton, M. Singh, S. Dalcanale, M. J. Uren, M. H. Wong, K. Sasaki, A. Kuramata, S. Yamakoshi, M. Higashiwaki, and M. Kuball, *IEEE Electron Device Lett.* **40**, 189 (2019).
47. S. Martin-Horcajo, A. Wang, M.-F. Romero, M. J. Tadjer, and F. Calle, *IEEE Trans. Electron Devices* **60**, 4105 (2013).

48. J. Pomeroy, M. Bernardoni, A. Sarua, A. Manoi, D. C. Dumka, D. M. Fanning, and M. Kuball, in *Proc. IEEE Compound Semiconductor Integrated Circuit Symposium*, Monterey, CA, USA, 13–16 Oct. 2013, pp. 1–4.
49. H. Sun, J. W. Pomeroy, R. B. Simon, D. Francis, F. Faili, D. J. Twitchen, and M. Kuball, *IEEE Electron Device Lett.* **37**, 621 (2016).
50. B. Chatterjee, A. Jayawardena, E. Heller, D. W. Snyder, S. Dhar, and S. Choi, *Rev. Sci. Instrum.* **89**, 114903 (2018).
51. N. A. Moser, J. P. McCandless, A. Crespo, K. D. Leedy, A. J. Green, E. R. Heller, K. D. Chabak, N. Peixoto, and G. H. Jessen, *Appl. Phys. Lett.* **110**, 143505 (2017).
52. C.-H. Lin, N. Hatta, K. Konishi, S. Watanabe, A. Kuramata, K. Yagi, and M. Higashiwaki, *Appl. Phys. Lett.* **114**, 032103 (2019).
53. A. L. Barry, B. Lehmann, D. Fritsch, and D. Bräunig, *IEEE Trans. Nucl. Sci.* **38**, 1111 (1991).
54. A. Ionascut-Nedelcescu, C. Carlone, A. Houdayer, H. J. von Bardeleben, J.-L. Cantin, and S. Raymond, *IEEE Trans. Nucl. Sci.* **49**, 2733 (2002).
55. D. C. Look, D. C. Reynolds, J. W. Hemsky, J. R. Sizelove, R. L. Jones, and R. J. Molnar, *Phys. Rev. Lett.* **79**, 2273 (1997).
56. J. C. Bourgoin and B. Massarani, *Phys. Rev. B* **14**, 3690 (1976).
57. J. Koike, D. M. Parkin, and T. E. Mitchell, *Appl. Phys. Lett.* **60**, 1450 (1992).
58. S. Ahn, Y.-H. Lin, F. Ren, S. Oh, Y. Jung, G. Yang, J. Kim, M. A. Mastro, J. K. Hite, C. R. Eddy, Jr., and S. J. Pearton, *J. Vac. Sci. Technol. B* **34**, 041213 (2016).
59. J. Yang, F. Ren, S. J. Pearton, G. Yang, J. Kim, and A. Kuramata, *J. Vac. Sci. Technol. B* **35**, 031208 (2017).
60. J. Yang, Z. Chen, F. Ren, S. J. Pearton, G. Yang, J. Kim, J. Lee, E. Flitsiyan, L. Chernyak, and A. Kuramata, *J. Vac. Sci. Technol. B* **36**, 011206 (2018).
61. J. Yang, C. Fares, Y. Guan, F. Ren, S. J. Pearton, J. Bae, J. Kim, and A. Kuramata, *J. Vac. Sci. Technol. B* **36**, 031205 (2018).
62. G. Yang, S. Jang, F. Ren, S. J. Pearton, and J. Kim, *ACS Appl. Mater. Interfaces* **9**, 40471 (2017).
63. M. H. Wong, A. Takeyama, T. Makino, T. Ohshima, K. Sasaki, A. Kuramata, S. Yamakoshi, and M. Higashiwaki, *Appl. Phys. Lett.* **112**, 023503 (2018).
64. Y. Tanaka, S. Onoda, A. Takatsuka, T. Ohshima, and T. Yatsuo, *Mat. Sci. Forum* **645–648**, 941 (2010).
65. K. Zeng, K. Sasaki, A. Kuramata, T. Masui, and U. Singisetti, in *Proc. 74th IEEE Device Research Conference*, Newark, DE, USA, 19–22 Jun. 2016, pp. 1–2.
66. M. H. Wong, Y. Nakata, A. Kuramata, S. Yamakoshi, and M. Higashiwaki, *Appl. Phys. Express* **10**, 041101 (2017).
67. T. Kamimura, Y. Nakata, M. H. Wong, and M. Higashiwaki, *IEEE Electron Device Lett.* **40**, 1064 (2019).
68. K. D. Chabak, J. P. McCandless, N. A. Moser, A. J. Green, K. Mahalingam, A. Crespo, N. Hendricks, B. M. Howe, S. E. Tetlak, K. Leedy, R. C. Fitch, D. Wakimoto, K. Sasaki, A. Kuramata, and G. H. Jessen, *IEEE Electron Device Lett.* **39**, 67 (2018).
69. H. Zhou, M. Si, S. Alghamdi, G. Qiu, L. Yang, and P. D. Ye, *IEEE Electron Device Lett.* **38**, 103 (2017).
70. H. Zhou, K. Maize, G. Qiu, A. Shakouri, and P. D. Ye, *Appl. Phys. Lett.* **111**, 092102 (2017).

71. K. D. Chabak, N. Moser, A. J. Green, D. E. Walker, Jr., S. E. Tetlak, E. Heller, A. Crespo, R. Fitch, J. P. McCandless, K. Leedy, M. Baldini, G. Wagner, Z. Galazka, X. Li, and G. Jessen, *Appl. Phys. Lett.* **109**, 213501 (2016).
72. A. Kyrtsos, M. Matsubara, and E. Bellotti, *Appl. Phys. Lett.* **112**, 032108 (2018).
73. J. L. Lyons, *Semicond. Sci. Technol.* **33**, 05LT02 (2018).
74. T. Gake, Y. Kumagai, and F. Oba, *Phys. Rev. Mater.* **3**, 044603 (2019).
75. K. Sasaki, Q. T. Thieu, D. Wakimoto, Y. Koishikawa, A. Kuramata, and S. Yamakoshi, *Appl. Phys. Express* **10**, 124201 (2017).
76. Z. Hu, K. Nomoto, W. Li, N. Tanen, K. Sasaki, A. Kuramata, T. Nakamura, D. Jena, and H. G. Xing, *IEEE Electron Device Lett.* **39**, 869 (2018).
77. Z. Hu, K. Nomoto, W. Li, Z. Zhang, N. Tanen, Q. T. Thieu, K. Sasaki, A. Kuramata, T. Nakamura, D. Jena, and H. G. Xing, *Appl. Phys. Lett.* **113**, 122103 (2018).
78. Z. Hu, K. Nomoto, W. Li, R. Jinno, T. Nakamura, D. Jena, H. G. Xing, in *Proc. 31st International Symposium on Power Semiconductor Devices and ICs*, Shanghai, China, 19–23 May 2019, pp. 483–486.
79. M. H. Wong, K. Goto. Y. Morikawa, A. Kuramata, S. Yamakoshi, H. Murakami, Y. Kumagai, and M. Higashiwaki, *Appl. Phys. Express* **11**, 064102 (2018).
80. M. H. Wong, K. Goto, H. Murakami, Y. Kumagai, and M. Higashiwaki, *IEEE Electron Device Lett.* **40**, 431 (2019).
81. J. N. Shenoy, J. A. Cooper, Jr., and M. R. Melloch, *IEEE Electron Device Lett.* **18**, 93 (1997).
82. T. Kimoto, *Jpn. J. Appl. Phys.* **54**, 040103 (2015).
83. M. Kanechika, M. Sugimoto, N. Soejima, H. Ueda, O. Ishiguro, M. Kodama, E. Hayashi, K. Itoh, T. Uesugi, and T. Kachi, *Jpn. J. Appl. Phys., Part 2* **46**, L503 (2007).
84. S. Chowdhury, B. L. Swenson, M. H. Wong, and U. K. Mishra, *Semicond. Sci. Technol.* **28**, 074014 (2013).
85. H. Nie, Q. Diduck, B. Alvarez, A. P. Edwards, B. M. Kayes, M. Zhang, G. Ye, T. Prunty, D. Bour, and I. C. Kizilyalli, *IEEE Electron Device Lett.* **35**, 939 (2014).
86. B. J. Baliga, *Fundamentals of Power Semiconductor Devices* (Springer, New York, 2008).
87. M. H. Wong, C.-H. Lin, A. Kuramata, S. Yamakoshi, H. Murakami, Y. Kumagai, and M. Higashiwaki, *Appl. Phys. Lett.* **113**, 102103 (2018).
88. H. Peelaers, J. L. Lyons, J. B. Varley, and C. G. Van de Walle, *APL Mater.* **7**, 022519 (2019).
89. K. Nomura, K. Goto, R. Togashi, H. Murakami, Y. Kumagai, A. Kuramata, S. Yamakoshi, and A. Koukitu, *J. Cryst. Growth* **405**, 19 (2014).
90. H. Murakami, K. Nomura, K. Goto, K. Sasaki, K. Kawara, Q. T. Thieu, R. Togashi, Y. Kumagai, M. Higashiwaki, A. Kuramata, S. Yamakoshi, B. Monemar, and A. Koukitu, *Appl. Phys. Express* **8**, 015503 (2015).
91. K. Goto, K. Konishi, H. Murakami, Y. Kumagai, B. Monemar, M. Higashiwaki, A. Kuramata, and S. Yamakoshi, *Thin Solid Films* **666**, 182 (2018).

High Efficiency AlN/GaN HEMTs for Q-Band Applications with an Improved Thermal Dissipation

Riad Kabouche*, Romain Pecheux, Kathia Harrouche, Etienne Okada, and Farid Medjdoub†

Institut d'Electronique, de Microélectronique et de Nanotechnologie, University Lille, Avenue Poincaré, Villeneuve d'Ascq, France
**riad.kabouche@ed.univ-lille1.fr*
†farid.medjdoub@univ-lille.fr

Joff Derluyn, Stefan Degroote, and Marianne Germain

EpiGaN company, Kempische Steenweg 293, Hasselt, Belgium

Filip Gucmann, Callum Middleton, James W. Pomeroy, and Martin Kuball

H. H. Wills Physics Laboratory, University of Bristol, BS8 1TL, Bristol, United Kingdom

In this paper, we demonstrate Q-band power performance of carbon doped AlN/GaN high electron mobility transistors (HEMTs) using a deep sub-micrometer gate length (120 nm). With a maximum drain current density I_D of 1.5 A/mm associated to a high electron confinement and an extrinsic transconductance g_m of 500 mS/mm, this structure shows excellent electrical characteristics. A maximum oscillation frequency f_{max} of 242 GHz has been observed As a result, a state-of-the-art combination at 40 GHz of output power density (P_{OUT} = 7 W/mm) and power added efficiency (PAE) of 52% up to V_{DS} = 25V has been obtained. The achievement of such outstanding performance is attributed to the reduced thermal resistance (R_{TH}) as compared to the commonly used double heterostructure by means of Raman thermography.

Keywords: High electron mobility transistors (HEMTs); double heterostructure field effect transistor (DHFET); GaN; Raman thermography; output power density (P_{OUT}); power added efficiency (PAE).

1. Introduction

The requirements for the emerging wireless communication systems such as 5G significantly increases the need for compact solid-state high power amplification based on gallium nitride (GaN) material. Achieving both high power-added-efficiency (PAE) and output-power-density (P_{OUT}) in the millimeter-wave range represents currently one of the key goals for the GaN technology. Indeed, higher PAE not only saves electrical power usage but also can reduce the size and cost of high power amplifiers (HPAs), due to the lower amount of heat dissipated. For instance, in space applications, the traveling wave tube amplifiers (TWTA) are still commonly used, because of the high PAE while delivering high P_{OUT}. Despite excellent PAE demonstrated so far with GaN devices up to Ka band [1]-[5], a limited set of data in the Q band (40 GHz) and above are available. As shown by HRL laboratory [6],[7] the double heterostructure field effect transistor (DHFET) using an AlGaN back barrier allowed combining a high electron confinement with high frequency performance together with low trapping effects [8]-[14]. However, AlGaN alloys have the drawback to provide much lower thermal conductivity than the binaries GaN or AlN [15], which in turn causes an increase in the peak channel temperature during operation [16]. Compared to Fe doped GaN buffers, carbon-doped HEMT structures show lower memory effects [17], lower risk of diffusion and the elimination of contamination risks when used in Si CMOS-based foundries. That is why, this structure is widely used for high voltage power applications. In this work, a carbon-doped HEMT structure [19], using a 0.12 µm gate technology has been evaluated and compared to a DHFET [20] with the aim of pushing the bias operation in the millimeter-wave range while maintaining a reduced junction temperature.

2. Device Fabrication

The AlN/GaN heterostructures were grown by metal organic chemical vapor deposition (MOCVD) on 4 in. SiC substrates. The HEMT structure consists of a transition layer to GaN, a 1 µm-thick carbon-doped GaN buffer layer followed by a 150 nm thick un-doped GaN channel, a 4 nm ultrathin AlN barrier layer and a 10-nm-thick in situ Si_3N_4 cap layer (Fig. 1). The in-situ SiN layer is used both as early passivation as well as to prevent strain relaxation [21]-[23]. The channel thickness has been chosen with respect to the trade-off between the electron confinement and the trapping effects. In the second structure called DHFET, the GaN buffer layer is replaced by a 1 µm-thick $Al_{0.08}Ga_{0.92}N$ layer as shown in Figure 1. Room temperature Hall measurements showed high electron sheet concentrations of 1.8×10^{13} and 1.6×10^{13} cm^{-2} with an electron mobility of about 1100 $cm^2V^{-1}s^{-1}$ in the HEMT and DHFET heterostructures, respectively. The device fabrication details can be found in [20]. Ohmic contact resistance (Rc) extracted from linear transmission line method (TLM) was as low as 0.3 Ω.mm for both heterostructures. A 0.12 µm Ni/Au T-gate length was defined by e-beam lithography (see Fig. 1). The gate-source and gate drain spacing were 0.3 and 2 µm, respectively, and the unit device width was 50 µm.

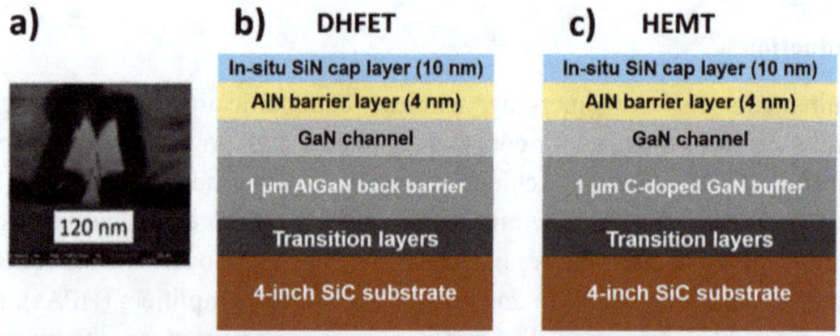

Fig. 1. a) FIB view of the 0.12 µm T-gate and schematic cross section of b) DHFET and c) C-doped HEMT.

3. DC and Small Signal Characterizations

DC characterizations have been performed with a Keysight A2902A parameter analyzer. Fig. 2 shows some typical I_D-V_{DS} characteristics for both structures. The gate source voltage was swept from −6V to +2V with a step of 0.5 V. For the DHFET, a maximum drain current density (I_{Dmax}) of 1.3 A/mm is observed. The devices based on the HEMT structure deliver a higher I_{Dmax} of 1.5 A/mm under the same conditions reflecting the higher carrier concentration.

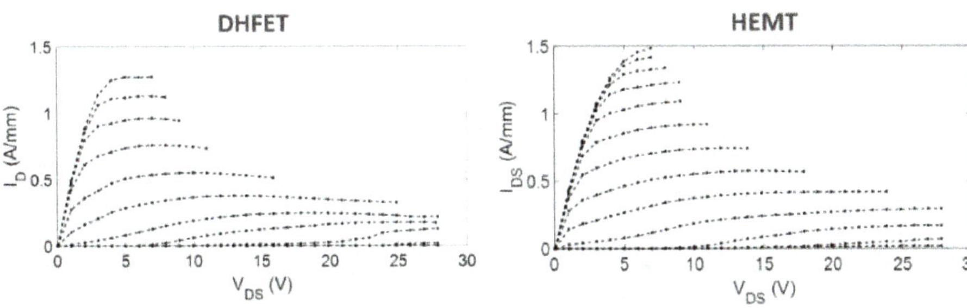

Fig. 2. Output characteristics of a 2×50 μm AlN/GaN DHFET and HEMT.

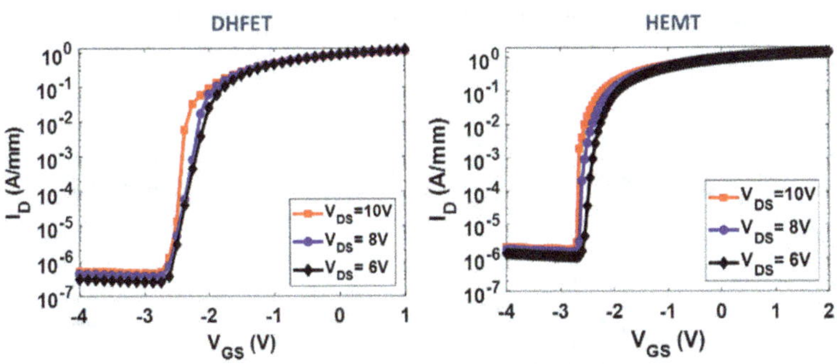

Fig. 3. Transfer characteristics at V_{DS} = 6, 8, 10 V of a 2×50 μm AlN/GaN DHFET and HEMT.

The transfer characteristics of both structures at V_{DS} = 6, 8 and 10 V appear in Fig. 3. Excellent device pinch-off behavior was obtained, which is illustrated by the low off-state leakage current below 10 μA/mm and the high breakdown voltages of 80V and 110V for the HEMT and DHFET structures, respectively. A low threshold voltage shift as a function of V_{DS} is also observed in both cases. This confirms that a good electron confinement can be also obtained without the insertion of an AlGaN back barrier, despite the use of short gate lengths (< 150 nm). In addition, a slightly higher extrinsic transconductance is observed for the HEMT structure with a g_m around 500 mS/mm at V_{DS} = 10V against 470 mS/mm at V_{DS} = 10V for the DHFET. The current-gain and power-gain cut-off frequencies are extracted from the scattering (S) parameters using Rhode and Schwarz ZVA67GHz network analyzer at V_{DS} = 20V. The quiescent current bias point has been varied around the peak transconductance so that the maximum frequency performances are extracted. The DHFET yields a $f_T = 45$ GHz and $f_{max} = 235$ GHz, while slightly better RF performances are achieved for the HEMT with $f_T = 60$ GHz and $f_{max} = 242$ GHz (see Fig. 4). This indicates that we may benefit from a shorter effective gate length with the HEMT structure at such high bias. Figure 5 show the pulsed I-V characteristics performed with a quiescent drain voltage ($V_{Q,DS}$) up to 25V at $V_{GS} = +1$V (all pulse conditions are described on Fig. 5). For both structures, a low gate-lag and a rather similar drain-lag of about 20% are observed.

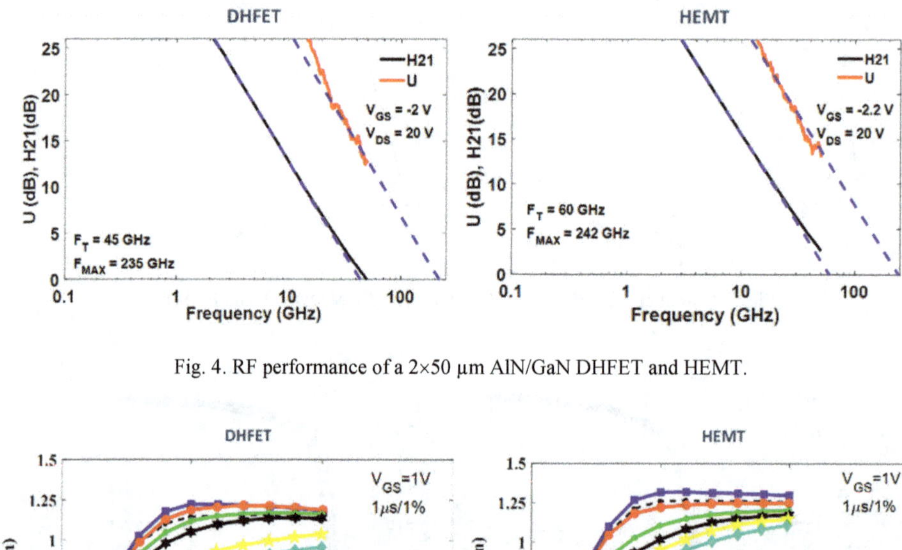

Fig. 4. RF performance of a 2×50 μm AlN/GaN DHFET and HEMT.

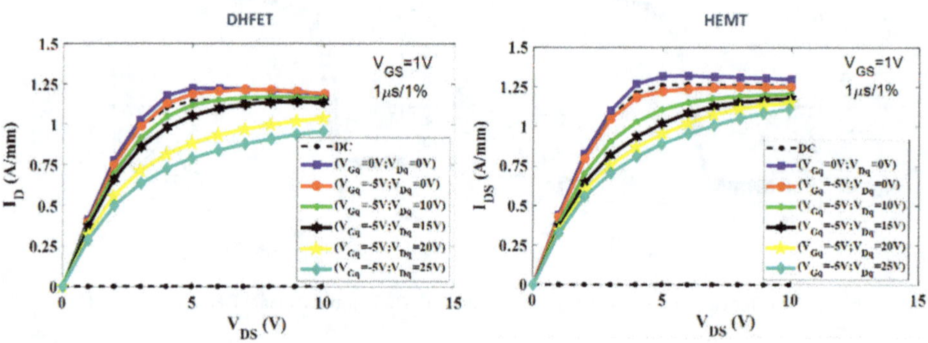

Fig. 5. Pulsed I-V characteristics of a 2×50 μm AlN/GaN DHFET and HEMT.

4. Large Signal Characterization at 40 GHz

Large signal characterizations have been realized on a nonlinear vector network analyzer system (Keysight Network Analyser PNA-X, N5245A-NVNA) capable of on-wafer large signal device characterization up to the Q-band in continuous and pulsed mode. Further details can be found in [24]. On both structures, load-pull measurements have been carried out in pulsed mode (1 μs width and 1% duty cycle). It can be stressed that the same measurement conditions with a drain current density of 100 mA/mm has been used for the HEMTs and DHFETs. Figure 6 shows the pulsed power performance of a 2×25 μm

AlN/GaN HEMT at 40 GHz with V_{DS} = 15V, 20V, and 25 V. The rather low gain is due to the biasing conditions in deep class AB. A high saturated P_{OUT} of 7 W/mm was achieved with a peak PAE of 52% (corresponding to a drain efficiency of 74%) associated to a linear power gain above 8 dB. Furthermore, at V_{DS} = 10 V a PAE as high as 56% (corresponding to a drain efficiency of 75%) combined with an output power density of 1.6 W/mm have been reached. Similarly, the pulsed power performance of a 2×25 μm DHFET structure at 40 GHz is shown in Fig. 7. The output power density evolves linearly as a function of the drain bias (see Fig. 8) up to V_{DS} = 25V for both structures reflecting the high material

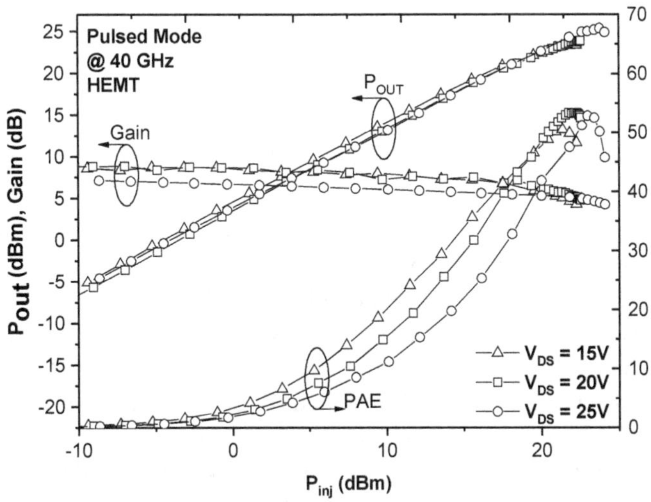

Fig. 6. Pulsed power performance of a 2×25 μm AlN/GaN HEMT at 40 GHz with V_{DS} = 15, 20, 25V.

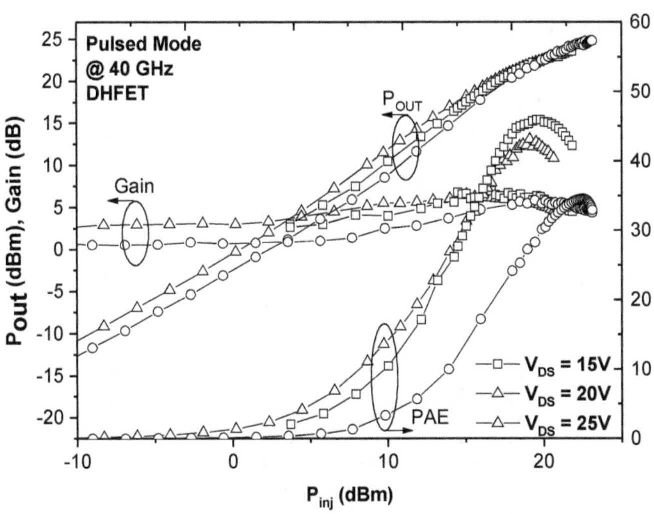

Fig. 7. Pulsed power performance of a 2×25 μm AlN/GaN DHFET at 40 GHz with V_{DS} = 15, 20, 25V.

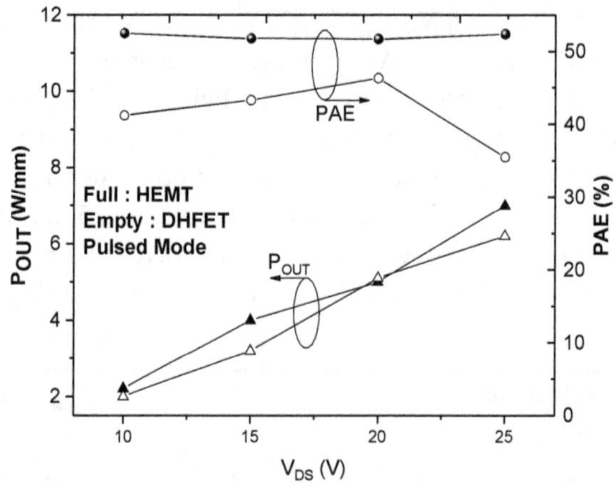

Fig. 8. Pulsed output power density (triangle) and PAE (circle) of both structures versus V_{DS} at 40 GHz.

quality and associated processing. It is worth noting that the HEMT structure shows the ability to deliver a PAE above 50% up to $V_{DS} = 25V$.

For the DHFET, an increasing evolution of the PAE up to $V_{DS} = 20V$ is observed, reaching more than 45% associated to an output power density of 5.1 W/mm at $V_{DS} = 20V$. Nevertheless, at $V_{DS} = 25V$ a strong degradation is observed with a PAE decreasing to 35%. It can be noticed that no device degradation (Fig. 9) is observed in both cases (such as an eventual gate leakage current increase) subsequent to the number of pulsed power sweeps. Thus, the significant drop of the PAE at $V_{DS} = 25V$ is attributed to the self-heating, despite the pulsed mode. In turn, the 1 µs pulse width is large enough to allow the increase of the junction temperature at such a high voltage within the DHFET structure. As can be seen from the benchmark in Fig. 10, the achieved PAE / P_{OUT} combination at 40 GHz compares favorably to the state-of-the-art, especially for power densities above 5 W/mm.

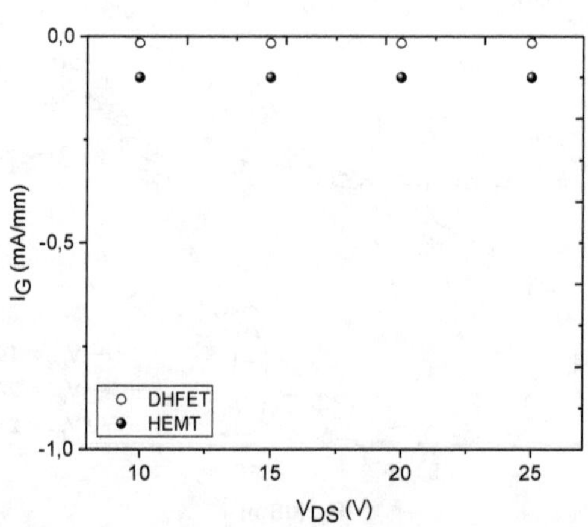

Fig. 9. Evolution of the gate leakage for the HEMTs (filled) and DHFETs (empty) at 40 GHz.

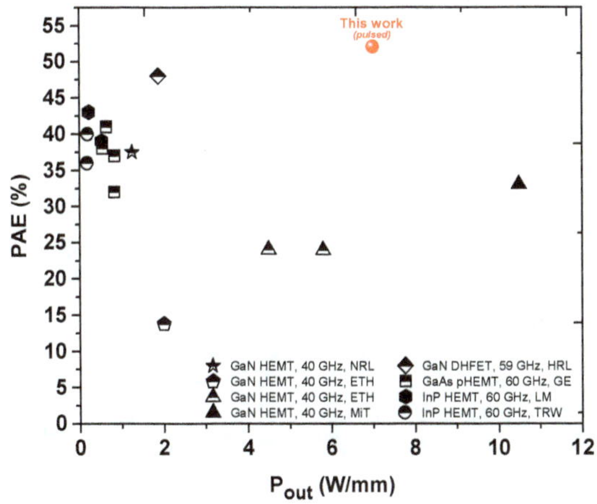

Fig. 10. Benchmark of peak PAE vs output RF power density for Q and V band [12], [15], [25]-[30].

5. Raman Thermography Measurement

The performance enhancement observed on the carbon-doped GaN HEMTs was attributed to a better thermal dissipation. To verify this statement, we assessed the channel temperature of these devices using Raman thermography [31]-[35]. These measurements were carried out on same dimensions 2×50 µm DHFET and HEMT GaN/SiC devices. Thermal shift of E_g Raman peak of TiO_2 nanoparticles (NP) [36] deposited on the device surface, with respect to a reference value at the device pinch-off, was used to determine device surface temperature. This was measured by confocal Renishaw InVia Raman microscope in backscattering configuration under 488 nm laser excitation. A schematic representation of Raman TiO_2 NP temperature measurement location, on top of the

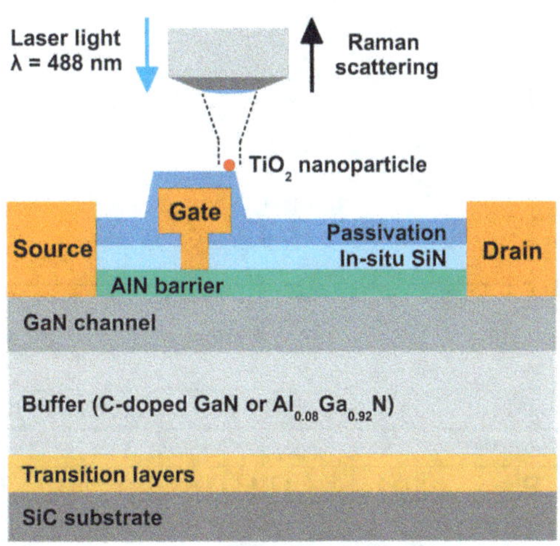

Fig. 11. Schematic representation of Raman TiO_2 NP temperature measurement.

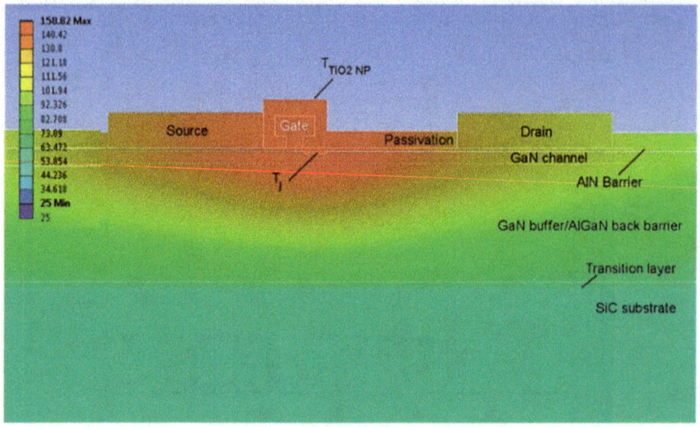

Fig. 12. Device thermal modeling fitted to the experimental data.

passivation at the drain edge of the gate is displayed in Fig. 11. A finite elements steady-state thermal model (Fig. 12) was subsequently calibrated with the measured surface temperature data and used for the junction temperature (T_J) extraction.

The determined peak junction temperatures are shown in Fig. 13. We note that the temperature rise difference between T_{TiO2} NP and T_J was ~3.5% and ~5.5% for the DHFET and the HEMT, respectively as the device surface is close to the highest temperature location in the device. As expected, the DHFET reached much higher T_J

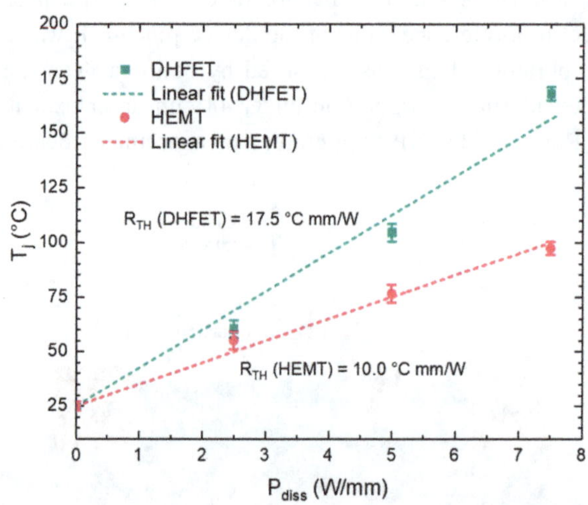

	DHFET				HEMT			
I_D (mA) @ V_{DS} = 10V	0	25	50	75	0	25	50	75
T_J (°C)	25	69.6	114.2	158.8	25	49.9	74.9	99.8

Fig. 13. Peak channel temperature of the HEMTs and DHFETs as a function of the dissipated power density. The table specifies the current used for each extracted junction temperature.

(approaching ~60%) than the HEMT for $P_{diss} > 5$ W/mm. The thermal resistance $R_{TH} = \Delta T/P_{diss}$ was extracted from the slope of fitted trend lines of T_J vs P_{diss}. Overall, the results showed that R_{TH} of the HEMT (10°C mm/W) was ~43% lower as compared to the DHFET (17.5°C mm/W) and comparable to reported values of GaN on SiC devices of similar dimensions [37], confirming the results we obtained from the electrical measurements.

6. Conclusion

This work shows that a careful architecture of buffer layers should be employed in order to perform high performance millimeter-wave GaN devices. The use of higher bias operation ($V_{DS} \geq 20$ V) is possible when using deep sub-micrometer gate lengths only if the thermal resistance induced by the buffer layers is reduced. In particular, it is shown that a thick AlGaN back barrier (DHFET structure) with 8% Al into the buffer results in a significant drop of the PAE at $V_{DS} > 20$V even in pulsed mode. This is illustrated by a much higher thermal resistance (obtained by Raman Thermography measurements) as compared to the HEMT structure. The optimized AlN/GaN HEMT structure enabled to deliver a state-of-the-art combination of PAE (> 50%) and an output power density of 7 W/mm at 40 GHz and $V_{DS} = 25$V.

Acknowledgments

This research was funded and supported by the French RENATECH network and the French Defense Procurement Agency (DGA) under Project EDA-EuGaNiC and Contract FUI-VeGaN.

References

1. P. Saad, H. M. Nemati, M. Thorsell, K. Andersson, and C. Fager, "An inverse class-F GaN HEMT power amplifier with 78% PAE at 3.5 GHz," Eur. Microw. Week 2009, EuMW 2009 Sci. Prog. Qual. Radiofreq. Conf. Proc. - 39th Eur. Microw. Conf. EuMC 2009, no. October, pp. 496–499, 2009, doi : 10.23919/EUMC.2009.5296560
2. J. S. Moon, R. Grabar, D. Brown, I. Alvarado-Rodriguez, D. Wong, A. Schmitz, H. Fung, P. Chen, J.-C. Kang, S. Kim, T. Oh, and C. Mcguire, ">70% power-added-efficiency dual-gate, cascode GaN HEMTs without harmonic tuning," *IEEE Electron Device Lett.*, vol. 37, no. 3, pp. 272–275, 2016, doi : 10.1109/LED.2016.2520488
3. B. Romanczyk, Steven Wienecke , Matthew Guidry, Haoran Li, Elaheh Ahmadi, Xun Zheng, Stacia Keller, and Umesh K. Mishra, "Demonstration of Constant 8 W/mm Power Density at 10, 30, and 94 GHz in State-of-the-Art Millimeter-Wave N-Polar GaN MISHEMTs," IEEE Trans. Electron Devices, vol. 65, no. 1, pp. 45–50, 2018, doi : 10.1109/TED.2017.2770087
4. K. Takagi, S. Takatsuka, Y. Kashiwabara, and S. Teramoto, "Ku-Band AlGaN/GaN-HEMT with over 30 % of PAE," *Engineering*, pp. 457–460, 2009.
5. J. S. Moon, J. S. Moon, D. Wong, M. Hu, P. Hashimoto, M. Antcliffe, C. McGuire, M. Micovic, and P. Willadson, "55% PAE and high power Ka-band GaN HEMTs with linearized transconductance via n+ GaN source contact ledge," *IEEE Electron Device Lett.*, vol. 29, no. 8, pp. 834–837, 2008, doi : 10.1109/LED.2008.2000792

6. A. Crespo, M. M. Bellot, K. D. Chabak, J. K. Gillespie, G. H. Jessen, V. Miller, M. Trejo, G. D. Via, D. E. Walker, Jr., B. W. Winningham, H. E. Smith, T. A. Cooper, X. Gao, and S. Guo, "High-power Ka-band performance of AlInN/GaN HEMT with 9.8-nm-thin barrier," *IEEE Electron Device Lett.*, vol. 31, no. 1, pp. 2–4, 2010, doi : 10.1109/LED.2009.2034875
7. M. Micovic, D. F. Brown, A. Kurdoghlian, D. Santos, B. Grabar, J. Magadia, I. Khalaf, H. Y. Tai, E. Prophet, S. D. Burnham, J. C. Wong, D. Regan, H. H. Fung, and Y. Tang, "GaN DHFETs Having 48% Power Added Efficiency and 57% Drain Efficiency at V-band," *IEEE Electron Device Lett.*, vol. 38, no. 12, pp. 1708–1711, 2017, doi : 10.1109/LED.2017.2763940
8. Y. Tang, K. Shinohara, D. Regan, A. Corrion, D. Brown, J. Wong, A. Schmitz, H. Fung, S. Kim, and M. Micovic., "Ultrahigh-Speed GaN High-Electron-Mobility Transistors With fT/fmax of 454/444 GHz," *IEEE Electron Device Lett.*, vol. 36, no. 6, pp. 549–551, 2015, doi : 10.1109/LED.2015.2421311
9. F. Medjdoub, M. Zegaoui, B. Grimbert, N. Rolland, and P.-A. Rolland, "Effects of AlGaN Back Barrier on AlN/GaN-on-Silicon High-Electron-Mobility Transistors," *Appl. Phys. Express*, vol. 4, no. 12, p. 124101, Nov. 2011, doi : 10.1143/APEX.4.124101
10. D. F. Brown, A. Williams, K. Shinohara, A. Kurdoghlian, I. Milosavljevic, P. Hashimoto, R. Grabar, S. Burnham, C. Butler, P. Willadsen, and M. Micovic, "W-Band Power Performance of AlGaN/GaN DHFETs with Regrown n+ GaN Ohmic Contacts by MBE," pp. 461–464, 2011, doi : 10.1109/IEDM.2011.6131584
11. F. Medjdoub, M. Zegaoui, B. Grimbert, D. Ducatteau, N. Rolland, and P. A. Rolland, "First demonstration of high-power GaN-on-silicon transistors at 40 GHz," *IEEE Electron Device Lett.*, vol. 33, no. 8, pp. 1168–1170, 2012, doi : 10.1109/LED.2012.2198192
12. E. Dogmus, R. Kabouche, A. Linge, E. Okada, M. Zegaoui, and F. Medjdoub, "High power, high PAE Q-band sub-10 barrier thickness AlN/GaN HEMTs ," *Phys. Status Solid (a)*, 2017, doi : 10.1002/pssa.201600797
13. T. Palacios, A. Chakraborty, S. Rajan, C. Poblenz, S. Keller, S. P. DenBaars, J. S. Speck, and U. K. Mishra, "High-power AlGaN/GaN HEMTs for Ka-band applications," *IEEE Electron Device Lett.*, vol. 26, no. 11, pp. 781–783, 2005, doi : 10.1109/LED.2005.857701
14. M. Micovic, P. Hashimoto, M. Hu, I. Milosavljevic, J. Duvall, P. J. Willadsen, W.- S. Wong, A. M. Conway, A. Kurdoghlian, P. W. Deelman, J.-S. Moon, A. Schmitz, and M. J. Delaney, "GaN Double Heterojunction Field Effect Transistor For Microwave and Millimeterwave Power Applications," *IEEE International Electron Devices Meeting* , pp. 807–810. 2004, doi : 10.1109/IEDM.2004.1419298
15. F. Medjdoub, B. Grimbert, D. Ducatteau, and N. Rolland, "Record Combination of Power-Gain Cut-Off Frequency and Three-Terminal Breakdown Voltage for GaN-on-Silicon Devices," *Appl. Phys. Express*, vol. 6, no. 4, p. 44001, Apr. 2013.
16. W. Liu and A. A. Balandin, "Thermal conduction in AlxGa1-xN alloys and thin films," *J. Appl. Phys.*, vol. 97, no. 7, pp. 1–6, 2005.
17. P. Moens, P. Vanmeerbeek, A. Banerjee , J. Guo , C. Liu , P. Coppens, A. Salih , M. Tack "On the Impact of Carbon-Doping on the Dynamic Ron and Off-state Leakage Current of 650V GaN Power Devices," *IEEE 27th International Symposium on Power Semiconductor Devices & IC's*, pp. 37–40, 2015, doi : 10.1109/ISPSD.2015.7123383
18. M. J. Uren, J. Moreke, and M. Kuball, "Buffer design to minimize current collapse in GaN/AlGaN HFETs," *IEEE Trans. Electron Devices*, vol. 59, no. 12, pp. 3327–3333, 2012, doi : 10.1109/TED.2012.2216535

19. M. Kuball, J. M. Hayes, M. J. Uren, T. Martin, J. C. H. Birbeck, R. S. Balmer, and B. T. Hughes, "Measurement of Temperature in Active High-Power AlGaN/GaN HFETs Using Raman Spectroscopy," *IEEE Electron Device Lett.*, vol. 23, no. 1, pp. 10–12, 2002, doi : 10.1109/55.974795
20. R. Pecheux, R. Kabouche, E. Okada, M. Zegaoui, and F. Medjdoub, "C-doped AlN/GaN HEMTs for High efficiency mmW applications," *International Workshop on Integrated Nonlinear Microwave and Millimetre-wave Circuits (INMMIC)* pp. 31–33, 2018. doi : 10.1109/INMMIC.2018.8430021
21. R. Pecheux, R. Kabouche, E. Dogmus, A. Linge, E. Okada, M. Zegaoui, and F. Medjdoub "Importance of buffer configuration in GaN HEMTs for high microwave performance and robustness," *ESSDERC*, pp. 4–7, 2017, doi : 10.1109/ESSDERC.2017.8066633
22. J. Derluyn, S. Boeykens, K. Cheng, R. Vandersmissen, J. Das, W. Ruythooren, S. Degroote, M. R. Leys, M. Germain, and G. Borghs "Improvement of AlGaN/GaN high electron mobility transistor structures by *in situ* deposition of a Si_3N_4 surface layer," *J. Appl. Phys.*, vol. 98, no. 5, p. 54501, 2005, doi : https://doi.org/10.1063/1.2008388
23. D. Marcon, M. V. Hove, D. Visalli, J. Derluyn, J. Das, F. Medjdoub, S. Degroote, M. Leys, K. Cheng and R. Mertens, "Excellent Stability of GaN-on-Si High Electron Mobility Transistors with 5 μm Gate–Drain Spacing Tested in Off-State at a Record Drain Voltage of 200 V and 200 °C," *Jpn. J. Appl. Phys.*, vol. 49, no. 4, p. 04DF07, Apr. 2010.
24. D. Marcon, F. Medjdoub, D. Visalli, M. V. Hove, J. Derluyn, J. Das, S. Degroote, M. Leys, K. Cheng, S. Decoutere, R. Mertens, M. Germain, and G. Borghs "High temperature on- and off-state stress of GaN-on-Si HEMTs with in-situ Si_3N_4 cap layer," *IEEE Int. Reliab. Phys. Symp. Proc.*, pp. 146–151, 2010, doi : 10.1109/IRPS.2010.5488836
25. R. Kabouche, E. Okada, E. Dogmus, A. Linge, M. Zegaoui, and F. Medjdoub, "Power Measurement Setup for On-Wafer Large Signal Characterization Up to Q-Band," IEEE Microw. Wirel. Components Lett., vol. 27, no. 4, pp. 419–421, 2017, doi : 10.1109/LMWC.2017.2678424
26. B. P. Downey, D. J. Meyer, D. S. Katzer, and J. A. Roussos, "SiN_x/InAlN/AlN/GaN MIS-HEMTs With 10.8 THz ·V Johnson Figure of Merit," *IEEE Electron Device Lett.*, vol. 35, no. 5, pp. 527–529, 2014, doi : 10.1109/LED.2014.2313023
27. D. Marti, S. Tirelli, A. R. Alt, J. Roberts, and C. R. Bolognesi, "150-GHz cutoff frequencies and 2-W/mm output power at 40 GHz in a millimeter-wave AlGaN/GaN HEMT technology on silicon," *IEEE Electron Device Lett.*, vol. 33, no. 10, pp. 1372–1374, 2012, doi : 10.1109/LED.2012.2204855
28. S. Tirelli, L. Lugani, D. Marti, J. F. Carlin, N. Grandjean, and C. R. Bolognesi, "AlInN-based HEMTs for large-signal operation at 40 GHz," *IEEE Trans. Electron Devices*, vol. 60, no. 10, pp. 3091–3098, 2013, doi : 10.1109/TED.2013.2262136
29. M. Kao, P. M. Smith, P. Ho,P.-C. Chao,K.H.G. Duh, A.A. Jabra and J.M. Ballingall "Very High Power-Added Efficiency and Low-Noise 0.15-μm Gate-Length Pseudomorphic HEMT's" *IEEE Electron Device Lett.*, vol. I, no. 12, pp. 580–582, 1989, doi : 10.1109/55.43146
30. W. M. T. Kong, Sujane C. Wang, Pane-Chane Chao, Der-Wei Tu, Kuichul Hwang, O. S. A. Tang, Shih-Ming Liu, Pin Ho, Kirby Nichols, and John Heaton "Very high efficiency V-band power InP HEMT MMICs," *IEEE Electron Device Lett.*, vol. 21, no. 11, pp. 521–523, 2000, doi : 10.1109/55.877196
31. R. Grundbacher, R. Lai, M. Nishimoto, T. P. Chin, Y. C. Chen, M. Barsky, T. Block, and D. Streit "Pseudomorphic InP HEMT's with dry-etched source vias having 190 mW output power

and 40% PAE at V-band," *IEEE Electron Device Lett.*, vol. 20, no. 10, pp. 517–519, 1999, doi : 10.1109/55.791928

32. M. Kuball, and J. W. Pomeroy, "A Review of Raman Thermography for Electronic and Opto-Electronic Device Measurement With Submicron Spatial and Nanosecond Temporal Resolution," *IEEE Transactions on Device and Materials Reliability,* vol. 16, no. 4, pp. 667–684, 2016, doi : 10.1109/TDMR.2016.2617458
33. D. M. Risbud, K. Pedrotti, M. Power, J. W. Pomeroy, and M. Kuball, "Thermal characterization of high voltage GaN-on-Si Schottky Barrier Diodes (SBD) for designing an on-chip thermal shutdown circuit for a power HEMT," *WiPDA*, pp. 156–161, 2015, doi : 10.1109/WiPDA.2015.7369293
34. J. Pomeroy, M. Bernardonia, A. Saruaa, A. Manoia, D.C. Dumkab, D.M. Fanningb and M. Kuballa "Achieving the Best Thermal Performance for GaN-on-Diamond," *2013 IEEE Compound Semiconductor Integrated Circuit Symposium (CSICS)*, pp. 1–4, 2013, doi : 10.1109/CSICS.2013.6659210
35. M. Power, James W. Pomeroy, Yohei Otoki, Takeshi Tanaka, Jiro Wada, Masaaki Kuzuhara, Wolfgang Jantz, Andrew Souzis, Martin Kuball "Measuring the thermal conductivity of the GaN buffer layer in AlGaN/GaN HEMTs," *Phys. Status Solidi*, 2015, vol. 212, no. 8, pp. 1742–1745, doi : 10.1002/pssa.201431788
36. Y.L. Du, Y. Deng, M .S. Zhang, " Variable-temperature Raman scattering study on anatase titanium dioxide nanocrystal, " *Journal of Physics and Chemistry of Solids,* Volume 67, Issue 11, 2006, Pages 2405-2408, ISSN 0022-3697
37. S Garcia, I Íñiguez-de-la-Torre, J Mateos, T González and S Pérez "Impact of substrate and thermal boundary resistance on the performance of AlGaN/GaN HEMTs analyzed by means of electro-thermal Monte Carlo simulations" *Semicond. Sci. Technol.* 2016 31 065005 doi : 10.1088/0268-1242/31/6/065005
38. J. Joh, J.A. del Alamo, U. Chowdhury, T. Chou, H. Tserng and J. L. Jimenez, "Measurement of Channel Temperature in GaN High-Electron Mobility Transistor, " *IEEE Transactions on Electron Devices,* Volume 56, Issue 12, 2009, Pages 2895-2901, doi : 10.1109/TED.2009.2032614

Recent Progress in Gallium Oxide and Diamond Based High Power and High-Frequency Electronics

Md Nazmul Hasan, Edward Swinnich, and Jung-Hun Seo*

Department of Materials Design and Innovation, University at Buffalo, Buffalo, NY 14260, USA
**junghuns@buffalo.edu*

In recent years, the emergence of the ultrawide-bandgap (UWBG) semiconductor materials that have an extremely large bandgap, exceeding 5eV including AlGaN/AlN, diamond, β-Ga_2O_3, and cubic BN, provides a new opportunity in myriad applications in electronic, optoelectronic and photonics with superior performance matrix than conventional WBG materials. In this review paper, we will focus on high power and high frequency devices based on two most promising UWBG semiconductors, β-Ga_2O_3 and diamond among various UWBG semiconductor devices. These two UWBG semiconductors have gained substantial attention in recent years due to breakthroughs in their growth technique as well as various device engineering efforts. Therefore, we will review recent advances in high power and high frequency devices based on β-Ga_2O_3 and diamond in terms of device performance metrics such as breakdown voltage, power gain, cut off frequency and maximum operating frequency.

Keywords: β-Ga_2O_3; diamond; high-power transistors; high-frequency transistors.

1. Introduction

The term wide bandgap (WBG) semiconductor is used to describe materials that have relatively large bandgaps of greater than 3eV, which is opposed to conventional narrow bandgap semiconductors, such as Si, Ge, and some III-V materials such as GaAs and InP. Since the 1990s, a series of technical breakthroughs were made that resulted in a surge of development in electronics and optoelectronics, such as high power radio-frequency (RF) transistors, blue light emitting diodes (LEDs), and UV photodetectors [1]. In recent years, the emergence of ultrawide-bandgap (UWBG) semiconductor materials, with bandgaps exceeding 5eV, including AlGaN/AlN, diamond, β-Ga_2O_3, and cubic BN, provides a new opportunity in myriad applications in electronics, optoelectronics, and photonics, with superior performance matrix than conventional WBG materials [2, 3]. For example, as most of the figures-of-merit (FOM) for device performance such as Johnson's FOM and Baliga FOM, increase in a highly non-linear manner as bandgap increases, these UWBG materials have the potential for far superior performance in high-frequency and high-power applications [4].

In this review paper, we will focus on high power and high frequency devices based on the two most promising UWBG semiconductors, β-Ga_2O_3 and diamond. These two UWBG semiconductors have gained substantial attention in recent years due to their excellent electrical properties among other UWBG semiconductors as displayed in Fig. 1. β-Ga_2O_3 has a distinct advantage, an ability to tune its bandgap by alloying with indium and aluminum, allowing for bandgap tuning from 3.6eV [5] to 8.6eV [6]. Thus, β-Ga_2O_3 possess an important position by possibly replacing current GaN technology markets. With a decent mobility and a high breakdown field, β-Ga_2O_3 can be an ideal material candidate for future high power and high frequency devices. On the other hand, diamond has been known as a promising UWBG semiconductor for a decade, as it shows excellent mobility and electric field (the highest in the UWBG semiconductor family) and, high thermal conductivity which are the most important parameters for the next generation high performance electronics [7]. There are still many difficulties to overcome in order to use diamond for commercial purposes, such as small substrate size and difficulty of growth, but the status of diamond as a next generation UWBG semiconductor remains unchanged.

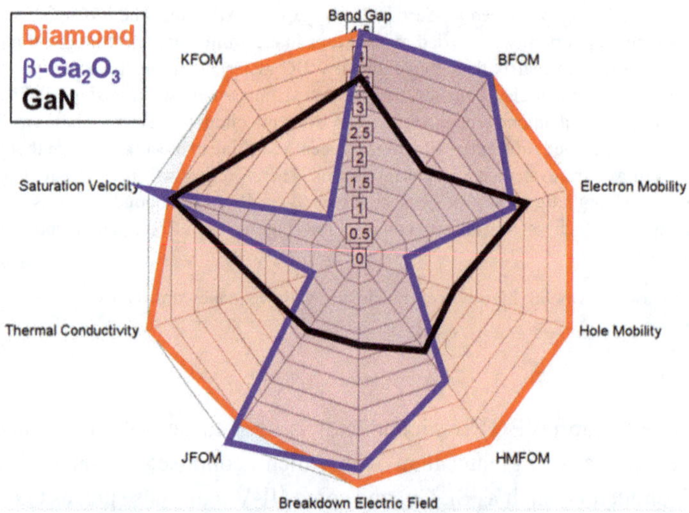

Material parameters and FOM	Si	GaN	Diamond	β-Ga_2O_3
Bandgap (eV)	1.1	3.4	5.5	4.9
Electron Mobility μ ($cm^2 Vs^{-1}$)	1400	1200	2000	300
Hole Mobility μ ($cm^2 Vs^{-1}$)	800	150	2800	0.001
Breakdown Field E_c (MV/cm)	0.3	3.3	10	8
Thermal conductivity σth (W/°C)	1.5	2.1	10	0.27
Saturation Velocity vs (cm/s) x 10^7	0.9	1.5	1	1.5
JFOM	5	83	167	200
KFOM	3	5	20	1
BFOM	7	6220	176282	24615
HMFOM	2	18	72	22

Fig. 1. A performance comparison of diamond and β-Ga_2O_3 vs. GaN.

2. β-Ga_2O_3 Electronics

A single crystal β-phase gallium oxide (β-Ga_2O_3) metal-semiconductor field effect transistor (MESFET) was first demonstrated by *Higashiwaki et al.* in 2012 [3]. A circular MESFET was fabricated on a Sn-doped Ga_2O_3 layer, grown on a semi-insulating β-Ga_2O_3 substrate by molecular beam epitaxy (MBE). The device showed an ideal transistor behavior with a maximum drain current of 15 mA/mm. The author also reported on β-Ga_2O_3 MOSFET which had Sn doped channel layer, Si ion implanted source-drain regions and a 20 nm Al_2O_3 gate dielectric material [8]. Figure 2(a) shows the schematic of the device cross-section of the MOSFET with a gate length of 2 μm provided extremely low off-state drain leakage current. The very low gate leakage current attributed to maximizing the drain current and also improved the on/off ratio of ten orders of magnitude. Figure 2(b) plots off-state breakdown characteristics at room temperature (RT) and at 250 °C under the pinch-off condition. Both off-state gate and drain current stayed below the detection limit before the device broke down at 404 V for RT and 378 V when the temperature was 250 °C. The breakdown scheme upto 250 °C indicated that the device performance can be observed without obvious permanent degradation.

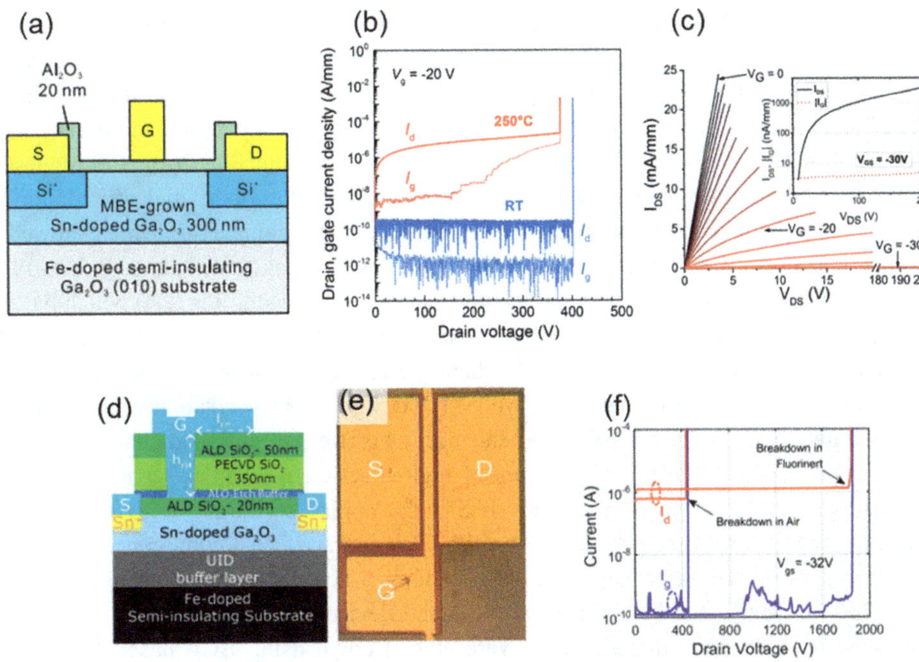

Fig. 2. (a) Schematic cross section of MOSFET from similar structure of very first demonstrated MESFET. (b) Breakdown voltage measurement of β-Ga_2O_3 MOSFET at RT and 250°C. (c) MOSFET family output plots and the inset I-V curves for drain current in pinch off condition. (d) and (e) Field plated β-Ga_2O_3 MOSFET device architecture cross section and optical image and (f) record 1.85 kV three terminal breakdown measurement (blue) in Fluorinert and compared with in air (red).

Green et al. reported on Sn doped β-Ga_2O_3 MOSFET with a gate to drain field strength of 3.8 MV/cm. This is the highest reported value so far and also one half of the theoretical limit [9]. It was achieved by applying a gate-drain voltage of 230V across a gate to drain spacing of 0.6 μm. The device was fabricated on Mg-doped (100) single crystal β-Ga_2O_3 substrate on which 200 nm Sn-doped β-Ga_2O_3 was grown homoepitaxially by metal organic vapor phase epitaxy (MOVPE). The carrier concentration was recorded 1.7×10^{18} cm^{-3} (as grown) which was higer than previously reported carrier concentration of 7×10^{17} cm^{-3} by MBE method [8]. Device mesa isolation was performed using BCl_3 gas in ICP (inductively coupled plasma)/RIE (reactive-ion etch) etching. The typical MOSFET family output curves are shown in Fig. 2(c). To avoid thermally induced degradation, the output curves were measured at limited power by applying a gate bias from 0 V to -30 V and a gate step of -2 V. The inset figure shows the gate and drain leakage current under the pinch-off condition.

Very recent, *Zhang et al.* reported a record high breakdown voltage of 1.85 kV for a field plated MOSFET measured in fluorinert, a liquid dielectric [10]. The device was fabricated on a Sn-doped epi-layer grown by MBE on a semi-insulating Fe-doped β-Ga_2O_3 substrate. The selective Sn+ spin on glass (SOG) doping was established under the S/D region to achieve low ohmic contact resistance, shown in Fig. 2(d). Peak electric field in the lateral MOSFET is typically near the drain-side edge of the gate in the gate-drain region. The higher breakdown voltage was obtained by depositing an oxide field plate that reduces the parasitic paths for catastrophic breakdown. The optimal length of depositing a thick, composite field oxide plate was halfway between the 20 μm gate and drain (L_{gd} 20 μm). The top view of final device architecture is shown in Fig. 2(e). The highest catastrophic breakdown >1800 V was recorded inside a fluorinert dielectric liquid. Another voltage breakdown test was carried out in the air for comparison which showed that the breakdown was extrinsic to the channel and device breakdown voltage was 440V only. The high breakdown voltage results have been summarized in Fig. 2(f) which pave the way for high voltage-power electronic applications based on β-Ga_2O_3 devices.

While there are several reports on reducing contact resistance and increasing the breakdown voltage using conventional MOSFET design, *S. Krishnamoorthy et al.* first demonstrated the delta-doped method to enhance the carrier mobility and charge density in the channel [11]. Thin (12 nm) layers of high sheet charge density channels were created using a shutter pulsing scheme of Si delta doping technique to avoid source oxidation and to obtain the flat doping profile. The average flat doping concentrations in the range of ~$4\text{x}10^{19}$ cm^{-3} led to a high conductivity contact and a high drain current (I_{Dmax} 236 mA/mm). A schematic of the designed device has been shown in Fig. 3(a) where the top two delta doped layers were etched down using BCl_3 based ICP/RIE chemistry to realize a channel with one delta-doped sheet. A gate dielectric (Al_2O_3) was deposited using the atomic layer deposition (ALD) method followed by deposition of Pt/Au/Ni gate metal stack. The transistor had a gate length of 1 μm and a source-drain spacing of 20 μm. Figure 3(b) shows the three terminal off-state breakdown behavior of the delta-doped FET devices. A breakdown voltage of 35 V was measured when the

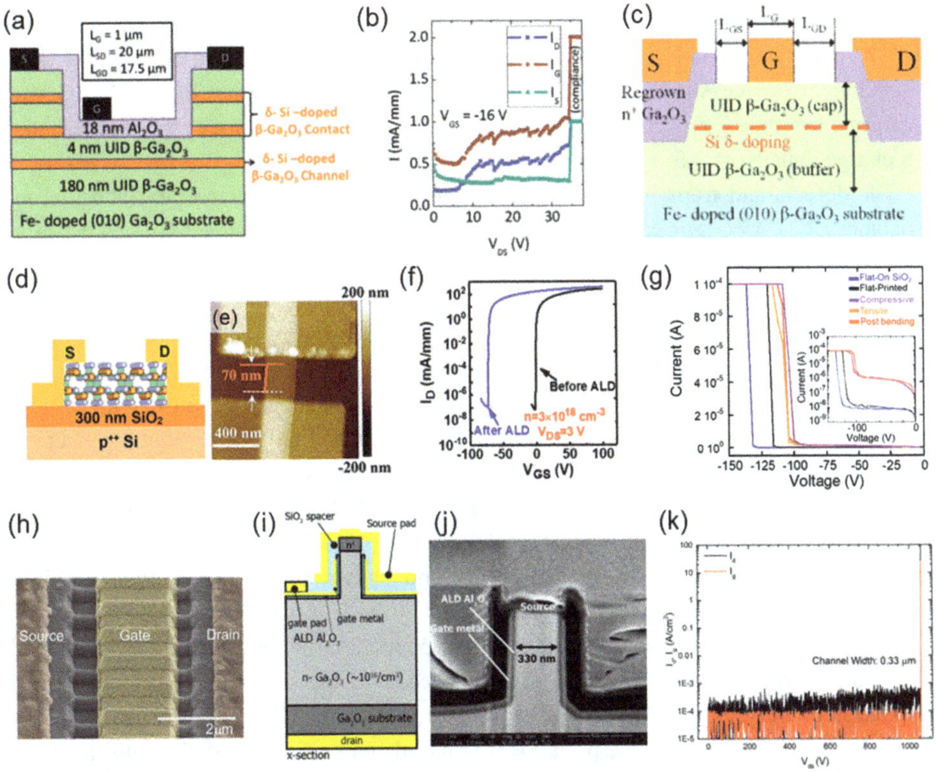

Fig. 3. (a) Schematic cross section of Si delta doped β-Ga_2O_3 MOSFET using shutter short pulsing technique. (b) Three terminal off state breakdown of Si delta doped β-Ga_2O_3 MOSFET. (c) Schematic cross section of delta doped β-Ga_2O_3 field effect transistors with regrown ohmic contacts. (d) Schematic of heavily doped β-Ga_2O_3 nanomembranes (NMs) based back gated transistors. (e) Atomic force microscopy (AFM) topography of the device top view. (f) Drain current after surface passivation of depositing 15nm Al_2O_3 using atomic layer deposition (ALD). (g) Breakdown voltage of the flexible SBD in flat and under bending conditions (inset shows logarithm scale I-V) (h) (i) and (j) shows the SEM images of wrap gate finFET, schematic cross-section of fin shaped vertical transistor and its SEM image. (k) First high voltage breakdown (>1 kV) of vertical finFET of the β-Ga_2O_3.

applied V_{GS} was -16V. The delta doping approach in β-Ga_2O_3 based FETs achieved higher electron mobility of 80 $cm^2 \cdot V^{-1} \cdot s^{-1}$ at the high charge density of 2.4 x 10^{14} cm^{-2}.

β-Ga_2O_3 delta doped FET structure with gate recess could produce low electron mobility, surface damage and remote impurity scattering associated with the etching process. To avoid gate recess related damages, *Xia et al.* presented a β-Ga_2O_3 based delta doped MESFET using regrown ohmic contacts [12]. A Si delta-doped region formed a 2-dimensional electron gas (2DEG) with higher sheet density channel. The device fabrication started with depositing 500 nm thick SiO_2 using PECVD followed by removing the S/D contact regions using CF_4 based ICP/RIE etching. Then a BCl_3 based ICP/RIE etch was performed to obtain a β-Ga_2O_3 trench down to ~40 nm for the regrown process. A patterned regrowth of heavily doped n-type β-Ga_2O_3 with a doping concentration of 2 x 10^{20} cm^{-3} was regrown in the selective patterned regions by Si shutter pulsing technique to obtain very low contact resistance for source and drain contacts. A schematic of device structure is shown in Fig. 3(c). The presence of the 2DEG was confirmed from the CV profile and a peak electron mobility of 95 $cm^2V^{-1}s^{-1}$ was measured at a charge density of 3.5×10^{12} cm^{-2}.

The defects and traps in a FET can change the threshold voltage (V_T) and the on-resistance (R_{on}). *McGlone et al.* reported on the trapping effects in Si delta-doped β-Ga_2O_3 MESFETs [13]. The threshold voltage instability was attributed to two distinct trap levels in the upper part of the bandgap. The empty trap level at E_C-0.7 eV and E_C-0.77 eV was shown to be the source of positive threshold voltage shift where the electrons were being trapped. The presence of trap levels also suggested that Fe doped substrate could possibly be causing Fe impurity trap close to the conduction band edge. To further investigate the type and the source of these traps, *Joishi et al.* published an article about the effect of an unintentionally doped (UID) buffer layer on the device performance of Si delta doped β-Ga_2O_3 FETs [14]. The effect of Fe substrate on the 2DEG was observed by varing the buffer layer thickness. It was found that the thicker (600 nm) buffer layer confirmed a negligible Fe concentration near the 2DEG and shown to have improved mobility resulting reduce DC-RF dispersion.

Similar to the other two-dimensional semiconductor materials like graphene and MoS_2, the larger lattice constant of β-Ga_2O_3 leads to the opportunity of mechanical exfoliation of free-standing nanomembranes (NMs) from the bulk substrate [15]. *Zhou et al.* reported a high performance β-Ga_2O_3 NM based field effect transistor with the drain current of 600 mA/mm and 450 mA/mm for depletion mode and enhancement mode operation respectively [16]. The enhancement mode FET was easily fabricated by controlling the thickness of β-Ga_2O_3 NMs which modulated the transistor threshold voltage (V_T). The device fabrication started after transfer printing the NMs onto a back gated Si substrate. The drain-source metals were deposited on top of the NMs after patterning the source drain regions by electron beam lithography (EBL) technique. The same research group published another article on β-Ga_2O_3 NM based FET with a record drain current exceeding 1.5 A/mm by using a heavily doped (8×10^{19} cm^{-3}) β-Ga_2O_3 NM acting as a channel [17]. The schematic of β-Ga_2O_3 NM based back-gated transistors and the atomic force microscopy (AFM) topography of the device are shown in Fig. 3(d-e). The depletion effect of the unpassivated transistor surface causes the shift of threshold voltage, V_T for different thickness of NMs. To passivate the top surface, 15 nm of Al_2O_3 was deposited by ALD. The negative shift of V_T more than 70 V shown in Fig. 3(f) confirms the existence of surface depletion on the unpassivated surface. *Kim et al.* also demonstrated a quasi 2D heterostructure based MISFET using mechanically exfoliated and transfer printed stacks of β-Ga_2O_3 nanobelts and h-BN flakes [18]. Apart from transistors based on β-Ga_2O_3 NMs on a rigid substrate such as Si, β-Ga_2O_3 NMs can also be transferred on a flexible plastic substrate to achieve high power switching behavior under different strain conditions. *Swinnich et al.* first demonstrated Schottky barrier diodes (SBDs) on a plastic substrate utilizing β-Ga_2O_3 NMs [19]. The first demonstration of the high breakdown voltage of β-Ga_2O_3 SDBs measured under flat and different bending conditions shows the potential of β-Ga_2O_3 NMs for high power flexible electronics.

A normally "off" type transistor is preferred to diminish off-state power dissipation and also for safe high voltage operation. Conventional MOSFETs require high doping concentration to achieve high current density which causes negative V_T. Nonplanar fin-shaped channels offer enhanced control of the channel charge by depleting it from the side walls in addistion to the top surface without sacrificing the doping level. This leads to the positive shift of V_T. *Chabak et al.* reported a finFET with a centered two fingers gate layout [20]. Each gate finger had 48 wrapping fins. The fin array was extended over 3 μm of the source to drain where the total source to drain distance was 4 μm. An SEM image of wrap gated fin channels has been depictred in Fig. 3(h).

A vertical fin-shaped enhancement mode (E-mode) MISFET shown in Fig. 3(i-j) was first reported by *Hu et al.* [21]. This transistor offered a breakdown voltage over 1 kV which did not require a field plate as previously stated as for β-Ga_2O_3 MOSFET. The top 50 nm of the drift layer was heavily doped by Si ion implantation to reduce the contact resistance. An ICP system was used to etch the vertical fin channels. This transistor offered E-mode, off state operation. The three-terminal off state for this vertical MISFET is shown in Fig. 3(k) at V_{GS}=0 V.

The high electron mobility of 100 cm^2/Vs as measured paves the way for β-Ga_2O_3 transistors to operate as RF amplifiers for power device applications [22]. *Green et al.* first reported the RF operation of β-Ga_2O_3 MOSFETs [23]. The device had a heavily doped capping layer to reduce the contact resistance in the drain-source region. A 0.7 μm gate recess was typically performed to the gated region to partially remove and deplete the channel for normally off operation. The typical small signal RF gain performance characteristics such as current gain (h_{21}), maximum stable gain/maximum available gain (MSG/MAG) and unilateral power gain (U-gain) have been plotted in Fig. 4(c). A cut off frequency (f_T) was 2.7 GHz and a maximum oscillating frequency (f_{MAX}) of 12.9 GHz was recorded at V_{DS}=40V and V_{GS} at -3.5 V.

Chabak et al. provided two approaches to device architecture shown in Fig. 4(a-b) for RF application [24]. In the first β-Ga_2O_3 FET design, a highly Si doped layer was deposited for capping to reduce the contact resistance in the source-drain region. A gate recess process was implemented to allow gate and channel device scaling that led the f_T and f_{MAX} of 3 GHz and 13 GHz respectively. The second approach was just using a highly doped thin channel and a T shaped gate patterned by EBL. Figure 4(d) shows the small signal RF gain performance obtained from the T gated and highly doped channel device. The f_T/f_{MAX} were boosted up to 5.1/17.1 GHz respectively.

A comparison study of continuous wave (CW) and pulsed large signal operation in β-Ga_2O_3 MOSFETs was demonstrated by *Singh et al.* [25]. The pulsed IV helped to prevent device heating. An Fe-doped semi-insulating β-Ga_2O_3 (010) substrate with 1.2 μm UID layer grown by MBE was used to start the fabrication. Figure 4(e) shows the schematic cross-section of the device. The S/D contacts were deposited followed by a heavily doped

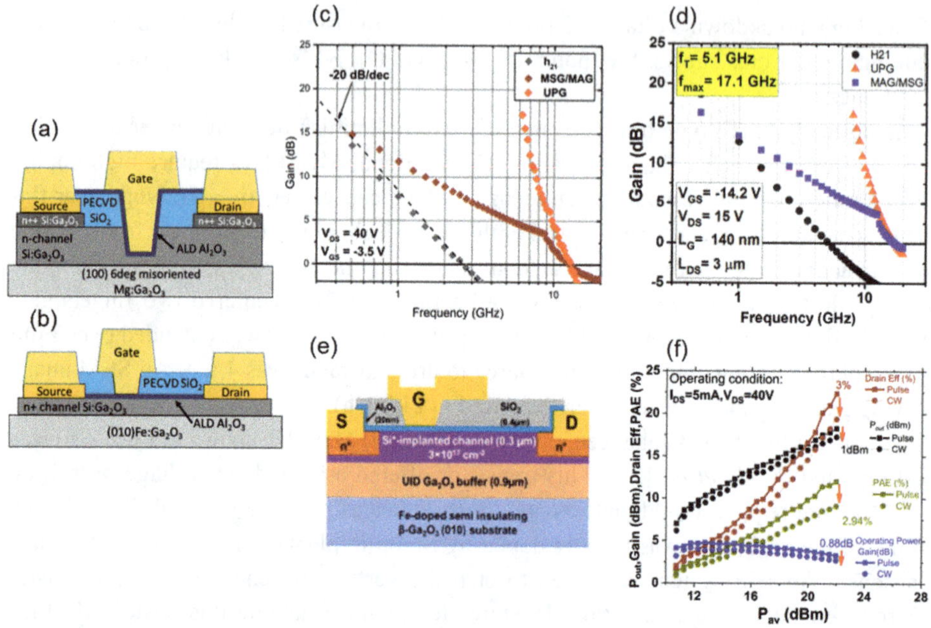

Fig. 4. Schematic cross-section of β-Ga_2O_3 RF MOSFET (a) gate recessed (b) heavily doped thin scaling channel. (c-d) Typical small signal RF gain performance. (e) Device architecture of a field plated MOSFET. (f) Comparison of CW and pulsed signal measurements.

layer by Si ion implantation. A 20 nm Al_2O_3 and a 0.4 μm SiO_2 gate dielectric combination was deposited and the gate recess was performed through the SiO_2 layer. The drain current improved to 150 mA/mm for the pulsed operation while the drain current was 58 mA/mm for the DC sweep. Figure 4(f) shows the measurements of output power, drain efficiency and power added efficiency (PAE) between CW and pulsed signal operation. A record PAE of 12% was achieved at 1 GHz for a 2 μm gate length.

3. Diamond Electronics

Figure 5(a) shows several material systems frequently used in the fabrication of devices for power electronics. Diamond is shown to have performance rivaling or surpassing mature technologies in SiC and AlGaN/GaN, especially in regard to the breakdown voltage and the breakdown electric field. Figure 5(b) shows the device structure of a typical lateral diamond MOSFET. It is very common to grow epitaxial layers of B-doped diamond on top of various substrates that are either undoped or include other dopants in order to precisely control the conductivity in the channel region [26–31]. Another approach is to form a two-dimensional hole gas (2DHG) layer in the channel region in order to increase the hole mobility and conductivity, which may be performed on an undoped substrate [32–36]. Common MOSFET structures used include vertical and lateral layouts with a trench gate, which is sometimes preferred in power MOSFET design due to the ability to shape the electric field in the channel, lower on-resistance, and better

(a)

Planar FET	Maximum Drain Current mA/mm	V_B Breakdown Voltage	L_{GD} Gate Drain Distance	V_B / L_{GD} Breakdown Field
SiC	90mA/mm	1600 V	20 μm	0.8 MV/cm
AlGaN/GaN	300mA/mm	200 - 1400 V	4 - 20 μm	1.0 MV/cm
AlGaN/AlGaN	200mA/mm	500 -1700 V	1 - 10 μm	1.7 MV/cm
Ga_2O_3	60mA/mm	400 V	8 μm	0.5 MV/cm
C-H Diamond p-FET	120mA/mm	350-1000 V 1000-1700V	1 - 9 μm 10-22 μm	1.0 MV/cm (Max 3.6 MV/cm) 0.8~1.0 MV/cm

(b)

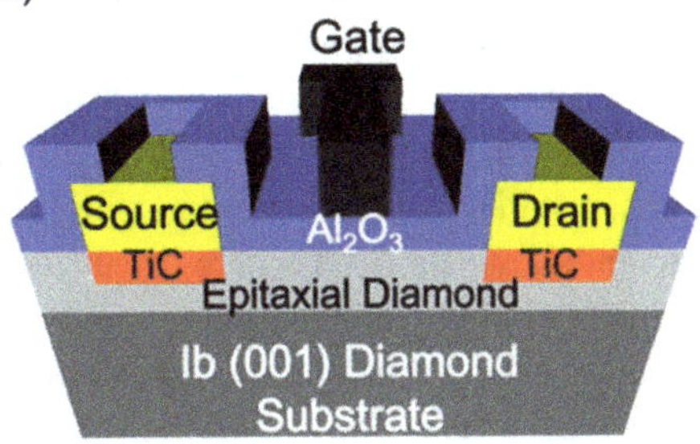

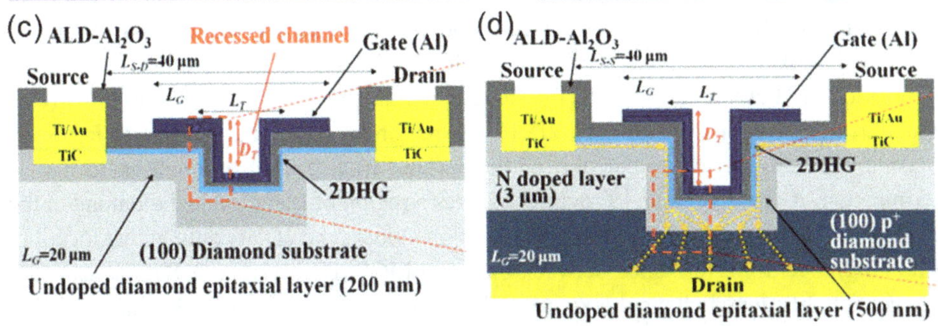

Fig. 5. (a) Table of various performance characteristics of several material systems commonly utilized in power electronics. (b) Lateral diamond MOSFET structure showing TiC regions in proximity to source/drain contacts. (c) Lateral and (d) vertical diamond MOSFET structures with a 2DHG layer and trench structure.

switching performance (Fig. 5(c-d)). This MOSFET structure has already been adapted for conventional wide bandgap semiconductors such as SiC and GaN, and now has become available for diamond as well [37]. Design and fabrication improvements have led to performance increases in diamond MOSFETs over time, such as substantial increases in transconductance from 200 μS/mm in the mid-1990s [27] to 200 mS/mm recently [38]. Typical maximum drain currents are on the order of ~100 mA/mm, but can increase up to 1.35 A/mm [38]. Polycrystalline substrates are occasionally used as well, and can be processed using the same techniques to form the 2DHG. Due to their polycrystalline structure, however, the MOSFET performance is limited, although high breakdown voltages (1.8 kV) are still possible [34, 39, 40].

A unique property of diamond is the ability to create a 2DHG layer near the substrate surface. This is achieved by covering the surface with either adsorbates or films that produce a negative charge [30]. The 2DHG layer results in hole mobility on the order of 100 cm^2/Vs, which when combined with a high carrier density of 10^{18} cm^{-3}, results in potential device performances that exceed other wide bandgap materials [41]. In order to achieve carrier densities of 10^{18} cm^{-3} or greater, which is desirable for power MOSFET operation, it is necessary to use a 2DHG formed from a H-terminated surface due to the lower activation energy of the 2DHG relative to boron, which is commonly used as a p-type dopant [41]. The H-termination approach is by far the most common for producing 2DHG layers [27–30, 32–34, 36, 38, 39, 42], and produces a 2DHG layer that is confined to within 10 nm of the surface [26]. If these 2DHG layers are used in a closely packed device structure, it is possible for the device performance to exceed that of other materials used for power electronics [41]. An illustration of this crystal structure is displayed in Fig. 6(a), where the surface of the bulk diamond structure has been terminated by H-atoms, resulting in C-H bonding at the surface. Figure 6(b) depicts the band diagram of a diamond/Al_2O_3 junction where the diamond has been H-terminated. The H-termination results in a buildup of negative charges on the diamond surface due to the H-atoms due to the unoccupied orbitals at the surface, which forms an electric in the bulk diamond just below the surface. The electric field will attract holes to the surface and is the origin of the holes in the 2DHG [41]. Al_2O_3 is a common material used for surface passivation and stabilization of the C-H bond layer [26, 28–30, 32, 33, 35, 38, 42], and can prevent the desorption of H-atoms from the surface at temperatures up to 550 °C [43]. Figure 6(c-d) illustrates the primary method of 2DHG formation by H_2 irradiation by microwave plasma, resulting in the conversion of the surface from a C-O termination to a C-H termination. A typical channel structure includes a region of H-terminated diamond in the channel, as shown in Fig. 6(e). This configuration induces the 2DHG only in that location. The difference in performance between diamond MOSFETs using a 2DHG layer and a conventional B-doped channel is quite stark, as indicated by Fig. 6(f). By implementing a 2DHG channel, it is possible to achieve current densities two to three orders of magnitude higher than by using a B-doped channel, which clearly indicates the superiority of this fabrication method for improving the device performance.

A particular advantage of diamond and other wide bandgap semiconductors is the ability to operate at high temperatures and voltages. Figure 7(a) shows the breakdown characteristics for a lateral diamond MOSFET with a breakdown voltage of -194 V, and after simulations with Silvaco Atlas, the breakdown field is estimated to be 4 MV/cm [44]. To understand the high-temperature operation, it is also necessary to investigate the current-voltage characteristics. The operation temperature can greatly affect the device performance, such as the on/off ratio, as indicated in Fig. 7(b). This device has an on/off ratio of 10^8 at room temperature that reduces to 10^3 at 400 °C as well as exhibits a breakdown voltage of 606 V at room temperature [28]. It is worth noting that several studies have been done on the reliability of the 2DHG layer at higher temperatures, where passivation layers deposited by ALD have been shown to limit the desorption of hydrogen or other adsorbates up to 550 °C [32, 43]. Higher breakdown voltages are possible as well, with upper limits being reported close to 2 kV, as shown in Fig. 7(c-f). Figure 7(c) presents the breakdown characteristics for a diamond MOSFET with a 2DHG layer, exhibiting a breakdown voltage of 1646 V [9]. The breakdown voltage can be modified by varying the gate-drain length and the gate oxide thickness for the same general device structure. By increasing the Al_2O_3 thickness from 200 nm to 400 nm, the breakdown voltage increases from 1646 V to 1700 V even as the gate-drain length decreases from 22 μm to 16 μm, while the breakdown electric field remains near 1 MV/cm [35]. If polycrystalline diamond is considered, the breakdown voltage is able to reach even higher levels. Figure 7(e) shows the breakdown characteristics of a black polycrystalline diamond MOSFET with a breakdown voltage of 1824 V [40]. Despite the increase in breakdown voltage, the breakdown field does not increase relative to single crystal diamond. Figure 7(f) shows that the breakdown fields for these devices are limited to approximately 1 MV/cm as well, similar to Fig. 7(d) which represents single crystal devices.

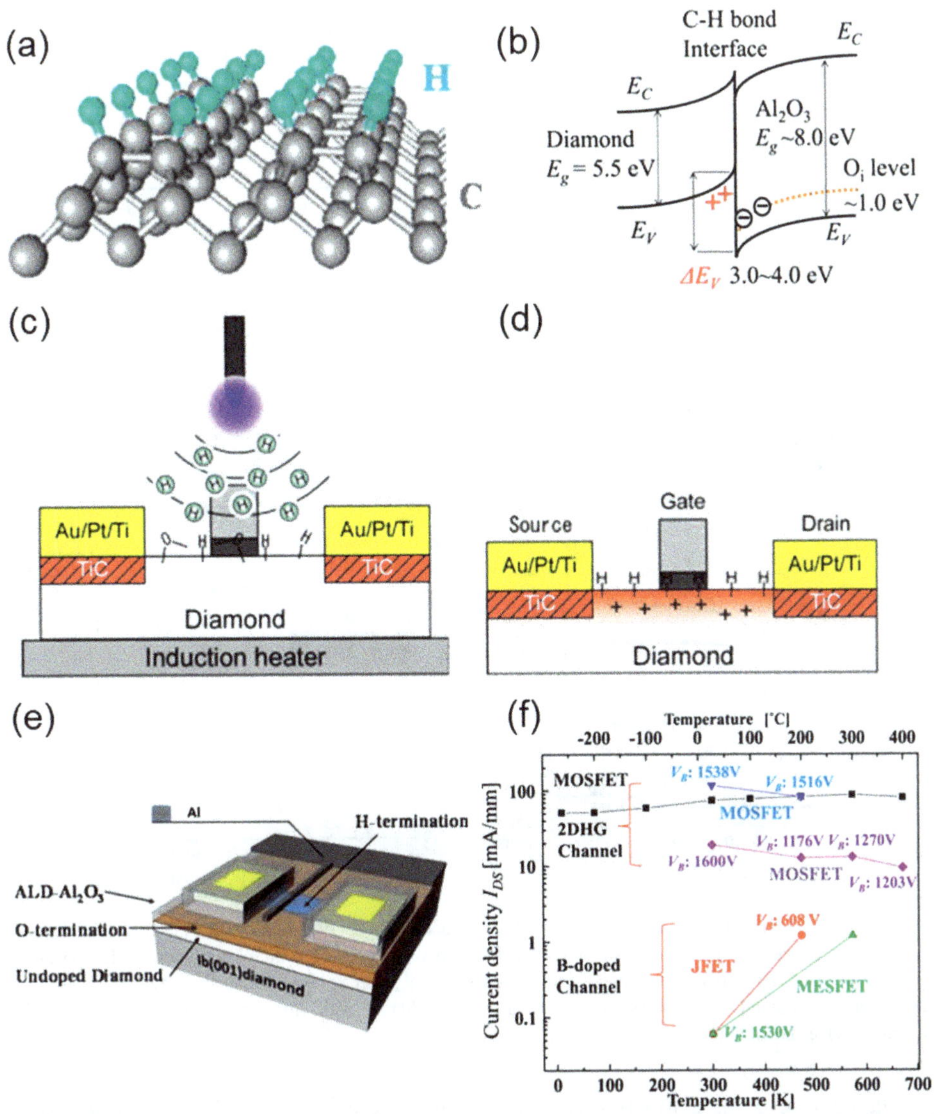

Fig. 6. (a) Illustration of diamond crystal structure showing H atoms bonded to C atoms at the surface, indicating H-termination of the surface. (b) Band diagram of a diamond/Al_2O_3 junction, depicting the effect of the H-termination at the interface by the inclusion of positive charges in the valence band. (c-d) Illustration of the H-termination process and the final H-terminated channel on a typical diamond MOSFET structure. (e) Device structure of a diamond MOSFET with H-terminated channel to produce a 2DHG layer. (f) Current density at various temperatures showing the performance effects of a 2DHG versus a B-doped channel.

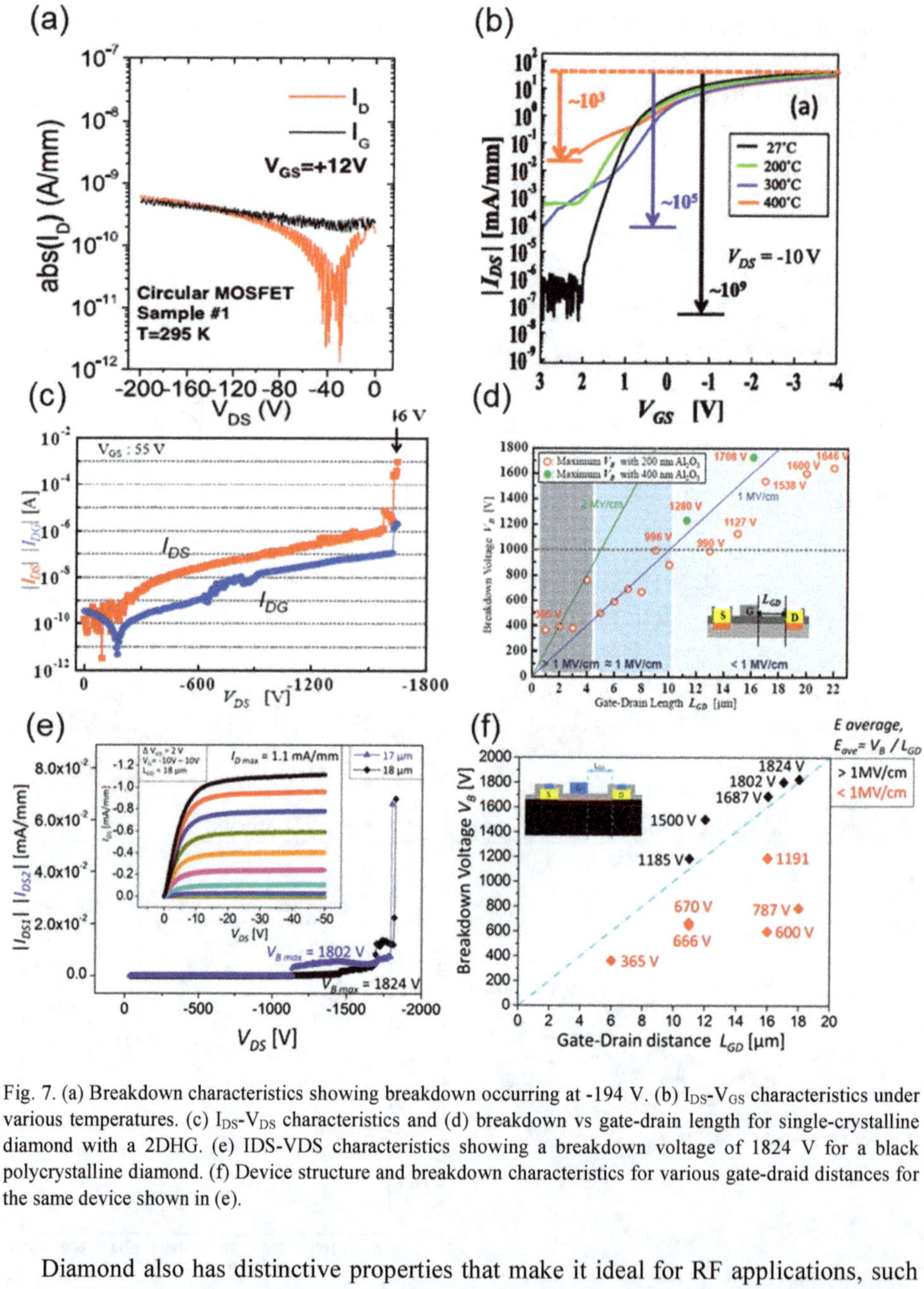

Fig. 7. (a) Breakdown characteristics showing breakdown occurring at -194 V. (b) I_{DS}-V_{GS} characteristics under various temperatures. (c) I_{DS}-V_{DS} characteristics and (d) breakdown vs gate-drain length for single-crystalline diamond with a 2DHG. (e) IDS-VDS characteristics showing a breakdown voltage of 1824 V for a black polycrystalline diamond. (f) Device structure and breakdown characteristics for various gate-draid distances for the same device shown in (e).

Diamond also has distinctive properties that make it ideal for RF applications, such as its high thermal conductivity (22 W/cm·K) and its high hole saturation velocity (1.1×10^7 cm/s) [38]. Figure 8 displays the selected RF characteristics of diamond MOSFETs. Figure 8(a-b) displays both small and large signal RF characteristics of a diamond MOSFET that has been treated with hydrogen plasma for H-terminated followed by NO_2 exposure. The NO_2 serves to increase the hole carrier density further than H-termination alone, and is passivated with Al_2O_3 to prohibit NO_2 desorption [38]. The cutoff frequencies of this device are 35 GHz and 70 GHz for f_T and f_{MAX}, and it achieves an output power density of 2 W/mm at 33% efficiency [38]. Figure 8(c) shows the cutoff frequency vs. gate length for various diamond MOSFETs. As indicated in the figure, most devices fail to reach the saturation carrier velocity of 1.1×10^7 cm/s of the diamond, instead reaching a maximum of 0.6×10^7 cm/s. Only recently, as shown by Imanishi *et al.* in 2019, have diamond RF MOSFETs nearly reached the saturation hole velocity, with this device achieving 1.0×10^7 cm/s. This device also achieves a maximum f_T of 31 GHz [33]. Figure 8(d) shows a comparison of large signal RF performance of diamond MOSFETs as well as those from other material systems. For diamond, the maximum reported output power density is 3.8 W/mm, which is higher than GaAs and LDMOS devices, but still lower than some SiC and AlGaN/GaN HEMT devices [33].

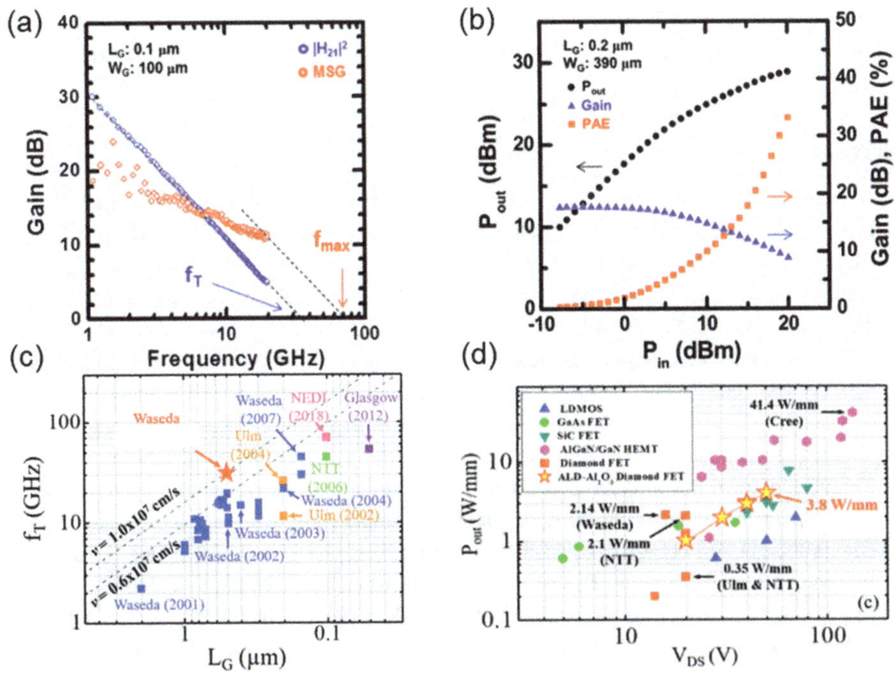

Fig. 8. (a) Small signal and (b) large signal RF characteristics of a diamond MOSFET. (c) Cutoff frequency versus gate length for various RF diamond MOSFETs. (d) Output power density versus V_{DS} for various RF diamond MOSFETs.

4. Conclusion

This paper presents an overview of recent developments in high-frequency devices based on two promising UWBG semiconductors, β-Ga_2O_3 and diamond. The unique UWBG semiconductor properties open the possibility for the next generation high power and high frequency devices where existing WBG systems show important limitations, such as high-temperature, high-voltage, and high-frequency operation. However, at present, the relatively small substrate size, especially in diamond, and unavailability of balanced dopants are major bottlenecks that β-Ga_2O_3 and diamond based device technologies must solve in order to be commercialized and compete with mature SiC and GaN technologies. As an alternative way, heterogeneous integration with other WBG or UWBG semiconductors can solve the above-mentioned technical issues. As an example, various research groups have recently attempted to create high-performance UWBG heterojunctions and related devices [45]. Although these attempts are all in the early stage, this is likely to be the easiest way to maximize the outstanding properties of UWBG semiconductors. With all such research, UWBG semiconductor devices will enable dramatic leaps not only in RF devices performance but also for other devices, such as power electronics and optoelectronics.

Acknowledgment

This work was supported by the National Science Foundation (Grant number: ECCS-1809077).

References

1. Jinwook W Chung, William E Hoke, Eduardo M Chumbes, and Tomás Palacios, IEEE Electron Device Letters **31** (3), 195 (2010); Yuanzheng Yue, Zongyang Hu, Jia Guo, Berardi Sensale-Rodriguez, Guowang Li, Ronghua Wang, Faiza Faria, Tian Fang, Bo Song, and Xiang Gao, IEEE Electron Device Letters **33** (7), 988 (2012); GY Xu, A Salvador, W Kim, Z Fan, C Lu, H Tang, H Morkoç, G Smith, M Estes, and B Goldenberg, Applied physics letters **71** (15), 2154 (1997); Xiaping Chen, Huili Zhu, Jiafa Cai, and Zhengyun Wu, Journal of Applied Physics **102** (2), 024505 (2007); Munho Kim, Jung-Hun Seo, Uttam Singisetti, and Zhenqiang Ma, Journal of Materials Chemistry C **5** (33), 8338 (2017).
2. Hideki Hirayama, Norimichi Noguchi, and Norihiko Kamata, Applied physics express **3** (3), 032102 (2010); Dong Liu, Sang June Cho, Jeongpil Park, Jiarui Gong, Jung-Hun Seo, Rafael Dalmau, Deyin Zhao, Kwangeun Kim, Munho Kim, and Akhil RK Kalapala, Applied Physics Letters **113** (1), 011111 (2018); D. Liu, S. J. Cho, J. Park, J. R. Gong, J. H. Seo, R. Dalmau, D. Y. Zhao, K. Kim, M. Kim, A. R. K. Kalapala, J. D. Albrecht, W. D. Zhou, B. Moody, and Z. Q. Ma, Applied Physics Letters **113** (1) (2018); Jan Isberg, Johan Hammersberg, Erik Johansson, Tobias Wikström, Daniel J Twitchen, Andrew J Whitehead, Steven E Coe, and Geoffrey A Scarsbrook, Science **297** (5587), 1670 (2002).
3. Masataka Higashiwaki, Kohei Sasaki, Akito Kuramata, Takekazu Masui, and Shigenobu Yamakoshi, Applied Physics Letters **100** (1), 013504 (2012).
4. JY Tsao, S Chowdhury, MA Hollis, D Jena, NM Johnson, KA Jones, RJ Kaplar, S Rajan, CG Van de Walle, and E Bellotti, Advanced Electronic Materials **4** (1), 1600501 (2018).
5. Caroline E. Knapp, Geoffrey Hyett, Ivan P. Parkin, and Claire J. Carmalt, Chemistry of Materials **23** (7), 1719 (2011).
6. Edward Swinnich, Yash Jayeshbhai Dave, E Bruce Pitman, Scott Broderick, Baishakhi Mazumder, and Jung-Hun Seo, Materials Discovery **11**, 1 (2018).
7. Munho Kim, Jung-Hun Seo, Uttam Singisetti, and Zhenqiang Ma, Journal of Materials Chemistry C **5** (33), 8338 (2017).

8. Masataka Higashiwaki, Kohei Sasaki, Takafumi Kamimura, Man Hoi Wong, Daivasigamani Krishnamurthy, Akito Kuramata, Takekazu Masui, and Shigenobu Yamakoshi, Applied Physics Letters **103** (12), 123511 (2013).
9. Andrew J Green, Kelson D Chabak, Eric R Heller, Robert C Fitch, Michele Baldini, Andreas Fiedler, Klaus Irmscher, Günter Wagner, Zbigniew Galazka, and Stephen E Tetlak, IEEE Electron Device Letters **37** (7), 902 (2016).
10. Ke Zeng, Abhishek Vaidya, and Uttam Singisetti, IEEE Electron Device Letters **39** (9), 1385 (2018).
11. Sriram Krishnamoorthy, Zhanbo Xia, Sanyam Bajaj, Mark Brenner, and Siddharth Rajan, Applied Physics Express **10** (5), 051102 (2017).
12. Zhanbo Xia, Chandan Joishi, Sriram Krishnamoorthy, Sanyam Bajaj, Yuewei Zhang, Mark Brenner, Saurabh Lodha, and Siddharth Rajan, IEEE Electron Device Letters **39** (4), 568 (2018).
13. Joe F McGlone, Zhanbo Xia, Yuewei Zhang, Chandan Joishi, Saurabh Lodha, Siddharth Rajan, Steven A Ringel, and Aaron R Arehart, IEEE Electron Device Letters **39** (7), 1042 (2018).
14. Chandan Joishi, Zhanbo Xia, Joe McGlone, Yuewei Zhang, Aaron R Arehart, Steven Ringel, Saurabh Lodha, and Siddharth Rajan, Applied Physics Letters **113** (12), 123501 (2018).
15. Wan Sik Hwang, Amit Verma, Hartwin Peelaers, Vladimir Protasenko, Sergei Rouvimov, Huili Xing, Alan Seabaugh, Wilfried Haensch, Chris Van de Walle, and Zbigniew Galazka, Applied Physics Letters **104** (20), 203111 (2014).
16. Hong Zhou, Mengwei Si, Sami Alghamdi, Gang Qiu, Lingming Yang, and D Ye Peide, IEEE Electron Device Letters **38** (1), 103 (2016).
17. Hong Zhou, Kerry Maize, Gang Qiu, Ali Shakouri, and Peide D. Ye, Applied Physics Letters **111** (9), 092102 (2017).
18. Janghyuk Kim, Michael A Mastro, Marko J Tadjer, and Jihyun Kim, ACS applied materials & interfaces **9** (25), 21322 (2017).
19. Edward Swinnich, Md Nazmul Hasan, Ke Zeng, Yash Dove, Uttam Singisetti, Baishakhi Mazumder, and Jung-Hun Seo, Advanced Electronic Materials **5** (3), 1800714 (2019).
20. Kelson D Chabak, Neil Moser, Andrew J Green, Dennis E Walker Jr, Stephen E Tetlak, Eric Heller, Antonio Crespo, Robert Fitch, Jonathan P McCandless, and Kevin Leedy, Applied Physics Letters **109** (21), 213501 (2016).
21. Zongyang Hu, Kazuki Nomoto, Wenshen Li, Nicholas Tanen, Kohei Sasaki, Akito Kuramata, Tohru Nakamura, Debdeep Jena, and Huili Grace Xing, IEEE Electron Device Letters **39** (6), 869 (2018).
22. Nan Ma, Nicholas Tanen, Amit Verma, Zhi Guo, Tengfei Luo, Huili Xing, and Debdeep Jena, Applied Physics Letters **109** (21), 212101 (2016).
23. Andrew Joseph Green, Kelson D Chabak, Michele Baldini, Neil Moser, Ryan Gilbert, Robert C Fitch, Günter Wagner, Zbigniew Galazka, Jonathan Mccandless, and Antonio Crespo, IEEE Electron Device Letters **38** (6), 790 (2017).
24. KD Chabak, DE Walker, AJ Green, A Crespo, M Lindquist, K Leedy, S Tetlak, R Gilbert, NA Moser, and G Jessen, presented at the 2018 IEEE MTT-S International Microwave Workshop Series on Advanced Materials and Processes for RF and THz Applications (IMWS-AMP), 2018 (unpublished).
25. Manikant Singh, Michael A Casbon, Michael J Uren, James W Pomeroy, Stefano Dalcanale, Serge Karboyan, Paul J Tasker, Man Hoi Wong, Kohei Sasaki, and Akito Kuramata, IEEE Electron Device Letters **39** (10), 1572 (2018).

26. Masafumi Inaba, Tsubasa Muta, Mikinori Kobayashi, Toshiki Saito, Masanobu Shibata, Daisuke Matsumura, Takuya Kudo, Atsushi Hiraiwa, and Hiroshi Kawarada, Applied Physics Letters **109** (3), 033503 (2016).
27. Hiroshi Kawarada, Makoto Aoki, and Masahiro Ito, Applied Physics Letters **65** (12), 1563 (1994).
28. H Kawarada, H Tsuboi, T Naruo, T Yamada, D Xu, A Daicho, T Saito, and A Hiraiwa, Applied Physics Letters **105** (1), 013510 (2014).
29. Hiroshi Kawarada, Tetsuya Yamada, Dechen Xu, Hidetoshi Tsuboi, Yuya Kitabayashi, Daisuke Matsumura, Masanobu Shibata, Takuya Kudo, Masafumi Inaba, and Atsushi Hiraiwa, Scientific reports **7**, 42368 (2017).
30. Nobutaka Oi, Masafumi Inaba, Satoshi Okubo, Ikuto Tsuyuzaki, Taisuke Kageura, Shinobu Onoda, Atsushi Hiraiwa, and Hiroshi Kawarada, Scientific Reports **8** (1), 10660 (2018).
31. Jung-Hun Seo, Henry Wu, Solomon Mikael, Hongyi Mi, James P Blanchard, Giri Venkataramanan, Weidong Zhou, Shaoqin Gong, Dane Morgan, and Zhenqiang Ma, Journal of Applied Physics **119** (20), 205703 (2016).
32. Akira Daicho, Tatsuya Saito, Shinichiro Kurihara, Atsushi Hiraiwa, and Hiroshi Kawarada, Journal of Applied Physics **115** (22), 223711 (2014).
33. Shoichiro Imanishi, Kiyotaka Horikawa, Nobutaka Oi, Satoshi Okubo, Taisuke Kageura, Atsushi Hiraiwa, and Hiroshi Kawarada, IEEE Electron Device Letters **40** (2), 279 (2019).
34. Yoshikatsu Jingu, Kazuyuki Hirama, and Hiroshi Kawarada, IEEE Transactions on Electron Devices **57** (5), 966 (2010).
35. H Kawarada, T Yamada, D Xu, Y Kitabayashi, M Shibata, D Matsumura, M Kobayashi, T Saito, T Kudo, and Masafumi Inaba, presented at the 2016 28th International Symposium on Power Semiconductor Devices and ICs (ISPSD), 2016 (unpublished).
36. Zongyou Yin, Moshe Tordjman, Alon Vardi, Rafi Kalish, and Jesús A del Alamo, IEEE Electron Device Letters **39** (4), 540 (2018).
37. Richard K Williams, Mohamed N Darwish, Richard A Blanchard, Ralf Siemieniec, Phil Rutter, and Yusuke Kawaguchi, IEEE Transactions on Electron Devices **64** (3), 674 (2017).
38. Makoto Kasu and Toshiyuki Oishi, presented at the 2014 9th European Microwave Integrated Circuit Conference, 2014 (unpublished).
39. Yu Fu, Ruimin Xu, Yuehang Xu, Jianjun Zhou, Qingzhi Wu, Yuechan Kong, Yong Zhang, Tangsheng Chen, and Bo Yan, IEEE Electron Device Letters **39** (11), 1704 (2018).
40. M Syamsul, Y Kitabayashi, D Matsumura, T Saito, Y Shintani, and H Kawarada, Applied Physics Letters **109** (20), 203504 (2016).
41. in *Power Electronics Device Applications of Diamond Semiconductors*, edited by Satoshi Koizumi, Hitoshi Umezawa, Julien Pernot et al. (Woodhead Publishing, 2018), pp. 295.
42. Takeyasu Saito, Kyung-ho Park, Kazuyuki Hirama, Hitoshi Umezawa, Mitsuya Satoh, Hiroshi Kawarada, Zhi-Quan Liu, Kazutaka Mitsuishi, Kazuo Furuya, and Hideyo Okushi, Journal of Electronic Materials **40** (3), 247 (2011).
43. Atsushi Hiraiwa, Akira Daicho, Shinichiro Kurihara, Yuki Yokoyama, and Hiroshi Kawarada, Journal of Applied Physics **112** (12), 124504 (2012).
44. TT Pham, J Pernot, C Masante, D Eon, E Gheeraert, G Chicot, F Udrea, and N Rouger, presented at the 2017 IEEE International Electron Devices Meeting (IEDM), 2017 (unpublished).
45. Adam W Saxler, (Google Patents, 2006).

Application of Atom Probe Tomography for Advancing GaN Based Technology

Olivia G. Licata and Baishakhi Mazumder*

Department of Materials Design and Innovation,
University at Buffalo, Buffalo, New York 14260, USA
**baishakh@buffalo.edu*

Atom Probe Tomography (APT) has emerged as a stand-alone technique for material characterization at the sub-nanometer level, with unrivaled spatial resolution, coupled with three-dimensional atomic mapping, and equal sensitivity of all elements. Over the past decade, APT has proven to be a valuable tool in advancing the understanding and design of GaN-based semiconductor technology, by revealing correlations between atomic-level structure chemistry and device performance. The uniqueness of APT is exemplified by its ability to directly analyze nanoscale features within a commercial device. In this review, the quantitative requirements for advancement in GaN-based device metrology are defined as accurate measurement of composition, structural inhomogeneities, elemental incorporation within thin interlayers, and quantification of dopants and impurities. These are bolstered by a review of recent advances in GaN-based devices, realized through APT. The rich compositional and spatial data provided by APT is necessary for the continued advancement of III-V semiconductor technology.

Keywords: Atom probe tomography; wide bandgap; GaN power electronics; compositional accuracy.

1. Introduction

The III–V semiconductors have enabled a diverse range of epitaxial heterostructures in comparison to elemental group IV semiconductors or SiC [1, 2]. The III-V system is adaptable and suitable for quantum wells, modulation-doped heterointerfaces, and heterojunction structures, allowing for expansion into new spectral regions for optical devices and new operation regimes for electronic devices [3]. The advancements of III-V materials and their alloys are mainly due to the exceptional crystal growth and the ability to grow good quality p-n junctions [4]. The pioneering work in compound semiconductors, was based on the cubic (Al, Ga, In)(P, As, Sb) system (now sometimes referred to as the "conventional" III-Vs) and the II-VI materials [5]. Wide band gap, III-V nitrides, such as GaN, AlN, InN, and their nitrides crystallize as hexagonal wurtzite or zinc-blende polytypes. The wurtzite structure is polar and results in interesting properties for GaN-based heterostructures, including large internal electric fields due to discontinuities in spontaneous and piezoelectric polarization [5]. The nitrides have hugely benefited from the materials and device developments in the III-Vs and most device concepts in the nitrides directly follow from the III-Vs – with the major caveat that polarization-related charges ($>10^{13}$ cm^{-2}) or electric fields ($>$ 1 MV/cm) in heterostructures are of unprecedented magnitude for any semiconductor [5, 6]. The AlGaN alloys are typically used as barrier layer devices [7–14]. The end component, AlN has received considerable attention for its insulating property [15]. As an example, AlN has also found important applications in III-V based high power and high frequency electronic devices [16–21]. The wide bandgap of AlN, large conduction band offset to GaN, and high polarization charges render it attractive as the charge-inducing barrier for high electron mobility transistors (HEMTs) [22]. Inserting an AlN interlayer between the AlGaN barrier and GaN channel of the AlGaN/GaN heterostructure effectively reduces alloy scattering to the 2-dimensional electron gas (2DEG) thereby improving electron mobility [23–27]. As the 2DEG in typical Ga-polar and N-polar GaN HEMT structures are confined at the AlN/GaN interface with a net positive fixed polarization charge, the roughness and abruptness of such interfaces determine the transport properties of the 2DEG and hence the overall device performance [28]. However, there are still some problems in the production of GaN based devices. Dispersion effects in the wide bandgap devices have restricted initial expectations. The presence of trapping centers in FETs based on III-V compounds associated with surface, material, and/or interface states have been considered to be the origin of these effects [29]. Understanding the microstructure of those interfaces may therefore provide physical insight into the nature of the interfacial trap state.

(Al, In)GaN-based light emitting diodes (LED) [30, 31] and laser diodes (LD) [32, 33] occupy a huge market segment in optoelectronic devices, which has been further growing with the possibility of LEDs for general lighting [34] and LDs for full color laser projection [35–37]. Feedback from full device characterization and growth is very important to achieve high performance LEDs and LDs. InGaN LDs for instance are devices that present a gain medium (or active region) fabricated with InGaN Quantum wells (QWs). Since high gain is necessary for high performance LD in many applications, one would like to characterize the gain medium of a LD. Several techniques can be used [38–41], however most of them rely on prerequisite knowledge of the confinement factor (how much the lasing optical mode overlaps with the active region) to have a good estimate of the material gain. Thus precise knowledge of the thickness and composition of the various layers of a LD structure is very important to quantify this parameter.

The intentional introduction of impurities (dopant) determines the semiconductive properties of materials and allows for modification of the conductivity [42]. The species and concentration of the dopant determines the conductivity type as p-type (holes are majority carriers) and n-type (electrons are majority carriers). The electronic characteristics and overall quality of semiconductor-based devices are greatly affected by the homogeneity of dopant distribution [42]. In magnesium-doped GaN, for optoelectronic devices, the clustering of dopants at structural defects has been widely reported, and can significantly affect p-type conductivity [43]. The localized segregation of dopants or impurities can reveal insight on the material microstructure and how it correlates with device performance.

In order to obtain a complete understanding of the internal physical and chemical properties of nanostructure devices, quantitative compositional analyses must be conducted at the atomic level. Transmission electron microscopy (TEM) is often utilized for its ability to resolve crystal structure through electron diffraction. TEM can provide accurate atomic scale structure in projection; however, without use of supporting techniques; such as electron energy loss spectroscopy (EELS) or energy-dispersive x-ray spectroscopy (EDS), it can only provide average composition [44, 45]. EDS and EELS are indirect methods in comparison to secondary ion mass spectroscopy (SIMS), which utilizes mass spectrometry to characterize the chemical composition [46]. Despite the improved compositional information, the sputtering process in SIMS limits the determination of an atom's planar (xy) position to the spot size of the primary ion beam (~100 nm) [44]. The conventional techniques do not allow for visualization of dopants or atomic segregation and cannot define in-plane interfacial chemistry within complex nanostructures. An ideal technique for identifying subnanometer properties; such as, dopant distribution, impurity diffusion, and interfacial heterogeneity, would incorporate both enhanced spatial resolution and compositional sensitivity. Atom Probe Tomography (APT) is a powerful characterization technique that provides direct 3-D visualization of complex structure chemistry at the atomic level. In the past decade, APT has emerged as one of the most powerful techniques for mapping the elemental distribution in semiconductors with near-atomic spatial resolution and analytical sensitivity as low as one atomic part per million [44, 45, 47–50].

2. Atom Probe Tomography

APT combines a time-of-flight mass spectrometer with a point projection microscope and unlike conventional optical or electron microscopes, the magnification effect is provided by a highly curved electric field [51–53]. APT characterization does not require prior knowledge of the analyzed material and its high sensitivity is equivalent for all elements [50]. This makes APT exceptionally useful in defining interfacial chemistry [28, 54–56], impurities [57–60], and solute clustering [43, 47, 61]. APT's unique capabilities derive from the acquisition process and reconstruction methodology [53, 62]. During an APT data acquisition, atoms are successively removed from the apex of a needle-shaped specimen through the process of field evaporation, as shown in the schematic in Fig. 1. In this process, surface atoms are first ionized and then subjected to an electric field force causing them to evaporate and accelerate towards a detector [63]. The process is initiated by the combined effects of a standing (DC) electrostatic field and either high voltage or laser pulses applied to the surface atoms. The time between the applied pulse and the atom reaching the detector defines the time-of-flight. The time-of-flight data is converted into the respective mass-to-charge ratios and displayed as a histogram, referred to as the mass spectrum. Mass-to-charge ratios in addition to their isotopic abundancies are unique to elemental species. The quality of the mass spectrum, and therefore the accuracy of the chemical identification, is assessed by the mass resolution [63]. Signal-to-noise-ratio also influences the quality of an experiment and so it is often optimized through background correction offered in CAMECA's Integrated Visualization and Analysis Software (IVAS) [62]. After obtaining the most accurate mass spectrum, the sequential 2D ion position data from the detector is converted to a 3-D atom map, corresponding to the original analyzed

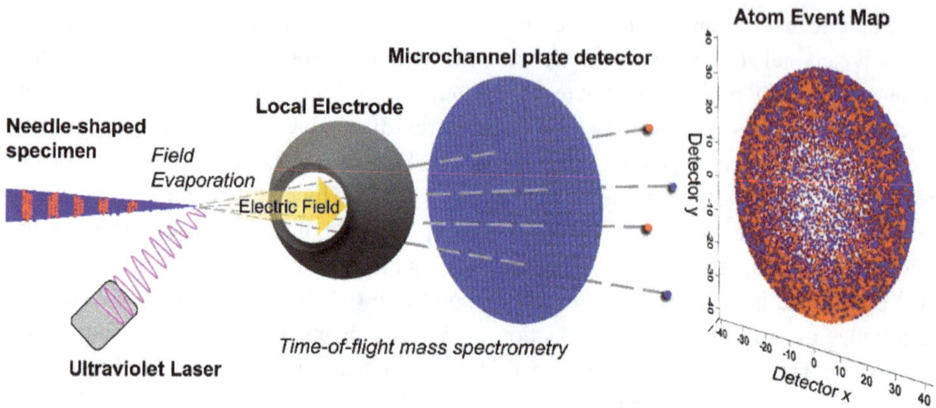

Fig. 1. Schematic of working principle of Atom probe tomography.

sample. The use of controlled field evaporation in APT captures individual ion positions in three dimensions, which facilitates quantification of elemental distribution in complex structures; such as, HEMTs, FETs, LEDs, and LDs.

The unique abilities of APT make it an asset in the optimization of semiconductor device design. APT continues to reveal microstructural and compositional attributes of light emitting diodes, laser diodes, and transistors, further improving our understanding towards optimized device design [45, 48, 64, 65]. Here we will highlight the major chemical and structural characteristics of GaN based heterostructures obtained through APT, including accuracy in compositional measurement, assessment of inhomogeneities, elemental incorporation and dopant distribution. We will review the corresponding influence on electronic properties; including, carrier transport (mobility), carrier localization in the energy landscape, radiative recombination, internal quantum efficiency, and optical efficiency.

2.1. *Composition Measurement Accuracy*

The design of a light emitting diode (LED) requires multiple interrelated components, with dimensional and compositional restrictions at the nanoscale. Light is emitted through the process of electroluminescence, when electrons and electron holes recombine. The wavelength of the light is determined by the energy bandgap of the neighboring materials and can therefore be tailored by the choice of semiconductor composition. Characterization of complex nanoscale components requires chemical imaging at the atomic level, which is enabled through atom probe tomography (APT). As discussed previously, APT time-of-flight data is converted into a histogram of mass-to-charge ratios known as the mass spectrum. Figure 2a shows the mass spectrum from the APT analysis of a commercial LED device, performed by Giddings *et al.* [66]. Here gallium (Ga) peaks appear in yellow at 69 and 71 Daltons (Da) with a distinct isotopic abundance ratio of 60:40. Ga also appears in the doubly charged state at 34.5 and 35.5 Da and the triply charged state at 23 and

23.67 Da. The range for each peak selection is defined by including at least the full width half maximum in addition to some of the peak tail, shown in Fig. 2a. The resulting 3-D atom map of the reconstructed data is shown in Fig. 2b [66]. APT distinguishes features at the atomic scale with high analytical sensitivity of ~10-100 atomic parts per million and with three dimensional sub-nanometer resolution [45]. The APT reconstruction, shown in Fig. 2b, defines the distinct features of the LED device, including the p-type (Mg-doped) layer, electron barrier layer (EBL), and InGaN quantum wells (QWs). Alternate metrology methods; such as, nanoscale secondary ion mass spectroscopy (nanoSIMS), with a lateral resolution of approximately 50 nm, cannot quantify these distinct features [51]. Transmission electron microscopy (TEM) offers high spatial resolution; however, it cannot provide the unique 3-D positions of atomic species derived from APT.

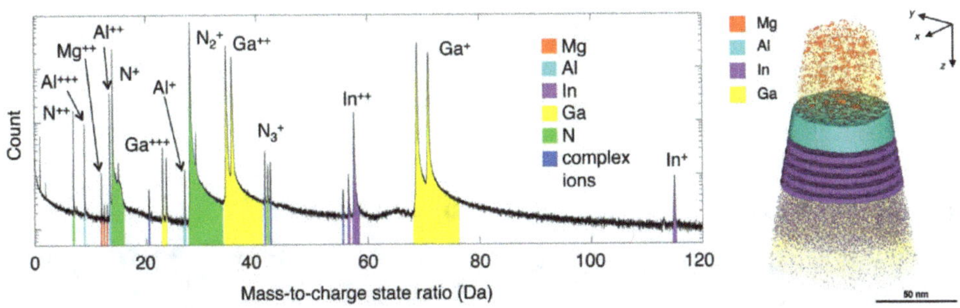

Fig. 2. (a) Mass spectrum for elemental and complex species identified from an APT analysis of a commercial LED device. (b) 3-D atom map of the analyzed APT specimen, cut from a commercial LED device. Reproduced from Reference 66 with permission of Cambridge University Press.
DOI: https://doi.org/10.1017/S1551929514000819

When analyzing compound semiconductor structures of type III-N in APT, it is necessary to consider the nitrogen loss mechanisms; such as, N_2 sublimation and neutral generation from molecular ion dissociation, which make it difficult to achieve accurate stoichiometry [67, 68]. Foronda *et al.* utilized composition measurements from APT to verify the lattice matching of $In_{0.18}Al_{0.82}N$ and GaN [69]. InAlN has a higher evaporation field than GaN, which may cause more nitrogen atoms to be evaporated as N_2^{2+}, as opposed to N^+. Since both $N_2^{2+}$and N^+ peaks are found at a mass-to-charge ratio of 14 Da, it is difficult to differentiate between the two species and obtain the accurate stoichiometry. For these reasons, the standard practice is to measure the ratio of group III site elements (In/In+Al and Al/In+Al), when reporting III-N material compositions in atom probe. Reporting III-site composition is the most reliable method and is not biased by the loss of nitrogen during evaporation.

Accurate compositional measurement through APT facilitates the investigation of electronic transport properties and optical efficiency of wide-bandgap semiconductors. Compositional fluctuations can impact carrier transport and consequently influence the overall device performance [70, 71]. Shivaraman *et al.* undertook an investigation to understand the influence of structure polarity on the energy landscape experienced by the charge carriers [71]. APT allowed for comparative compositional analysis of $20\bar{2}1$ and $20\bar{2}\bar{1}$ semipolar InGaN LED structures, shown in the upper portions of Figs. 3a and 3b [71]. Polarization discontinuity occurs in both the $20\bar{2}1$ and $20\bar{2}\bar{1}$ orientation; however, the direction is opposite with respect to the c plane and depletion field. The magnitude of the polarization discontinuity is influenced by the indium content within the active region of the LED [70]. Here Shivaraman *et al.* relied on APT for accurate compositional analysis and found that Indium composition was greater in the $20\bar{2}\bar{1}$ single quantum well (SQW) than the $20\bar{2}1$ SQW [71]. This is demonstrated by the 1-D composition profiles, from APT analysis, shown in the lower portion of Figs. 3a and 3b. This work exemplified the influence of the polarization fields on the energy landscape of charge carriers and correlated composition results from APT with optical emission wavelength. Semiconductor devices designed with components at the nanoscale require accurate metrology solutions for quantifying composition. However, a bulk compositional assessment, without indication of local distribution, such as that obtained through high resolution x-ray diffraction (HRXRD) or SIMS, will not provide sufficient characterization for nanostructured devices.

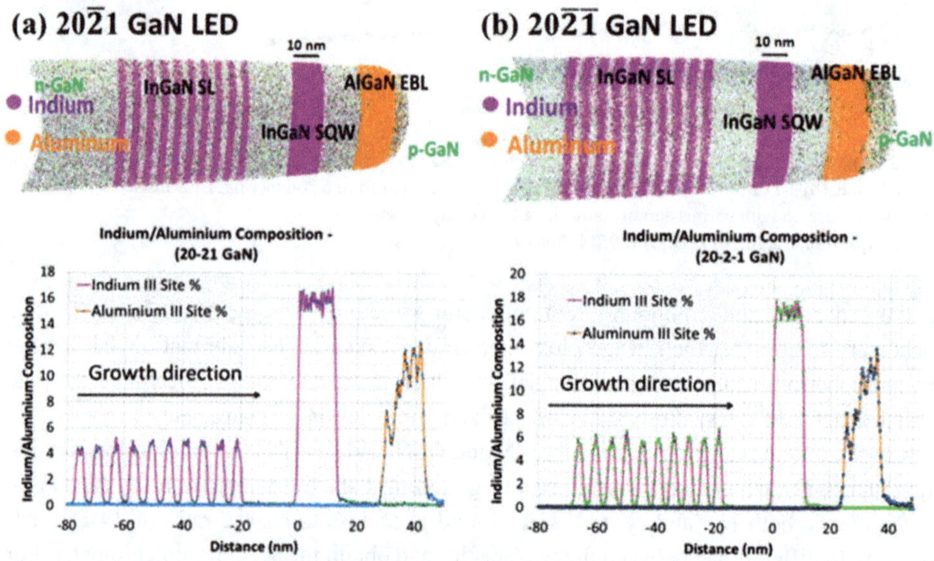

Fig. 3. (a) 3-D atom map of the LED device grown on $20\bar{2}1$ GaN with corresponding 1-D composition profile along the growth direction. (b) Corresponding 3-D atom map and 1-D composition profile for the same LED device on $20\bar{2}\bar{1}$ GaN. Reproduced from Reference 71, with permission of AIP Publishing.
DOI: https://doi.org/10.1063/1.4812363

2.2. *Structure Inhomogeneities and the Energy Landscape*

Investigation and quantification of chemical inhomogeneities and QW thickness in LED structures, using APT, have revealed insights on carrier localization [72–79]. Structural inhomogeneities found at the atomic level can be influenced by intrinsic material properties as well as the applied growth conditions and process parameters. In order to achieve optimal device performance, it is important to correlate structural and chemical phenomena with the corresponding growth condition or parameter of influence. Ren *et al.* and several other groups have investigated the impact of introducing hydrogen to the carrier gas in metal organic chemical vapor deposition (MOCVD) during growth of the GaN spacer in blue LEDs [80–83]. The introduction of hydrogen resulted in increased internal quantum efficiency (IQE), for Ren *et al.*, in addition to nm-scale discontinuities in the QWs [83]. Metrology methods such as HRXRD cannot capture, nor precisely locate, these discontinuities because they rely on the average value taken across indium-rich regions and gaps, resulting in inaccurate compositional information. APT measurements, obtained by Ren *et al.*, created a basis for a simulation of carrier distribution to understand how discontinuities impact the rate of radiative recombination and therefore the IQE [83].

In another study, Rigutti *et al.* hypothesized that radiative recombination takes place through exciton states, localized in In-rich regions. APT revealed compositional inhomogeneities within InGaN QW planes which were subsequently correlated with microphotoluminescence spectroscopy [73]. On the other hand, when studying nominally lattice-matched InAlN/GaN, Choi *et al.* found segregation of 10 nm wide Al-rich cores in all experiments using plasma-assisted molecular beam epitaxy (PAMBE), despite systematic modifications in growth condition, material orientation, and InN mole fraction [84]. Here, APT revealed large continuous inhomogeneities in the microstructure. Subsequent investigations performed in other groups demonstrated the sensitivity of APT to discrete fluctuations within several nm. Wu *et al.* and Yang *et al.* used APT to define nanoscale fluctuations in indium composition and distribution in a commercial grade c-plane LED structure [75, 76]. Results from APT were incorporated into a simulation of carrier transport to show the influence of indium fluctuations [75, 76]. The resulting simulated values were found to be in closer agreement to observed behaviors of emission spectra, bias voltages, and droop when small nanoscale fluctuations of indium composition were included [76]. Browne *et al.* used APT to examine the effect of random alloy fluctuations on electron transport through QW active regions in a unipolar LED-like structure, shown in Fig. 4a [72]. APT quantified well thicknesses as low as 1.5 nm and showed lateral compositional fluctuations within several nm, shown in Figs. 4b and 4c, respectively [72]. As shown in Fig. 4c, APT results revealed significant compositional fluctuations ranging from 8-20% [72]. 2D drift diffusion simulations were performed, including the magnitude and length scale of indium fluctuations obtained from APT [72]. These understandings will aid in assessing performance and improving future device design.

APT has also been utilized to define compositional and spatial variation resulting from sample polarity. Tang *et al.* demonstrated the impact of sample orientation on indium fluctuations in InGaN QWs of non-polar a-plane and polar c-plane samples [74]. The cross-sectional 3-D atom maps, obtained from APT, are shown in Figs. 5a (a-plane) and 5b (c-plane) [74]. As indicated with black arrows, there are regions of potential indium segregation in the a-plane sample [74]. The distribution in the a-plane sample is noticeably less random than the c-plane sample which appears completely homogenous. Frequency

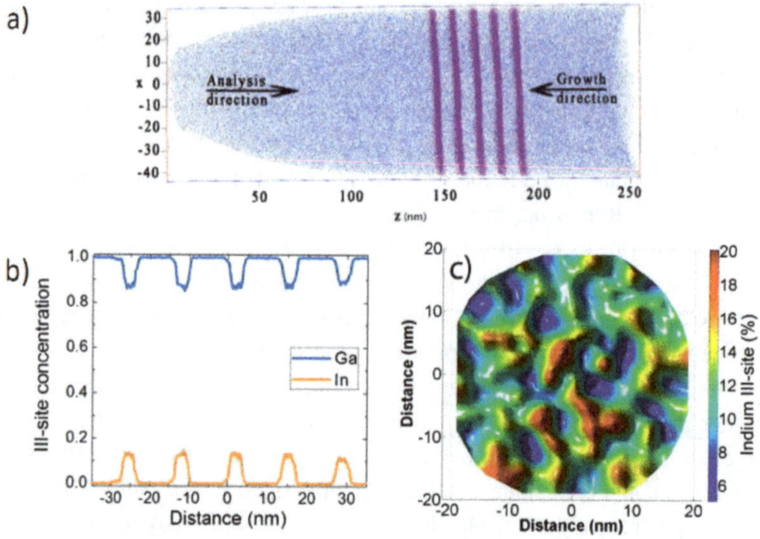

Fig. 4. APT analysis of a 5 × 3.0 nm quantum well sample. (a) 3-D atom map of analyzed tip showing the presence of the five QWs. (b) 1-D composition profile of group III-site occupation by gallium and indium through QWs. (c) 2D plot of the lateral indium composition within the central well. Reproduced from Reference 72, with permission of AIP Publishing. DOI: https://doi.org/10.1063/1.4919750

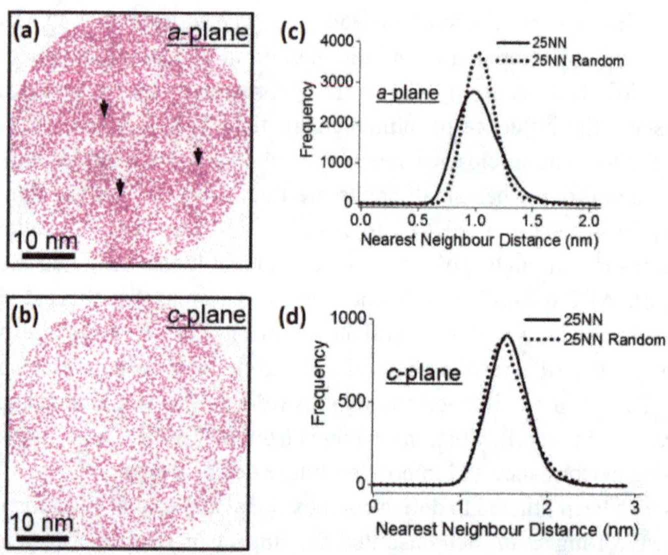

Fig. 5. (a) Cross section of xy plane (perpendicular to the analysis direction) of the a-plane sample orientation 3-D atom map. Black arrows indicate regions of potential segregation of In. (b) Cross section of c-plane 3-D atom map. Twenty five nearest neighbor distributions for (c) a-plane and (d) c-plane. The dotted lines indicate a calculated random binomial distribution. Reproduced from Reference 74, with permission of AIP Publishing. DOI: https://doi.org/10.1063/1.4909514

distribution and nearest neighbor analyses were performed in order to further qualify the presence of indium clustering [74]. Figures 5c and 5d show the observed distribution for the twenty five nearest neighbors as a solid line along with the expected binomial distribution considering purely random fluctuations as a dotted line [74]. The variation in the a-plane distribution from the random distribution is a strong indicator for clustering of indium atoms. The two sample orientations also showed variations in optoelectronic properties, which were assessed through photoluminescence emission wavelength [74]. In-plane structural variations and chemical inhomogeneity can be directly observed at the atomic scale by APT and these findings can reveal pertinent internal physics and chemistry for the optimization of power electronic and optoelectronic devices. This technique can be extended to complex multi-layer structures and define the layer thickness, elemental homogeneity, and chemical purity of thin interlayers.

2.3. *Elemental Incorporation*

AlGaN/GaN heterostructures are vital to high-voltage, high-power, and high-temperature microwave electron devices. Including an AlN interlayer at the AlGaN/GaN hetero-interface has shown improved mobility, sheet concentration, and resistance in the 2DEG [23]. These benefits are realized with increased purity of the AlN interlayer, which necessitates the prevention of Ga incorporation during the structure growth process. Atom probe tomography has been rigorously utilized to quantify the chemical structure of interlayers of various chemistries to aid in optimization of growth parameters for improved structure design [60, 85–88]. Mazumder *et al.* utilized APT to demonstrate the purity of AlN interlayers, at the AlGaN/GaN heterointerface, grown in plasma-assisted molecular

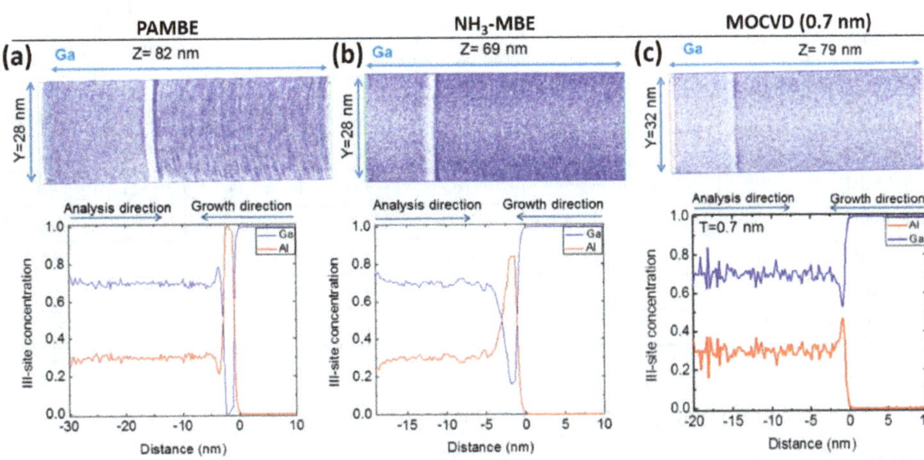

Fig. 6. Three dimensional atom maps for AlGaN/GaN 2 nm thick heterointerfaces grown by (a) PAMBE, (b) NH_3-MBE, and a 0.7 nm heterointerface grown by (c) MOCVD. Gallium atoms are shown in blue and Al atoms are omitted to demonstrate the interfacial abruptness. III site concentration profiles of Al (red) and Ga (blue) are shown below their respective 3-D atom maps for each growth method. Reproduced from Reference 87, with permission of AIP Publishing. DOI: https://doi.org/10.1063/1.4798249

beam epitaxy (PAMBE), ammonia-based molecular beam epitaxy (NH_3-MBE), and metal-organic chemical vapor deposition (MOCVD) [87]. Utilizing APT allowed for quantification of Ga incorporation within 2 nm thick AlN interlayers. The APT results can be visualized in 3-D, as shown in Fig. 6a-c, for PAMBE, NH_3-MBE, and MOCVD, respectively [87]. The 3-D atom maps show the Ga atoms only, in order to demonstrate the abruptness of the pure AlN interlayer. The results are further projected as concentration profiles along the analysis/growth direction, in the lower portion of Fig. 6a-c, for PAMBE, NH_3-MBE, and MOCVD, respectively [87]. The concentration profiles plot the III-site concentration of Al and Ga along the analysis/growth direction. PAMBE results, shown in Fig. 6a, indicate that the Ga concentration drops to 0 within the AlN interlayer [87].

In another work, Kaun *et al.* investigated metal-polar InAlN barriers grown in PAMBE under N-rich conditions [88]. They investigated a range of structure designs consisting of AlN and GaN interlayers of varying thicknesses in order to optimize the 2DEG [88]. The Hall measurements for each structure design conducted by Mazumder *et al.* and Kaun *et al.* are summarized in Table 1 [87, 88]. Mazumder *et al.* investigated both 0.7 nm and 2 nm thick AlN layers, grown in MOCVD, and found the composition to be independent of interlayer thickness [87]. Both samples grown in MOCVD showed relatively high (~0.53) Ga incorporation which was quantified through APT analysis [87]. This investigation helped elucidate what design parameters were relevant and irrelevant to the resulting level of Ga incorporated into the pure AlN interlayers, which is necessary for future device design optimization.

Table 1. Hall Measurements taken at room temperature (RT) and 77K under various growth conditions. Hall measurements include 2DEG sheet density (n_s), mobility (μ), and sheet resistivity (R_s) [87, 88].

Growth Method	Interlayer (nm)		n_s (10^{13} cm^{-2})		μ ($cm^2/V{\cdot}s$)		R_s (Ω/sq)	
	AlN	GaN	RT	77 K	RT	77 K	RT	77 K
MOCVD	2	0	1.15	1.10	1280	3316	425	171
NH_3-MBE	2	0	1.41	1.45	2107	8740	209	49
PAMBE	2	0	1.65	1.65	2030	9400	188	42
PAMBE	0	0	0.3090	0.3163	394	374	5321	5476
PAMBE	1.5	0	1.358	1.372	822	1437	559	316
PAMBE	3	0	1.308	1.350	1975	8376	241	55
PAMBE	3	2	2.274	2.298	1684	5885	163	46
PAMBE	3	4	2.029	2.081	1693	5592	181	53

2.4. *Detecting and Quantifying Dopants and Impurities*

GaN-based structures are utilized in a diverse range of device applications, requiring specific material characteristics. A dopant material is often introduced to intentionally modify the material properties; however, the level of dopant required to achieve a particular behavior is very constrained. The dopant interaction and incorporation within the microstructure can lead to unwanted effects that reduce the device efficiency. The need for controlled spatial distribution of dopant is intensified for compound semiconductor microstructures and heterostructures [42]. For this reason, it is necessary to define the composition of dopant achieved within the microstructure as well as the distribution throughout. This is applicable in alternate situations where the 'dopant' is an unintentional impurity, as well. Impurities can be introduced during material growth and processing steps and therefore reveal insights on how the processing methods affect the device performance.

In order to achieve p-type conductivity in GaN-based LEDs and lasers, a dopant must be introduced to increase the hole concentration in the p-type layer. The local distribution of dopant impacts the energy landscape and therefore the carrier transport and current across semiconductor structures. Magnesium (Mg) is the standard electron acceptor element, utilized as a dopant in GaN. In order to compensate for the low doping efficiency of Mg and enhance hole concentration by up to 10 times, an AlGaN/GaN superlattice is implemented [89]. TEM can reveal structural information; however, it cannot provide dopant distribution of light elements like Mg. SIMS can detect impurities; however, local variation or dopant solubility information cannot be obtained with this tool. APT allows chemical visualization of elements at the atomic level and provides quantification of dopant with local chemistry information.

Bennett *et al.* showed, for the first time, 3-D distribution of Mg clusters in AlGaN layers of an AlGaN/GaN superlattice using APT [43]. APT analysis allowed for identification and 3-D visualization of Mg clusters resulting from doping. The respective AlGaN layers and Mg clusters are shown in Fig. 7 [43]. The observed distribution of clusters was assessed and compared to a binomial distribution to validate that they were not a result of random alloy fluctuations [43]. This assessment is beneficial when assessing clustering in an APT dataset. Later, Giddings *et al.* quantified Mg clusters in a commercially available

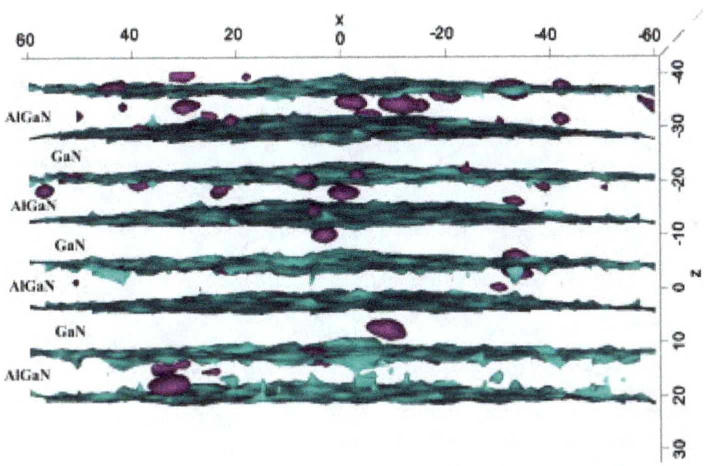

Fig. 7. APT reconstruction including flat, turquoise isoconcentration surfaces with 5.5 at. % Al at the upper and lower interface of each AlGaN layer. The AlGaN and GaN layers are labelled for clarity. The areas of increased Mg content are indicated by round, violet isoconcentration surfaces with 1.5 at. % Mg. Reproduced from Reference 43, with permission of Elsevier Publishing. DOI: https://doi.org/10.1016/j.ultramic.2010.11.028

LED device [66]. Giddings *et al.* also reports on Mg clusters found in the layers below the p-type layer, including the electron blocking layer (EBL) and a 7nm thick layer of GaN preceding the first InGaN QW [66]. These results have implications on the electrical properties of the material and can therefore offer useful insight in the improved design of LED structures.

Khromov *et al.* demonstrated that Mg cluster formation was dependent on the applied dopant concentration [90]. The highly doped sample ($5x10^{20}$ cm^{-3}) resulted in Mg clustering, while the low doping ($5x10^{19}$ cm^{-3}) sample demonstrated a homogenous distribution of Mg [90]. TEM was utilized as a complementary technique to APT, together revealing that the higher dopant concentration directly contributed to a higher density of basal plane stacking faults (SFs) [90]. Due to optical inactivity found in the highly doped samples, Khromov *et al.* posited that clustering of Mg atoms around SFs may have a screening (repulsive) effect on the holes, destroying the localization for excitons at the SFs [90]. These findings aid in the efforts towards optimization of applied dopant concentration and demonstrate the value of APT in detecting low concentrations of light elements. These benefits extend to scenarios in which a dopant is not intentional, such as, impurities introduced during processing procedures.

Oxygen is a common unintentional n-type dopant. Young *et al.* characterized impurities at the regrowth interface of a p-n tunnel junction using APT [91]. The concentration profile obtained from APT data, shown in Fig. 8, demonstrated that the p-type dopant (Mg) and the unintentional dopant (oxygen) overlapped at the regrowth interface, suggesting potential formation of an Mg-O complex [91]. This donor-acceptor complex would prevent Mg from incorporating into the n-side of the tunnel junction. Young *et al.* attributed the resulting lower forward operating voltage and higher efficiency to enhanced tunneling from O incorporation at the regrowth interface [91]. APT's ability to characterize the thin oxygen layer was pertinent in understanding how the oxygen impurities impacted tunnel junction device performance.

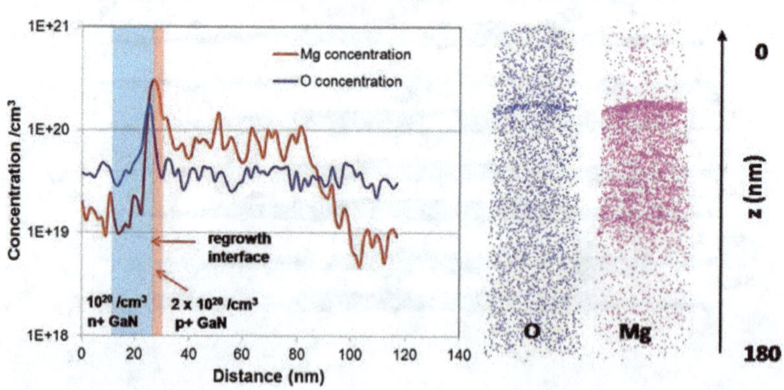

Fig. 8. Composition profile for O and Mg taken perpendicular to the regrowth interface (left). The respective 3-D atom maps for the distribution of O and Mg across the APT specimen are also shown (right). Reproduced from Reference 91, with permission from IOP Publishing © 2016 The Japan Society of Applied Physics. DOI: https://doi.org/10.7567/APEX.9.022102

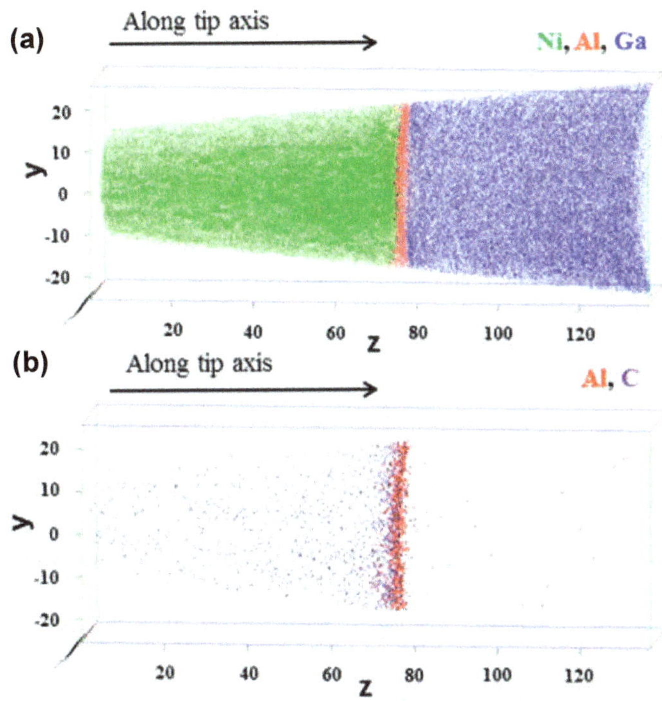

Fig. 9. (a) Three dimensional atom map of the Ni/Al_2O_3/GaN structure, analyzed in APT, with Ni in green, Al in red, and Ga in blue and (b) the same structure highlighting only the Al (red) and C (violet) atoms. Reproduced from Reference 92, with permission of AIP Publishing. DOI: https://doi.org/10.1063/1.4824211

Semiconductor device structures have specific chemical and physical design parameters, sensitive to changes at the nanometer scale. Trace amounts of impurities can have serious influence on the device performance. Mazumder *et al.* investigated the Al_2O_3 dielectric/GaN system interface through atom probe tomography to quantify the composition and assess incorporation of possible impurities [92]. The 3-D atom map of the analyzed Ni/Al_2O_3/GaN structure is shown in Fig. 9a [92]. APT analysis clearly showed the presence of carbon (C) impurities in the dielectric layer, shown in Fig. 9b [92]. Capacitance-voltage measurements allowed for correlation between estimated trap density and carbon concentration, determined by APT [92]. The estimated trap states were found to be on the same order of magnitude as the concentration of carbon impurities [92].

In another work, Mazumder *et al.* analyzed Al_2O_3 dielectrics in order to optimize the growth method and conditions for APT analysis [93]. Atom probe revealed an amorphous, intermixed interface, including C impurity levels at 2×10^{19} cm^{-3} in the ALD film [93]. APT allowed for accurate quantification of impurity chemistry without prior knowledge of the exact composition and quantified light elements at low concentrations (APT limit is above 10^{17} cm^{-3}) [50, 93, 94]. The understandings gained from APT analysis help inform improved growth method protocols and structural chemistry parameters for the evolution of semiconductor device design.

3. Conclusion

APT offers unique metrological capabilities ideal for the optimization of power electronic and optoelectronic device structures, including near-atomic spatial resolution and 3-D elemental distribution. These attributes surpass the limitations of alternate characterization techniques that can only provide bulk composition measurements or planar distribution of elements. The enhanced insight gained from APT allows for visualization and quantification of dopant distribution, layer interfaces and unintended impurities. The application of APT in GaN-based wide bandgap structures is reviewed to demonstrate the depth of chemical and physical properties revealed and how they correlate with device performance. The reviewed works make evident the role of APT in the investigation and quantification of atomic composition, structural inhomogeneities, elemental incorporation, dopants, and impurities to understand the respective influence on carrier transport and localization in the energy landscape, radiative recombination, internal quantum efficiency, and optical efficiency. The revealed correlations between material properties and performance are key for future advancement in GaN-based technology. APT will continue to serve as a necessary and ideal tool in gaining chemical and microstructural understanding towards optimization in wide bandgap structures.

References

1. J. A. del Alamo, Nature **479**, 317 (2011).
2. H. Amano, Y. Baines, E. Beam, M. Borga, T. Bouchet, P. R. Chalker, M. Charles, K. J. Chen, N. Chowdhury, R. M. Chu, C. De Santi, M. M. De Souza, S. Decoutere, L. Di Cioccio, B. Eckardt, T. Egawa, P. Fay, J. J. Freedsman, L. Guido, O. Haberlen, G. Haynes, T. Heckel, D. Hemakumara, P. Houston, J. Hu, M. Y. Hua, Q. Y. Huang, A. Huang, S. Jiang, H. Kawai, D. Kinzer, M. Kuball, A. Kumar, K. B. Lee, X. Li, D. Marcon, M. Marz, R. McCarthy, G. Meneghesso, M. Meneghini, E. Morvan, A. Nakajima, E. M. S. Narayanan, S. Oliver, T. Palacios, D. Piedra, M. Plissonnier, R. Reddy, M. Sun, I. Thayne, A. Torres, N. Trivellin, V. Unni, M. J. Uren, M. Van Hove, D. J. Wallis, J. Wang, J. Xie, S. Yagi, S. Yang, C. Youtsey, R. Y. Yu, E. Zanoni, S. Zeltner and Y. H. Zhang, J Phys D Appl Phys **51** (16) (2018).
3. S. J. Pearton, F. Ren, A. P. Zhang and K. P. Lee, Mat Sci Eng R **30** (3-6), 55-212 (2000).
4. V. Dmitriev, T. P. Chow, S. P. DenBaars, M. S. Shur and M. G. Spencer, 2000.
5. J. S. Speck and S. F. Chichibu, MRS Bull **34** (5), 304-312 (2009).
6. J. A. Garrido, J. L. Sanchez-Rojas, A. Jimenez, E. Munoz, F. Omnes and P. Gibart, Appl Phys Lett **75** (16), 2407-2409 (1999).
7. Y. Y. Zhang and G. R. Yao, J Appl Phys **110** (9) (2011).
8. G. J. Ding, L. W. Guo, Z. G. Xing, Y. Chen, P. Q. Xu, H. Q. Jia, J. M. Zhou and H. Chen, Acta Phys Sin-Ch Ed **59** (8), 5724-5729 (2010).
9. H. Hirayama, Y. Tsukada, T. Maeda and N. Kamata, Appl Phys Express **3** (3) (2010).
10. P. C. Chang, K. H. Lee, S. J. Chang, Y. K. Su and C. H. Liu, Semicond Sci Tech **24** (10) (2009).
11. H. C. Lee, S. Y. Hyun, H. I. Cho, C. Ostermaier, K. W. Kim, S. I. Ahn, K. I. Na, J. B. Ha, D. H. Kwon, C. K. Hahn, S. H. Hahm, H. C. Choi and J. H. Lee, Jpn J Appl Phys **47** (4), 2824-2827 (2008).
12. J. Zhao, Z. Lin, T. D. Corrigan, Z. Wang, Z. You and Z. Wang, Appl Phys Lett **91** (17) (2007).

13. Y. Yang, Y. Hao, J. C. Zhang, C. Wang and Q. Feng, Sci China Ser E **49** (4), 393-399 (2006).
14. X. F. Cai, S. P. Li and J. Y. Kang, Superlattice Microst **97**, 1-7 (2016).
15. Y. Uzawa, Z. Wang, A. Kawakami and B. Komiyama, Appl Phys Lett **66** (15), 1992-1994 (1995).
16. J. Y. Lee, B. R. Park, H. Kim, J. Kim and H. Y. Cha, Electron Mater Lett **10** (6), 1115-1120 (2014).
17. T. Palacios, Y. Dora, A. Chakraborty, C. Sanabria, S. Keller, S. P. DenBaars and U. K. Mishra, Phys Status Solidi A **203** (7), 1845-1850 (2006).
18. N. Maeda, K. Tsubaki, T. Saitoh, T. Tawara and N. Kobayashi, Opt Mater **23** (1-2), 211-217 (2003).
19. O. Ambacher, K. Chu, B. Green, V. Tilak, N. Weimann, J. Smark, J. R. Shealy and L. F. Eastman, Inst Phys Conf Ser (166), 483-487 (2000).
20. S. J. Pearton and F. Ren, Adv Mater **12** (21), 1571-+ (2000).
21. B. J. Baliga, Semicond Sci Tech **28** (7) (2013).
22. T. Zimmermann, D. Deen, Y. Cao, J. Simon, P. Fay, D. Jena and H. G. Xing, Ieee Electr Device L **29** (7), 661-664 (2008).
23. I. P. Smorchkova, L. Chen, T. Mates, L. Shen, S. Heikman, B. Moran, S. Keller, S. P. DenBaars, J. S. Speck and U. K. Mishra, J Appl Phys **90** (10), 5196-5201 (2001).
24. L. Shen, S. Heikman, B. Moran, R. Coffie, N. Q. Zhang, D. Buttari, I. P. Smorchkova, S. Keller, S. P. DenBaars and U. K. Mishra, Ieee Electr Device L **22** (10), 457-459 (2001).
25. S. Keller, S. Heikman, L. Shen, I. P. Smorchkova, S. P. DenBaars and U. K. Mishra, Appl Phys Lett **80** (23), 4387-4389 (2002).
26. Y. Cao and D. Jena, Appl Phys Lett **90** (18) (2007).
27. Y. Cordier, J. C. Moreno, N. Baron, E. Frayssinet, S. Chenot, B. Damilano and F. Semond, Ieee Electr Device L **29** (11), 1187-1189 (2008).
28. B. Mazumder, M. H. Wong, C. A. Hurni, J. Y. Zhang, U. K. Mishra and J. S. Speck, Appl Phys Lett **101** (22) (2012).
29. S. C. Binari, P. B. Klein and T. E. Kazior, P Ieee **90** (6), 1048-1058 (2002).
30. Y. P. Hsu, S. J. Chang, Y. K. Su, C. S. Chang, S. C. Shei, Y. C. Lin, C. H. Kuo, L. W. Wu and S. C. Chen, J Electron Mater **32** (5), 403-406 (2003).
31. E. Matioli, E. Rangel, M. Iza, B. Fleury, N. Pfaff, J. Speck, E. Hu and C. Weisbuch, Appl Phys Lett **96** (3) (2010).
32. S. Nakamura, M. Senoh, S. Nagahama, N. Iwasa, T. Yamada, T. Matsushita, Y. Sugimoto and H. Kiyoku, Appl Phys Lett **69** (20), 3034-3036 (1996).
33. K. M. Kelchner, Y. D. Lin, M. T. Hardy, C. Y. Huang, P. S. Hsu, R. M. Farrell, D. A. Haeger, H. C. Kuo, F. Wu, K. Fujito, D. A. Cohen, A. Chakraborty, H. Ohta, J. S. Speck, S. Nakamura and S. P. DenBaars, Appl Phys Express **2** (7) (2009).
34. K. Eichhorn, presented at the Light-Emitting Diodes: Research, Manufacturing, and Applications X, 2006 (unpublished).
35. J. W. Raring, M. C. Schmidt, C. Poblenz, Y. C. Chang, M. J. Mondry, B. Li, J. Iveland, B. Walters, M. R. Krames, R. Craig, P. Rudy, J. S. Speck, S. P. DenBaars and S. Nakamura, Appl Phys Express **3** (11) (2010).
36. U. Strauss, S. Bruninghoff, M. Schillgalies, C. Vierheilig, N. Gmeinwieser, V. Kummler, G. Bruderl, S. Lutgen, A. Avramescu, D. Queren, D. Dini, C. Eichler, A. Lell and U. T. Schwarz, Gallium Nitride Materials and Devices Iii **6894** (2008).

37. A. Avramescu, T. Lermer, J. Muller, C. Eichler, G. Bruederl, M. Sabathil, S. Lutgen and U. Strauss, Appl Phys Express **3** (6) (2010).
38. B. W. Hakki and T. L. Paoli, J Appl Phys **44** (9), 4113-4119 (1973).
39. K. L. Shaklee and R. F. Leheny, Appl Phys Lett **18** (11), 475-& (1971).
40. G. Frankowsky, F. Steuber, V. Harle, F. Scholz and A. Hangleiter, Appl Phys Lett **68** (26), 3746-3748 (1996).
41. S. Heppel, A. Hangleiter, S. Bader, G. Bruderl, A. Weimar, V. Kummler, A. Lell, V. Harle, J. Off, B. Kuhn and F. Scholz, Physica Status Solidi a-Applied Research **188** (1), 59-63 (2001).
42. E. F. Schubert, *Doping in III-V Semiconductors*. (E. Fred Schubert, 2015).
43. S. E. Bennett, R. M. Ulfig, P. H. Clifton, M. J. Kappers, J. S. Barnard, C. J. Humphreys and R. A. Oliver, Ultramicroscopy **111** (3), 207-211 (2011).
44. A. A. Gribb and T. F. Kelly, Adv Mater Process **162** (2), 31-34 (2004).
45. M. A. Khan, S. P. Ringer and R. K. Zheng, Adv Mater Interfaces **3** (12) (2016).
46. S. J. Pearton, in *Wide Bandgap Semiconductors - Growth, Processing and Applications* (William Andrew Publishing/Noyes).
47. P.-P. Choi, O. Cojocaru-Mirédin, D. Abou-Ras, R. Caballero, D. Raabe, V. S. Smentkowski, C. G. Park, G. H. Gu, B. Mazumder, M. H. Wong, Y.-L. Hu, T. P. Melo and J. S. Speck, Microscopy Today **20** (3), 18-24 (2012).
48. L. Rigutti, B. Bonef, J. Speck, F. Tang and R. A. Oliver, Scripta Mater **148**, 75-81 (2018).
49. A. Devaraj, D. E. Perea, J. Liu, L. M. Gordon, T. J. Prosa, P. Parikh, D. R. Diercks, S. Meher, R. P. Kolli and Y. S. Meng, Int Mater Rev **63** (2), 68-101 (2018).
50. T. F. Kelly and D. J. Larson, Annu Rev Mater Res **42**, 1-31 (2012).
51. A. Cerezo, P. H. Clifton, M. J. Galtrey, C. J. Humphreys, T. F. Kelly, D. J. Larson, S. Lozano-Perez, E. A. Marquis, R. A. Oliver, G. Sha, K. Thompson, M. Zandbergen and R. L. Alvis, Mater Today **10** (12), 36-42 (2007).
52. E. W. Muller, J. A. Panitz and S. B. Mclane, Rev Sci Instrum **39** (1), 83-& (1968).
53. M. K. Miller and R. G. Forbes, *Atom-Probe Tomography: The Local Electrode Atom Probe*. (Springer US, 2014).
54. M. J. Galtrey, R. A. Oliver, M. J. Kappers, C. J. Humphreys, P. H. Clifton, D. Larson, D. W. Saxey and A. Cerezo, J Appl Phys **104** (1) (2008).
55. F. Liu, L. Huang, R. F. Davis, L. M. Porter, D. K. Schreiber, S. V. N. T. Kuchibatla, V. Shutthanandan, S. Thevuthasan, E. A. Preble, T. Paskova and K. R. Evans, J Vac Sci Technol B **32** (5) (2014).
56. T. J. Prosa, P. H. Clifton, H. Zhong, A. Tyagi, R. Shivaraman, S. P. DenBaars, S. Nakamura and J. S. Speck, Appl Phys Lett **98** (23) (2011).
57. O. Cojocaru-Miredin, T. Schwarz and D. Abou-Ras, Scripta Mater **148**, 106-114 (2018).
58. S. Koelling, A. Li, A. Cavalli, S. Assali, D. Car, S. Gazibegovic, E. P. A. M. Bakkers and P. M. Koenraad, Nano Letters **17** (2), 599-605 (2017).
59. F. Tang, D. S. Gianola, M. P. Moody, K. J. Hemker and J. M. Cairney, Acta Mater **60** (3), 1038-1047 (2012).
60. H. R. Li, B. Mazumder, B. Bonef, S. Keller, S. Wienecke, J. S. Speck, S. P. Denbaars and U. K. Mishra, Semicond Sci Tech **32** (11) (2017).
61. M. J. Galtrey, R. A. Oliver, M. J. Kappers, C. J. Humphreys, D. J. Stokes, P. H. Clifton and A. Cerezo, Appl Phys Lett **90** (6) (2007).
62. D. J. Larson, T. J. Prosa, R. M. Ulfig, B. P. Geiser and T. F. Kelly, in *Local Electrode Atom Probe Tomography: A User's Guide* (Springer, New York, NY, USA, 2013).

63. B. Gault, M. P. Moody, J. M. Cairney and S. P. Ringer, in *Atom Probe Microscopy*, edited by R. Hull, C. Jagadish, R. M. Osgood Jr., J. Parisi and Z. M. Wang (Springer, New York, NY, USA, 2012).
64. D. J. Larson, D. Lawrence, W. Lefebvre, D. Olson, T. J. Prosa, D. A. Reinhard, R. M. Ulfig, P. H. Clifton, J. H. Bunton, D. Lenz, J. D. Olson, L. Renaud, I. Martin and T. F. Kelly, 17th International Conference on Microscopy of Semiconducting Materials 2011 **326** (2011).
65. S. E. Bennett, T. M. Smeeton, D. W. Saxey, G. D. W. Smith, S. E. Hooper, J. Heffernan, C. J. Humphreys and R. A. Oliver, J Appl Phys **111** (5) (2012).
66. A. D. Giddings, T. J. Prosa, D. Olson, P. H. Clifton and D. J. Larson, Microscopy Today **22** (5), 12-19 (2014).
67. R. J. H. Morris, R. Cuduvally, D. Melkonyan, C. Fleischmann, M. Zhao, L. Arnoldi, P. van der Heide and W. Vandervorst, J Vac Sci Technol B **36** (3) (2018).
68. N. A. Sanford, P. T. Blanchard, M. Brubaker, K. A. Bertness, A. Roshko, J. B. Schlager, R. Kirchhofer, D. R. Diercks and B. Gorman, Physica Status Solidi C: Current Topics in Solid State Physics, Vol 11, No 3-4 **11** (3-4), 608-612 (2014).
69. H. M. Foronda, B. Mazumder, E. C. Young, M. A. Laurent, Y. L. Li, S. P. DenBaars and J. S. Speck, J Cryst Growth **475**, 127-135 (2017).
70. A. E. Romanov, T. J. Baker, S. Nakamura and J. S. Speck, J Appl Phys **100** (2) (2006).
71. R. Shivaraman, Y. Kawaguchi, S. Tanaka, S. P. DenBaars, S. Nakamura and J. S. Speck, Appl Phys Lett **102** (25) (2013).
72. D. A. Browne, B. Mazumder, Y. R. Wu and J. S. Speck, J Appl Phys **117** (18) (2015).
73. L. Rigutti, I. Blum, D. Shinde, D. Hernandez-Maldonado, W. Lefebvre, J. Houard, F. Vurpillot, A. Vella, M. Tchernycheva, C. Durand, J. Eymery and B. Deconihout, Nano Lett **14** (1), 107-114 (2014).
74. F. Z. Tang, T. T. Zhu, F. Oehler, W. Y. Fu, J. T. Griffiths, F. C. P. Massabuau, M. J. Kappers, T. L. Martin, P. A. J. Bagot, M. P. Moody and R. A. Oliver, Appl Phys Lett **106** (7) (2015).
75. Y. R. Wu, R. Shivaraman, K. C. Wang and J. S. Speck, Appl Phys Lett **101** (8) (2012).
76. T. J. Yang, J. S. Speck and Y. R. Wu, Gallium Nitride Materials and Devices Ix **8986** (2014).
77. D. Watson-Parris, M. J. Godfrey, P. Dawson, R. A. Oliver, M. J. Galtrey, M. J. Kappers and C. J. Humphreys, Phys Rev B **83** (11) (2011).
78. L. Rigutti, L. Mancini, W. Lefebvre, J. Houard, D. Hernandez-Maldonado, E. Di Russo, E. Giraud, R. Butte, J. F. Carlin, N. Grandjean, D. Blavette and F. Vurpillot, Semicond Sci Tech **31** (9) (2016).
79. L. Mancini, F. Moyon, D. Hernandez-Maldonado, I. Blum, J. Houard, W. Lefebvre, F. Vurpillot, A. Das, E. Monroy and L. Rigutti, Nano Letters **17** (7), 4261-4269 (2017).
80. R. Czernecki, E. Grzanka, J. Smalc-Koziorowska, S. Grzanka, D. Schiavon, G. Targowski, J. Plesiewicz, P. Prystawko, T. Suski, P. Perlin and M. Leszczynski, J Cryst Growth **414**, 38-41 (2015).
81. Y. T. Moon, D. J. Kim, K. M. Song, D. W. Kim, M. S. Yi, D. Y. Noh and S. J. Park, J Vac Sci Technol B **18** (6), 2631-2634 (2000).
82. A. F. Tsatsulnikov, W. V. Lundin, E. E. Zavarin, A. E. Nikolaev, A. V. Sakharov, V. S. Sizov, S. O. Usov, Y. G. Musikhin and D. Gerthsen, Semiconductors+ **45** (2), 271-276 (2011).
83. X. C. Ren, J. R. Riley, D. D. Koleske and L. J. Lauhon, Appl Phys Lett **107** (2) (2015).
84. S. Choi, F. Wu, R. Shivaraman, E. C. Young and J. S. Speck, Appl Phys Lett **101** (4) (2012).
85. M. N. Fireman, D. A. Browne, B. Mazumder, J. S. Speck and U. K. Mishra, Appl Phys Lett **106** (20) (2015).

86. S. W. Kaun, B. Mazumder, M. N. Fireman, E. C. H. Kyle, U. K. Mishra and J. S. Speck, Semicond Sci Tech **30** (5) (2015).
87. B. Mazumder, S. W. Kaun, J. Lu, S. Keller, U. K. Mishra and J. S. Speck, Appl Phys Lett **102** (11) (2013).
88. S. W. Kaun, E. Ahmadi, B. Mazumder, F. Wu, E. C. H. Kyle, P. G. Burke, U. K. Mishra and J. S. Speck, Semicond Sci Tech **29** (4) (2014).
89. P. Kozodoy, M. Hansen, S. P. DenBaars and U. K. Mishra, Appl Phys Lett **74** (24), 3681-3683 (1999).
90. S. Khromov, D. Gregorius, R. Schiller, J. Losch, M. Wahl, M. Kopnarski, H. Amano, B. Monemar, L. Hultman and G. Pozina, Nanotechnology **25** (27) (2014).
91. E. C. Young, B. P. Yonkee, F. Wu, S. H. Oh, S. P. DenBaars, S. Nakamura and J. S. Speck, Appl Phys Express **9** (2) (2016).
92. B. Mazumder, M. Esposto, T. H. Hung, T. Mates, S. Rajan and J. S. Speck, Appl Phys Lett **103** (15) (2013).
93. B. Mazumder, X. Liu, R. Yeluri, F. Wu, U. K. Mishra and J. S. Speck, J Appl Phys **116** (13) (2014).
94. L. J. Lauhon, P. Adusumilli, P. Ronsheim, P. L. Flaitz and D. Lawrence, MRS Bull **34** (10), 738-743 (2009).

β-$(Al,Ga)_2O_3$ for High Power Applications — A Review on Material Growth and Device Fabrication

Ashley (Zhe) Jian*, Kamruzzaman Khan*,‡ and Elaheh Ahmadi†

Department of Electrical Engineering and Computer Science, University of Michigan, Ann Arbor, MI 48109, USA
†*eahmadi@umich.edu*

‡*Department of Materials Science and Engineering, University of Michigan, Ann Arbor, MI 48109, USA*

Gallium Oxide has attracted a great deal of attention for power electronics due to its large bandgap (~4.8 eV) as well as availability of cost effective, large area, high quality substrates. In this article, we will discuss advancements in homoepitaxial and heteroepitaxial growth of β-$(Al,Ga)_2O_3$ and their heterostructures via different growth techniques with the emphasis on metal organic chemical vapor deposition, molecular beam epitaxy, and halide vapor phase epitaxy. We will also review various device structures demonstrated for high power applications.

Keywords: Ga_2O_3; high power electronics; MBE; MOCVD; HVPE; mist-CVD.

1. Introduction

Wide Bandgap based RF and power electronics are going to define many crucial "more-electric" transportation systems and communication systems including RADAR. Underlying this success is the enhanced efficiency in conversion processes afforded by wideband gap materials from kHz to GHz. The need is currently being served by SiC and GaN but the requirements are even beyond the performance of these systems. Nonetheless, producing large scale, cost-effective, and high quality GaN and (to a lesser extent) SiC substrates remains the major challenge in the development roadmap of power electronics based on these wide bandgap materials.

Low on-resistance, high breakdown voltage, and small leakage current in the off-state are the key parameters needed for a low-loss, high-power, and high-frequency switch. To realize these key parameters, the semiconductor of choice requires to have high breakdown field, large bandgap and high electron mobility. Wide bandgap semiconductors, including β-Ga_2O_3, allow a maximum electric field breakdown strength over ten times larger and on-state resistance (R_{ON}) more than four hundred times lower than their conventional Si counterparts at a given voltage [1–7].

* These authors have equally contributed to this manuscript.

Gallium Oxide is the most promising material to provide the next step in performance based on cost effective, large area, high quality substrates. Ga_2O_3 can be found in five different polymorphs including α, β, δ, ε, and γ [8]. Among the five commonly identified polymorphs of Ga_2O_3, the β-Ga_2O_3 has recently attracted a great deal of attention due to its wide bandgap of ~4.8 eV, and availability of melt growth techniques to produce high quality substrates cost-effectively [9, 10, 11, 12]. Floating Zone (FZ) [13], Czochralski [10], and edge-defined film-fed growth (EFG) [14] methods have been successfully employed to produce β-Ga_2O_3 substrates.

Since 2012, when the first β-Ga_2O_3 field effect transistor (FET) was demonstrated, the field has been expanded and a lot of progress has been made. The quality of β-Ga_2O_3 substrates has improved remarkably and both insulating, and n-type doped substrates are now commercially available in different crystal orientations including ($\bar{2}01$), (010), (001), and (100). Epitaxial growth of β-$(Al,Ga)_2O_3$ thin films and n-type doping have been demonstrated via various growth techniques. Additionally, Studies on dry etching of β-Ga_2O_3, Ohmic contact, Schottky contact, and dielectrics have been reported by several groups. Furthermore, various β-Ga_2O_3-based electronic device structures, including Schottky [15–17] and PN diodes [18, 19], FETs [9, 20], metal-oxide-semiconductor FETs (MOSFETs) [21, 22], modulation-doped FETs (MODFETs) [23–25], and current aperture vertical MOSFET (CAVETs) [26, 27] have been realized.

Nevertheless, looking at the success roadmap of Si, GaN, and SiC, in order to realize the full potential of β-Ga_2O_3, various fundamental challenges need to be addressed. Some of these challenges include high-quality bulk and epitaxial growth of material, characterization of defects and understanding their impact on device performance, efficient and reliable n-type/p-type doping, low-resistance Ohmic contacts, low-leakage Schottky contacts, high-quality dielectrics, high electrical and thermal conductivity, high etch-rate and low damage dry/wet etching, and last but not least, identifying proper device design based on the application. Although significant progress has been made in addressing many of these challenges, attempts for p-type doping of β-Ga_2O_3 have been unsuccessful and it is widely believed to be impossible (at least via conventional techniques). In addition to a shallow acceptor not being foreseen, due to the relatively flat valence band of β-Ga_2O_3, the hole mobility is expected to be close to zero in this material. Furthermore, a relatively low electron mobility (limited by phonon scattering) and low thermal conductivity remain the two main obstacles that need to be overcome for β-Ga_2O_3 technology to succeed.

In this article, we will discuss advancements in homoepitaxial and heteroepitaxial growth of β-$(Al,Ga)_2O_3$ and their heterostructures via different growth techniques with the emphasis on metal organic chemical vapor deposition (MOCVD), molecular beam epitaxy (MBE), and halide vapor phase epitaxy (HVPE). We will then review various device structures demonstrated for high power applications.

2. Epitaxial Growth

Since the development of high-quality β-Ga_2O_3 substrate is still under progress and the technology is yet to be matured, the epitaxial growth on as-received β-Ga_2O_3 substrates typically leads to various crystal defects and a poor surface morphology. Therefore, proper surface pre-treatment prior to the homoepitaxial growth is of crucial importance. In-situ oxygen plasma treatment in the MBE chamber followed by Ga-etching has been shown to be effective to remove absorbates residues from the substrate surface and improve surface morphology [28–32]. Oshima *et al.* [29] showed that exposing β-Ga_2O_3 substrate to Ga flux in an MBE chamber in the absence of oxygen led to etching β-Ga_2O_3 surface, and the etching rate increased proportional to the Ga-flux. This behavior was observed on different planes. The root-mean-square roughness of the homo-epitaxially grown film was markedly improved when Ga-etching was performed prior to the growth. This enhancement in the surface morphology was attributed to removing chemical-mechanical polishing (CMP) related damage residue and other substrate contaminations. Figure 1 shows atomic force microscopy (AFM) images of a β-Ga_2O_3 substrate before and after Ga etching. It was later shown by the same group [32, 33] that Ga etching also reduces the unintentional

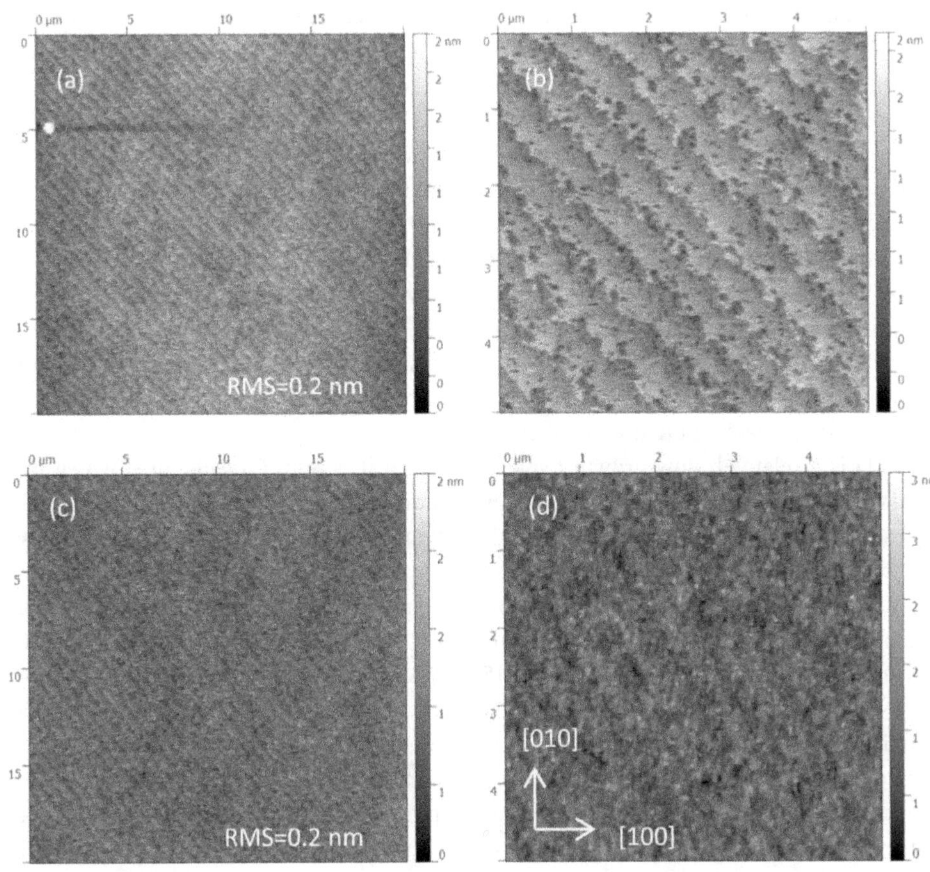

Fig. 1. AFM images of a (001) β-Ga_2O_3 substrate. (a), (b): virgin substrates, (c), (d): after Ga-etching (−29 nm) [29]. Copyright 2018 IOP publishing.

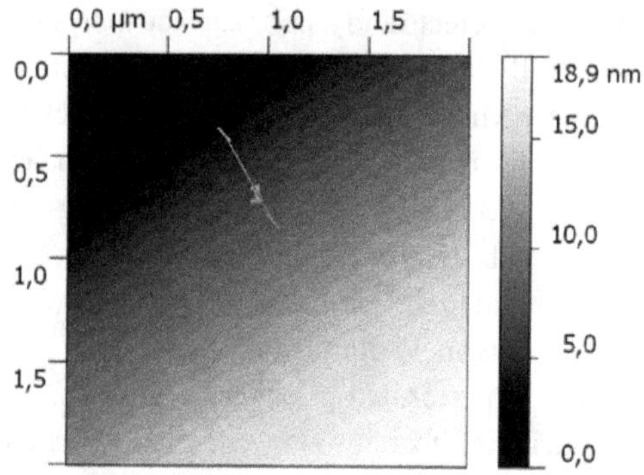

Fig. 2. AFM image of a β-Ga_2O_3 (100) surface after 1 h annealing at 1000 °C in oxygen-containing environment [35]. Copyright 2014 John Wiley and Sons.

doping concentration of Ge and Si at the growth interface. High-temperature annealing of β-Ga_2O_3 substrates at 850 °C - 1000 °C for 30-60 min in air/O_2 environment has been also proven to help obtaining a damage-free surface and resulting in a terrace-like surface morphology [34–36]. Figure 2 shows AFM image of a β-Ga_2O_3 substrate after annealing in O_2 environment at 1000 °C for 1min in an O_2 environment.

Various techniques have been used for epitaxial growth of β-Ga_2O_3, including sol-gel process [37], HVPE [38], Mist-CVD [39, 40], pulsed laser deposition (PLD) [41], molecular beam epitaxy (MBE) [42], and metal organic chemical vapor deposition (MOCVD) [43]. Substrates used for epitaxial growth of β-Ga_2O_3, growth rate, as well as electron concentration and mobility in β-Ga_2O_3 thin films grown via these growth techniques are summarized in Table 1. Figure 3 also depicts electron mobility vs electron concentrations in n-type doped β-Ga_2O_3 films grown via different growth techniques gathering all the data reported in the literature.

PLD is a relatively low growth temperature (300 °C - 800 °C) technique, which has been widely used before to study thin film growth of oxide [41, 44–51] and nitride [52, 53] [54] semiconductors. However, due to small growth rates and non-uniformity of thickness and composition on large scale substrates, it may not be very attractive for industrial applications. Hetero-epitaxial and homoepitaxial growth of β-Ga_2O_3 have been reported. Although β-Ga_2O_3 films grown hetero-epitaxially on sapphire [41, 45, 46] and MgAlO substrates suffer from poor crystal quality [50], β-Ga_2O_3 films grown homo-epitaxially on β-Ga_2O_3 substrates exhibited single crystal structure with relatively smooth surface morphology [46–48]. Si doping of β-Ga_2O_3 have been also demonstrated by PLD. However, due to oxidation of Si charge in the oxide environment of the PLD system, it has been challenging to achieve low/medium level of doping concentration via this growth technique. Zhang *et al.* [47] addressed this issue by using different β-Ga_2O_3:Si targets with various amount of Si concentration. However, this approach only enables Si doping concentrations close to the Si concentration of the target and does not allow for an arbitrary Si concentration.

Table 1. Summary of substrates used for epitaxial growth of β-Ga_2O_3, growth rate, as well as electron concentration and mobility in β-Ga_2O_3 thin films grown via different growth techniques.

Method	**Substrate**	**Dopants**	**Carrier concentration (cm^{-3})**	**Mobility range ($cm^2V^{-1}s^{-1}$) (carrier conc. for highest mobility)**	**Growth rate (μm/h)**	**Ref.**
MOCVD	β-Ga_2O_3 (010):Fe			176 ($7.4x10^{15}$)	0.8	[83]
MOCVD	β-Ga_2O_3 (010)	Sn	$2x10^{17}$-10^{19}	32-100 ($2x10^{17}$)	0.15-.20	[82]
MOCVD	β-Ga_2O_3 (010)	Si	10^{17}-$8x10^{19}$	50-130 (10^{17})	0.15-.20	[82]
MOCVD	β-Ga_2O_3 (010)	Sn	$6x10^{17}$-$7.1x10^{17}$	10-13.7 ($7.1x10^{17}$)	0.6	[34]
MOCVD	β-Ga_2O_3 (010)				0.018-0.048	[35]
MOCVD	a-sapphire and c-sapphire				0.15	[74]
MOCVD	c-sapphire (001)				10	[75]
MOCVD	c-sapphire (001)				0.01-0.48	[43]
MOCVD	β-Ga_2O_3 (001) 6 ° miscut toward [001]		$(1.1\pm0.2)x10^{18}$	80±10	0.12	[85]
MOCVD	β-Ga_2O_3 (001) 6 ° miscut toward [$00\bar{1}$]		$(1.6\pm0.2)x10^{18}$	8±1	0.12	[85]
MBE	β-Ga_2O_3 (010):Mg	Sn	$4x10^{16}$-10^{19}	35-130 ($4x10^{16}$)	0.125	[89]
MBE	β-Ga_2O_3:Sn (010)	Sn	10^{17}-10^{18}	10-80 (10^{17})	0.03-0.132	[28]
MBE	β-Ga_2O (010):Fe	Ge	$1.6x10^{18}$-$1.4x10^{19}$	39-97 ($1.6x10^{18}$)		[96]
MBE	β-Ga_2O_3 (010):Fe	Sn	$5x10^{16}$-10^{20}	39-120 ($5x10^{16}$)		[96]
MBE	β-Ga_2O_3 (001)	Ge	$6.7x10^{17}$-$9.5x10^{18}$	20.3-26.3 ($6.7x10^{17}$)	0.055	[33]
MBE	β-Ga_2O_3 (001)	Sn		25.3 ($1.1x10^{18}$)		[33]
MBE	β-Ga_2O_3 (010)	Si	$6.8x10^{19}$-$1.7x10^{20}$	77-81 ($6.8x10^{19}$)	0.155-0.24	[95]
MBE	c-sapphire				0.33-0.1	[86]
MBE	β-Ga_2O_3:Si (010)				0.5-0.6	[90]
MBE	c-sapphire				0.014	[94]
MBE	β-Ga_2O_3 (010)				0.1-.2	[29]
MBE	β-Ga_2O_3 (001)				0.05	[29]
MBE	β-Ga_2O_3 (100)				0.027	[29]
MBE	β-Ga_2O_3 (100)				0.01-0.046	[93]
PLD	c-sapphire	Si	$1x10^{15}$-$2.4x10^{20}$	0.03-0.3 ($1x10^{15}$)		[47]
PLD	β-Ga_2O_3 (010):Fe	Si	$2.61x10^{19}$-$1.74x10^{20}$	18.6-26.5 ($2.61x10^{19}$)		[36]
PLD	β-Ga_2O_3 (010)				0.005-0.2	[48]
PLD	c-sapphire				0.07-0.22	[41]
HVPE	β-Ga_2O_3 (001):Fe	Si	$3.18x10^{15}$-$1.18x10^{18}$	88-149 ($3.18x10^{18}$)	5-10.0	[115]
HVPE	β-Ga_2O_3 (001)			123 (1.1. $x10^{17}$)	25.0	[110]
HVPE	β-Ga_2O_3 (001)	Si	$2x10^{16}$-$1x10^{19}$	50-100 ($2x10^{16}$)	25.0	[110]
HVPE	c-sapphire				1-25	[111]
HVPE	β-Ga_2O_3 (001)				8.0	[108]

(Continued)

Table 1. (*Continued*)

Method	Substrate	Dopants	Carrier concentration (cm^{-3})	Mobility range ($cm^2V^{-1}s^{-1}$) (carrier conc. for highest mobility)	Growth rate (μm/h)	Ref.
HVPE	β-Ga_2O_3 (001):Sn				5.0	[38]
HVPE	Off axis c-sapphire				1-250	[112]
LPCVD	β-Ga_2O_3 (010)	Si		73 (1.2 x10^{18})	1.9	[69]
LPCVD	β-Ga_2O_3 (001)	Si		42 (9.5 x10^{17})	1.2	[69]
LPCVD	β-Ga_2O_3 (010)	Si		100 (2.5x10^{17})	2.0	[71]
LPCVD	c-sapphire	Si	9x10^{16}-3x10^{19}	8-42 (1.32x10^{18})	1.7	[68]
LPCVD	Off axis c-sapphire	Si	4x10^{17}-7x10^{18}	10-65 (4x10^{17})	1-30	[67]
LPCVD	Off axis c-sapphire		10^{17}-4x10^{19}	106.6 (4.83x10^{17})	0.5-10	[65]
LPCVD	c-sapphire				1.1	[64]
LPCVD	β-Ga_2O_3 (010)				1.3	[66]
LPCVD	β-Ga_2O_3 (010)	Si			6.0	[70]
LPCVD	Sapphire ($11\bar{2}0$)				1.8	[64]
LPCVD	Sapphire ($1\bar{1}02$)				0.4	[64]
Mist-CVD	β-Ga_2O_3 (010):Sn				0.12-0.9	[40]
Mist-CVD	β-Ga_2O_3 (010)	Sn	10^{18}-5x10^{20}	10-45(10^{18})		[40]

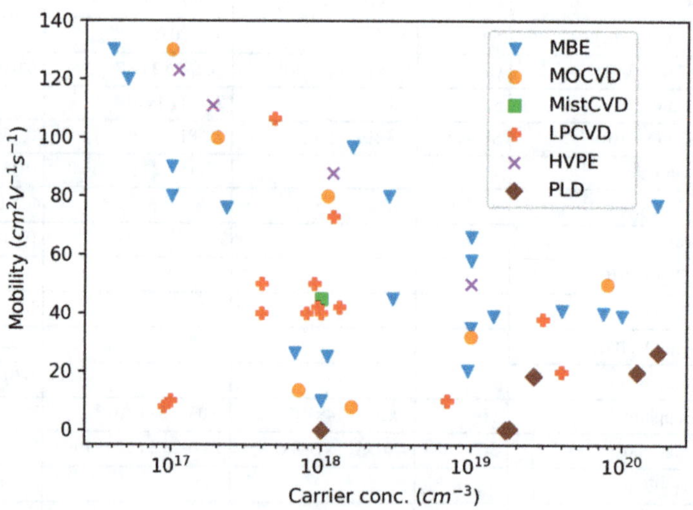

Fig. 3. Electron mobility vs. electron concentrations in n-type doped β-Ga_2O_3 films grown via different growth techniques with all the data reported in the literature.

In the past, low pressure chemical vapor deposition (LPCVD) has been widely employed for deposition of thin films such as polycrystalline silicon [55–57], Si_xN_y [58–60], SiO_2 [61, 62] and high-k dielectrics [63]. LPCVD is considered a relatively simple and low-cost growth technique with high reproducibility and scalability. Recently, Zhao and her group members have developed this deposition technique for epitaxial growth of Ga_2O_3 on sapphire [64, 65] and β-Ga_2O_3 [66] substrates. High purity gallium pallets and oxygen were used as the precursors for gallium and oxygen, respectively, and Ar was used as the carrier gas. They have studied the impacts of substrate plane, substrate miscut, and growth conditions on β-Ga_2O_3 crystal quality [64, 66, 67]. Epitaxial growth of pure β-phase Ga_2O_3 ($\bar{2}01$) has been reported on c-plane sapphire via this deposition technique. Si doping of β-Ga_2O_3 have been also demonstrated using $SiCl_4$ as the precursor [68–71] with a wide range of carrier concentrations ($9x10^{16}$ - $4x10^{19}$ cm^{-3}) with room-temperature mobilities ranging from 32 cm^2/Vs to 72 cm^2/Vs. Very recently, Joishi *et al.* [71] demonstrated room temperature electron mobility of ~100 cm^2V^{-1} s^{-1} for a carrier density of $2.5x10^{17}$ cm^{-3}, which is comparable with that achieved via more common growth techniques such as MOCVD and MBE.

Mist-CVD is considered a low-cost environment-friendly growth technique, in which the liquid solution is atomized by an ultrasonic transducer and the mist particles are transferred by the carrier gas to the substrate. For mist-CVD of β-Ga_2O_3, Ga solution consists of gallium acetylacetonate [$Ga(C_5H_8O_3)_3$: .05 mol/L] dissolved in deionized water with addition of a slight amount of hydrochloric acid (HCl) and O_2 is employed as the oxidant. The practical utility of mist-CVD is evident in the recent establishment of Flosfia, a start-up based in Japan. Flosfia has demonstrated power device operation of Schottky barrier diodes (SBDs) made from α-Ga_2O_3 deposited by mist-CVD, which have record low on-resistance of 0.1mΩcm^2, and when package mounted, have a thermal conductivity comparable to SiC commercial devices [72]. However, there are only a few reports [39, 40, 73] on using this technique for epitaxial growth of β-Ga_2O_3. Sn doping of β-Ga_2O_3 has been demonstrated with a reasonable carrier concentration ranging from 10^{18} cm^{-3} to $5x10^{20}$ cm^{-3} with a maximum electron mobility of 45 $cm^2/(V.s)$ at a carrier concentration of 10^{18} cm^{-3}.

Although LPCVD and mist-CVD are both promising deposition techniques for epitaxial growth of β-Ga_2O_3, MOCVD, MBE, and HVPE have been more commonly employed, so far, for studying this material system. Therefore, in the following sections we will review the progress made via these growth techniques in more detail. The effect of growth conditions and substrate miscut on film quality will be highlighted and n-type doping will be discussed.

Metal-Organic Chemical Vapor Deposition: Epitaxial growth of β-Ga_2O_3 by MOCVD also called metal organic vapor pressure epitaxy (MOVPE) has been reported on sapphire [34, 43, 74–76], β-Ga_2O_3 [34, 35, 77], MgO [44, 78, 79], GaAs [80] and $MgAl_2O_4$ [81]. Oxygen and water vapor have been used as oxidants in this growth technique. Although the attempt of Wagner *et al.* [35] to grow thin film β-Ga_2O_3 (100) by oxygen was unsuccessful, other groups have demonstrated high quality β-Ga_2O_3 films on b-plane and a-plane as well as on sapphire by oxygen as the oxidant [75, 76]. On the other hand, Wagner *et al.* [35] demonstrated smooth homoepitaxial growth of β-Ga_2O_3 films with water as an oxidant in MOCVD. They used Trimethylgallium (TMG) as Ga precursor and high purity Ar as the carrier gas in this work.

TMG has been typically used as the Ga precursor to grow β-Ga_2O_3 by MOCVD while Baldini *et al.* [82] and Zhang *et al.* [83] used Triethylgallium (TEG) as Ga precursor. Moreover, Alema *et al.* [75] have demonstrated successful growth of β-Ga_2O_3 on c-plane sapphire using TMG, TEG and Ga $(DPM)_3$ as Ga precursor. They used a close coupled showerhead MOCVD (CCS-MOCVD) reactor for this study to prevent the premature oxidation of the MO precursors [75]. A record growth rate of 10 μm/hour was achieved using TMG as Ga precursor due to TMG's higher pyrolyzing temperature.

Alema *et al.* [75] have also done an extensive study of the growth conditions on growth rate of β-Ga_2O_3. They showed that, when TEG was used as the Ga precursor, the β-Ga_2O_3 growth rate decreased with increasing the substrate temperature. In contrast, when TMG was used as the precursor, the growth rate remained relatively constant at substrate temperatures ranging from 850 °C to 950 °C. On the other hand, Sbrockey *et al.* [43] showed an increase in growth rate with an increase in substrate temperature from 500 °C to 700 °C when TMG was used as the precursor. This is because TEG pyrolyzes at a lower temperature (~300 °C) compared with that for TMG (500 °C), which leads to TEG being more susceptible to the gas phase parasitic reaction [75]. Additionally, high-temperature growth may cause decomposition of grown β-Ga_2O_3 layer and desorption of Ga from the β-Ga_2O_3 surface which would lead to lower growth rates. On the other hand, when Ga $(DPM)_3$ was employed as the Ga precursor, the growth rate of β-Ga_2O_3 increased with increasing the substrate temperature from 750 °C to 900 °C due to an increase in the efficiency of Ga $(DPM)_3$ pyrolysis. The growth rate reportedly decreased [43, 75] as the reactor pressure increased from 40 Torr to 120 Torr independent of reactor configuration (CCS vs vertical rotating disc (VRD)) or Ga precursor.

Si and Sn doping of β-Ga_2O_3 have been demonstrated by MOCVD using Tetraethylorthosilicate (TEOS) and Tetraethyltin (TESn) as metal-organic precursors for Si [82, 83, 84], and Sn [34], respectively. A uniform doping profile has been demonstrated using either Si or Sn as n-type dopants. A wide range of electron concentrations have been achieved with Si as the dopant ($1x10^{17}$ cm^{-3} - $8x10^{19}$ cm^{-3}). Carrier concentrations as high as $1x10^{19}$ cm^{-3} have been also demonstrated using Sn. The surface morphology was not in general affected by Sn or Si doping and carrier density increased linearly by increasing the doping concentration, except for Sn doping concentrations beyond $1x10^{19}$ cm^{-3}, which led to elongated blocks due to step bunching. Additionally, electron density of only $2x10^{18}$ cm^{-3} was measured on a film with $8x10^{19}$ cm^{-3} Sn concentration, which may be due to self-compensation of dopants or generation of point defects which behave as compensating centers. Moreover, unlike Si doping, a severe memory effect was observed when high flow rates of TESn was used in the MOCVD reactor which is due to higher vapor pressure of TESn compared with that of TEOS in addition to larger atomic mass of Sn. Consequently, Sn concentrations as high as $4x10^{17}$ cm^{-3} were measured on unintentionally doped β-Ga_2O_3 films which were grown right after growth of a highly Sn-doped film in the same reactor. Similar range of electron mobilities (50 cm^2/Vs - 130 cm^2/Vs) were measured on Sn and Si doped β-Ga_2O_3 films grown by MOCVD. Zhang *et al.* [83] have reported an electron mobility of 176 cm^2/V.s and 3481 cm^2/V.s at room temperature and 54 K, respectively, on an unintentionally-doped β-Ga_2O_3 film grown by MOCVD with an intrinsic carrier concentration of $7.4x10^{15}$ cm^{-3}. Very recently, record electron mobilities of 184 cm^2 /V s at room temperature and 4984 cm^2/Vs at low temperature (45 K) were measured for β-Ga_2O_3 thin films with room-temperature doping concentrations of $2.5x10^{16}$ and $2.75x10^{16}$ cm^{-3}, respectively. It is worth mentioning that these two record mobilities belong to two different samples, and have not been simultaneously measured on the same sample [84].

Schewski *et al.* [85] studied the impact of miscut orientation on the structural and electronic properties of homoepitaxial growth of β-Ga_2O_3 (100) by MOCVD. They showed that, after annealing, substrates with miscuts toward [00$\bar{1}$] exhibited monolayer steps whereas both monolayer and bilayer (mainly bilayer) steps were observed for miscuts toward [001]. Epitaxial growth on both substrates occurred in step-flow mode, and films on substrates with a miscut toward [00$\bar{1}$] were free of structural defects. In contrast, growth on a substrate with a miscut toward [001] exhibited three dimensional islands indicating rough homoepitaxial layers and steep edges due to step bunching. They achieved an electron mobility of about 90 $cm^2V^{-1}s^{-1}$ at electron concentrations in the range of $1\text{-}2\times10^{18}$ cm^{-3} on films grown on (100) substrates with 6° miscut toward the [00$\bar{1}$]. However, layers grown under identical conditions but with 6° miscut toward the [001] direction exhibited low electron mobilities of around 10 cm^2 $V^{-1}s^{-1}$.

Molecular Beam Epitaxy: MBE have been utilized for hetero-epitaxial growth of β-Ga_2O_3 on sapphire substrates [86, 87], and GaN templates [88] as well as homo-epitaxial growth [28, 42, 89–91, 92]. RF plasma source has been commonly used to supply activated oxygen [23, 86, 87, 88, 91] while Sasaki *et al.* [42, 89, 90] used a combination of ozone (5%) and oxygen (95%) to supply activated oxygen.

The substrate plane has significant impact on the β-Ga_2O_3 growth rate, when grown by MBE. This is in clear contrast with MOCVD by which growth rates of β-Ga_2O_3 on different planes remain on the same order. Sasaki *et al.* [42, 89] showed that the growth rate on non-cleavage planes ((310) and (010) plane) is more than ten times higher than that on cleavage planes ((100) and (001) plane) due to low adhesion energy of cleavage planes (Fig. 4). Growth rate of β-Ga_2O_3 has also been studied as a function of Ga flux [29, 93]. Oshima *et al.* [29] observed that the growth rate with an increase in Ga/O flux was dependent on substrate crystal planes e.g., (001), (100) and (010). The growth rate on (001) at the plateau was ~0.056 μm/h which was ~1/4 of that on (010) and ~twice of that on (100). Tsai *et al.* [93] showed that with increasing Ga/O flux ratio the surface morphology changed from three dimensional columnar to terrace-like surface morphology. Okumura *et al.* [28] observed the growth rate increased with increasing growth temperature for temperatures below 650 °C due to excess Ga adlayer preventing oxygen atoms from incorporating into Ga_2O_3 films or formation of volatile gallium sub-oxide (Ga_2O). The growth rate was constant for a substrate temperature between 650-750 °C due to desorption of excess Ga from the surface without the formation of Ga_2O, then the growth rate decreased with a temperature above 750 °C for Ga rich condition. They also observed an improvement in the surface morphology from three-dimensional island features to a uniform surface with

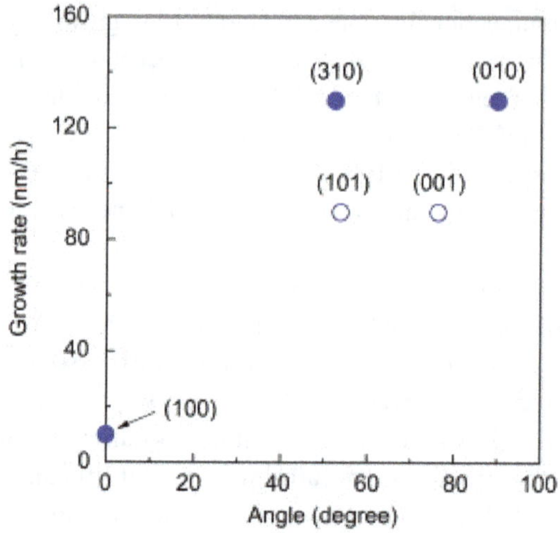

Fig. 4. Relationship between surface orientation of β-Ga_2O_3 substrate and homoepitaxial growth rate. The horizontal axis is the angle between the substrate surface and (100) plane [90]. Copyright 2012 The Japan Society of Applied Physics.

an increase in temperature due to increasing adatoms mobility [89, 42]. Han *et al.* [33] found that higher growth temperatures were required to achieve smooth surface morphology and high-quality crystal on c-plane compared with that on b-plane.

Vogt *et al.* [94] performed a thorough investigation of the reaction kinetics and growth window of β-$Ga_2O_3(\bar{2}01)$ on sapphire by PAMBE. The growth rate as a function of gallium flux was measured in situ by laser reflectometry at different growth temperatures and gallium-to-oxygen ratios. The suboxide (Ga_2O) flux desorbing off the growth surface (Φ_{Ga2O}) was also measured by a line-of-sight quadrupole mass spectrometer (QMS). It was shown that by increasing substrate temperature from 500 °C to 700 °C, the maximum possible growth rate reduces (Fig. 5). At each growth temperature, by increasing Ga flux, the growth rate reached a plateau and then reduced by further increasing the Ga flux. A growth diagram (Fig. 6) was developed describing the growth window with three distinct growth regimes: 1) Ga transport limited, 2) Ga_2O desorption limited, 3) O transport limited.

Sn, Si, and Ge have been used for n-type doping of β-Ga_2O_3 by MBE. Although a wide range of doping concentrations has been demonstrated for Sn and Ge, it has been proven to be difficult to achieve low and medium level of Si concentrations in β-Ga_2O_3. This is due to oxidization of Si metal source in the oxide environment of the MBE and formation of a SiO_2 cap layer on the metal which cannot be easily desorbed. Therefore, achieving a uniform Si doping concentration in thick films as well as obtaining a consistent doping concentration using same Si cell temperature is very difficult, if not impossible. Krishnamoorthy *et al.* [95] tried to address this issue by pulsing the Si shutter to avoid source oxidation. They obtained a flat doping profile with carrier concentrations ranging from 4×10^{17} to 1.7×10^{20} cm^{-3}. It is important to note that by changing Si cell temperature from 850 °C to 875 °C, the Si concentration jumped from 4×10^{17} cm^{-3} to 2×10^{19} cm^{-3}. On

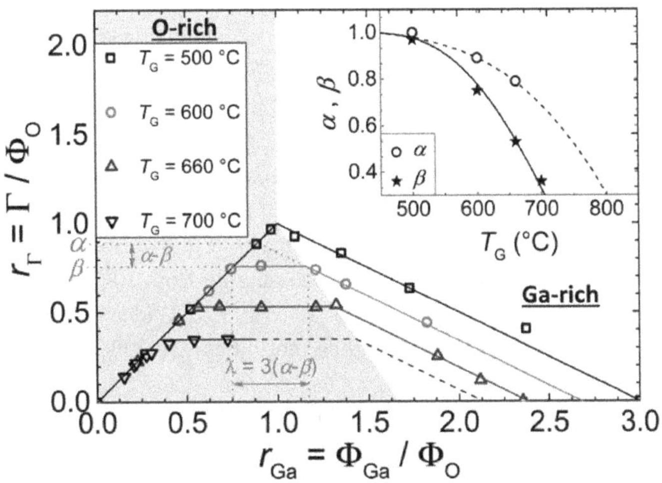

Fig. 5. Growth rates normalized by the O flux ($r_\Gamma = \Gamma/\Phi_O$) as a function of the Ga-to-O ratio (r_{Ga}) for four different growth temperatures. The supplied active O flux was 1.7 A°/s for T_G = 500; 600; 660 °C and 5.1 A°/s for 700 °C. The growth rate evolution is parametrized by the maximum growth rate (β) and the O flux (α) that may react with Ga (to form Ga_2O_3 or Ga_2O), normalized by the active O flux. Inset: the influence of T_G, α and β. The lines are guides for the eye [94]. Copyright 2016 AIP Publishing LLC.

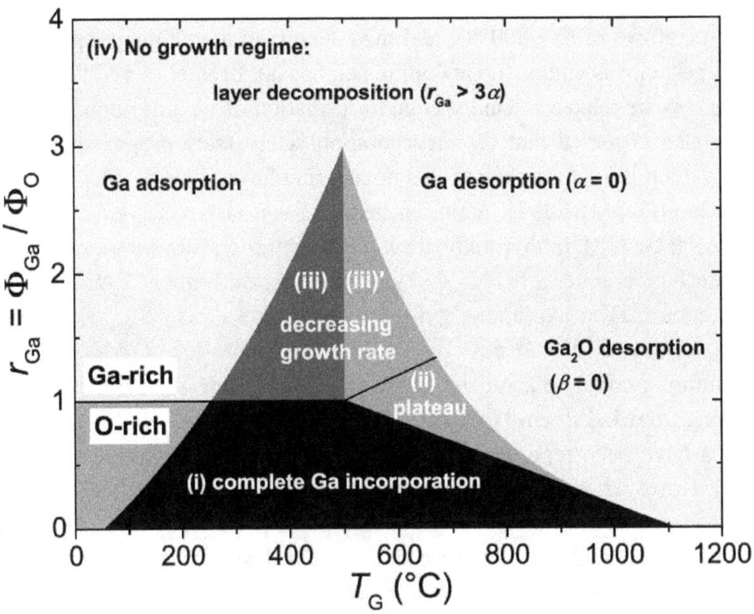

Fig. 6. A growth diagram (GD) for the Ga_2O_3 MBE growth including all experimental parameters (Φ_{Ga}; Φ_O) and T_G. It is divided into two major regimes: O-rich and Ga-rich. The GD illustrates regimes of complete (i), partial (ii, iii, iii), and no Ga-incorporation (iv) as a function of T_G and r_{Ga} [94]. Copyright 2016 AIP Publishing LLC.

the other hand, Si concentration remains relatively unaltered when Si cell temperature is increased from 875 °C to 900 °C. These observations suggest poor control over Si concentration in β-Ga_2O_3 films grown by PAMBE. Sn doping also suffers from oxidation of Sn metal source, not as severely as Si though, but enough to make it challenging to achieve consistent doping concentration on a daily basis. Sasaki *et al.* [42,89] used SnO_2 powder as Sn source in an ozone MBE to address this issue. They achieved carrier concentrations in the range of 4×10^{16} cm^{-3} - 10^{19} cm^{-3} with electron mobilities ranging from 130 cm^2/V.s to 35 cm^2/V.s, respectively. Okumura *et al.* [28] also studied Sn doping by plasma assisted MBE with carrier concentrations ranging from 10^{17} cm^{-3} to 10^{18} cm^{-3} and electron mobilities between 10 cm^2/V.s to 80 cm^2/V.s. Wider range of carrier concentrations (5×10^{16} - 10^{20} cm^{-3}) with higher mobilities (39-120 cm^2/(V.s)) were later demonstrated by PAMBE [96].

Ahmadi *et al.* [96] studied Ge doping of β-Ga_2O_3 (010). As opposed to SiO_2 and SnO_2, GeO_2 is more volatile. Therefore, although oxygen environment of the MBE chamber oxidizes Ge, the oxide can be easily desorbed by heating up the cell to 850 °C. Removing the oxide from source material allows consistent and uniform Ge doping in an oxide MBE. The impact of substrate temperature, Ge cell temperature, and growth regime (O-rich vs Ga-rich) on Ge doping concentration was studied. They observed that for a constant Ge flux, Ge incorporation is inversely proportional to gallium flux which may be related to Ge suboxide formation and competition between Ga and Ge. Moreover, Ge incorporation increased with increasing Ge cell temperature [29]. However, Ge incorporation saturated over cell temperatures of 625-800 °C, and then decreased at cell temperature higher than 800 °C. This behavior is unique to Ge doping and has not been observed for Sn/ Si doping [42,89]. This may be related to Ge loss due to formation of volatile suboxides at high Ge fluxes. They also observed that Ge incorporation is inversely proportional to substrate temperature, which is again unique to this dopant and has not been observed for Si or Sn doping of β-Ga_2O_3 by MBE [97]. A similar behavior was also observed on (001) β-Ga_2O_3 when doped with Ge [32]. In this study, the Ge concentration showed a strong dependence on the growth temperature, whereas the Sn concentration remained independent of the growth temperature. The maximum growth temperature at which a wide range of Ge concentrations (from 10^{17} to 10^{20} cm^{-3}) could be achieved was 675 °C while the same range of Sn concentration could be achieved at growth temperature of 750 °C. Carrier concentrations ranging from 2×10^{16} cm^{-3} to 1.4×10^{19} cm^{-3} with mobilities ranging from 39 cm^2/V.s to 97 cm^2/V.s have been reported on b-plane β-Ga_2O_3 using Ge as dopant by PAMBE [96, 98–100]. However, much lower electron mobility was obtained (26.3 cm^2 V^{-1} s^{-1} at the electron concentration of 6.7×10^{17} cm^{-3}) on c-plane β-Ga_2O_3 by this growth technique.

Alloying β-Ga_2O_3 with Al is of great interest for design and fabrication of β-Ga_2O_3 based devices for high power applications. However, Al_2O_3 naturally occurs in corundum crystal structure, whereas β-Ga_2O_3 is most stable in monoclinic crystal structure (β-phase). Moreover, the solubility limit of Al_2O_3 in β-Ga_2O_3 is shown to be ~65% for temperatures ranging from 850 °C to 1950 °C [101], and this solubility is much lower at lower temperatures. Therefore, growth of high Al content β-$(Al_xGa_{1-x})_2O_3$ films via MBE is challenging as MBE is a relatively low temperature growth technique (600-800 °C). Kaun *et al.* [102] studied β-$(Al_xGa_{1-x})_2O_3$ films by PAMBE for the first time. They were able to achieve an Al_2O_3 mole fraction up to 0.18 at 600 °C. However, the surface became rougher when Al content in the film increased due to low surface mobility of Al adatoms at 600 °C. The surface morphology improved and the solubility limit of Al in β-Ga_2O_3 enhanced with increasing substrate temperature.

Vogt *et al.* [103] demonstrated that supplying indium can drastically increase the growth rate and expand the possible growth domain of β-Ga_2O_3 [104] and β-$(Al_xGa_{1-x})_2O_3$ [103] films by PAMBE. They attributed this enhancement to rapid formation of In_2O_3, followed by a transformation of In_2O_3 to Ga_2O_3 due to an inter-atomic exchange. They called this technique metal-oxide catalytic epitaxy (MOCATAXY). Figure 7 shows the growth rate of β-$(Al_xGa_{1-x})_2O_3$ as a function of substrate temperature in the absence of In (squares) and in the presence of In (circles). MOCATAXY resulted in higher quality β-$(Al_xGa_{1-x})_2O_3$ surface morphology and higher growth rate compared with conventional PAMBE.

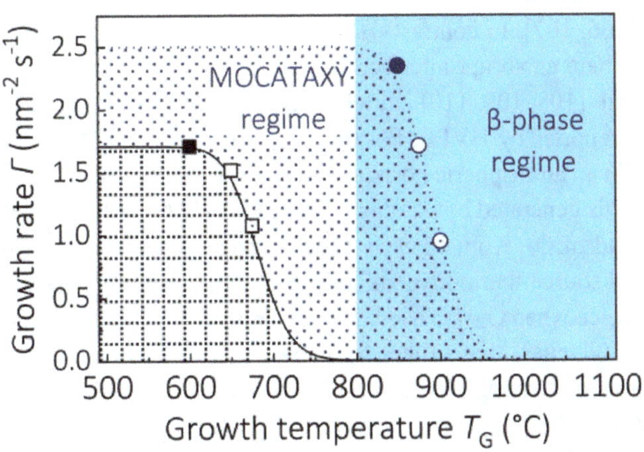

Fig. 7. Growth rate Γ as a function of T_G. The data plotted as squares and discs show Γ of β-$(Al_xGa_{1-x})_2O_3$ (010) in the absence of In (i.e. for $\phi_{In} = 0$) and in the presence of In (i.e. for $\phi_{In} > 0$), respectively. The lines are guides to the eye. The growth domains of "conventional" PAMBE and MOCATAXY are depicted by gridlines and points, respectively. The blue highlighted area shows the regime of high possible Al incorporation into β-$(Al_xGa_{1-x})_2O_3$, as suggested by the Al_2O_3-Ga_2O_3 phase diagram [103]. Copyright 2012 The Japan Society of Applied Physics.

Ge [96, 32] and Sn doping [105] of β-$(Al_xGa_{1-x})_2O_3$ films have been demonstrated. Ahmadi *et al.* showed that electron concentration reduced significantly from $1x10^{19}$ cm^{-3} to $5x10^{17}$ cm^{-3}, when Al content in β-$(Al_xGa_{1-x})_2O_3$ films increased from 0% to 12%. Table 2 shows the electron concentrations and mobilities in these films grown at 600 °C by PAMBE. Successful modulation doping of β-$(Al_xGa_{1-x})_2O_3$/ β-Ga_2O_3 heterostructure was separately demonstrated by Ahmadi *et al.* [32] and Krishnamoorthy *et al.* [105], using Ge and Si doping, respectively. These heterostructures have been employed to design and

Table 2. Electron concentrations and mobilities in the films grown at 600 °C with increasing Al content in β-$(Al_xGa_{1-x})_2O_3$ by PAMBE.

Al%	n_s (cm^{-3})	μ (cm^2/vs)
12	$5.3x10^{17}$	11
6	$1.4x10^{18}$	77
2	$2.5x10^{18}$	77
0	$1.4x10^{19}$	66

fabricate modulation-doped field effect transistors (MODFETs) which will be discussed in more details later in the device section.

Halide Vapor Phase Epitaxy: The growth rates achieved by Halide Vapor Phase Epitaxy (HVPE) (1-250 μm/h) are typically much higher than that achieved by the other epitaxial growth techniques discussed earlier. This makes HVPE attractive for growth of thick films such as the drift region in vertical MOSFETs or Schottky diodes. This growth technique is also used widely in nitride industry, in particular for producing free-standing GaN substrates [106, 107]. In contrast, HVPE is not suitable for growth of thin layers or heterostructures where an abrupt interface is desired.

Homo-epitaxial [108, 109, 110] and hetero-epitaxial [111, 112, 113] growth of β-Ga_2O_3 have been reported by HVPE. For this purpose, GaCl and O_2 are typically utilized as the precursors in an atmospheric horizontal reactor, while N_2, He or Ar has been used as a carrier gas. GaCl is generated by passing Cl2 or HCl over Ga metal. Nikolaev *et al.* [111] used dry air instead of O_2. Konishi *et al.* [108] compared the impact of using O_2 versus H_2O as an oxygen source thermodynamically and experimentally. They showed that the growth rate was twice when O_2 was used. Moreover, using H_2O led to higher unintentional Si impurities ($2x10^{16}$ cm^{-3}) due to the decomposition of the quartz glass reactor in the presence of hydrogen. Nonetheless, using H_2O resulted in a smoother surface compared with when O_2 was used.

The first homoepitaxial growth of β-Ga_2O_3 (001) by HVPE was demonstrated by Murakami *et al.* [109]. In this study, the growth rate remained relatively unchanged (5μm/h) at growth temperatures ranging from 800 °C to 1050 °C. The growth rate of films grown at 1000 °C increased linearly with increasing the GaCl partial pressure while the VI/III ratio was fixed at 10 for all the growths, suggesting that the growth rate at and above 900 °C depended on the mass transportation as was predicted by thermodynamic analysis [114]. The surface morphology of the β-Ga_2O_3 (001) also improved remarkably by increasing substrate temperature from 800 °C to 1050 °C. Leach *et al.* [110] studied the impact of substrate miscut on surface morphology of β-Ga_2O_3 grown homoepitaxially by HVPE. They compared films grown on on-axis substrate with those grown on vicinal substrates with 2° miscut in [100] and [001] directions and observed the most smooth surface morphology on the substrate with 2° miscut in [100].

Goto *et al.* [115, 116] studied Si-doping of c-plane β-Ga_2O_3 using $SiCl_4$ gas as the precursor for Si. The HVPE of β-Ga_2O_3 films was followed by chemical-mechanical polishing which removed 2 μm of the films and then annealing at 1150 °C in N_2 environment for an hour to activate Si dopants. The n-type carrier density at room temperature, extracted from Hall measurements, was almost equal to the doped silicon concentration, measured by SIMS, which was controlled in the range of 10^{15} cm^{-3} and 10^{18} cm^{-3}. The Si doping concentration in the unintentionally doped film was below the detection level of SIMS ($5x10^{15}$ cm^{-3}) and the electron density was measured to be $3x10^{15}$ cm^{-3}. A low background doping is desirable in power switching applications, in particular for achieving large breakdown voltage.

3. Ga_2O_3-Based Field-Effect Transistors

The first Ga_2O_3 FET was reported by Higashiwaki *et al.* and was fabricated on Sn-doped Ga_2O_3 (010) film grown homoepitaxially on a Mg-doped β-Ga_2O_3 substrate by MBE [9]. A circular FET pattern which was used since a device isolation technique was not yet developed. Although this transistor suffered from a poor Ohmic contact and a large gate leakage current, breakdown voltage more than 250 V and an on/off drain current ratio of ~10^4 was obtained. They later developed Si-implantation in the source/drain region area followed by an activation annealing at 925 °C in N_2 environment to address the high contact resistance [117, 118]. They also introduced atomic layer deposition (ALD) Al_2O_3 dielectric to reduce high gate current leakage previously observed on their devices. A high drain current on/off ratio of over ten orders of magnitude was achieved with an extremely low off-state drain leakage of a few pA/mm. This device exhibited a breakdown voltage as high as 370 V in the off-state. Moreover, stable transistor operation was sustained at temperatures up to 250 °C, although the drain leakage current increased six orders of magnitude at this high temperature. Later, development of gate-connected field plates allowed a breakdown voltage up to 750 V on Ga_2O_3 devices [119]. In contrast to previous devices reported by this group, device fabrication began with a 1.2-μm UID Ga_2O_3 epilayer grown on an Fe-doped semi-insulating β-Ga_2O_3 (010) substrate by ozone MBE [120]. The MOSFET channels were then defined by selective-area 300 nm-deep Si implanted with a uniform concentration of 3×10^{17} cm^{-3}. This allowed a rectangular device pattern, as opposed to circular patterns on earlier devices, since devices were isolated by UID Ga_2O_3. Room temperature channel mobilities of 90–100 cm^2V^{-1} s^{-1} were measured on these devices [120]. A comparison between pulsed and CW large signal RF performance of these devices [121] showed a reduction in self-heating when pulsed which led to a power added efficiency of 12%, drain efficiency of 22.4%, output power density of 0.13 W/mm, and maximum gain up to 4.8 dB at 1 GHz for a 2-μm gate length device. Gamma-ray tolerance of these devices was also studied. The MOSFETs' on-current, on-resistance (R_{ON}), and threshold voltage (V_T) remained stable and hysteresis in the transfer characteristics remained negligible after exposure to the highest dose (230 kGy). It was found that the overall radiation resistance of these devices [122] was limited by radiation-induced degradations in the gate insulation and surface passivation, perhaps due to dielectric damage and interface trap generation.

The development of chlorine-based dry etching of Ga_2O_3 has enabled device isolation through mesa etch [123–125]. The group at AFRL have fabricated MOSFETs on Sn-doped β-Ga_2O_3 channels grown by MBE [126] and MOCVD [127] on Fe-doped β-Ga_2O_3 (010) and Mg-doped β-Ga_2O_3 (100) substrates, respectively. They reported the gate-to-drain electric field of 3.8 MV/cm, which is the highest reported for any transistor and surpassing bulk GaN and SiC theoretical limits. Moser *et al.* [126] fabricated MOSFETs with ALD HfO_2 as the gate dielectric. A pulsed current density of >450 mA/mm was observed. The same group demonstrated the first MOSFET with Ge-doped channel grown by molecular beam epitaxy on (010) Fe-doped semi-insulating substrates [127]. The Ge-doped channel devices performed similarly to previously reported devices with Sn- and Si-doped channels with the drain current ON/OFF ratios of $>10^8$ and the saturated drain current of >75 mA/mm at VG = 0 V. Hall effect measurements showed a high carrier mobility of 111 $cm^2/(V\cdot s)$ with 4×10^{17} cm^{-3} active carriers. They were also first to report the RF performance of β-Ga_2O_3 MOSFETs [128, 129]. To enable RF performance, the contact resistance was reduced using a highly doped cap layer. A gate recess design was used, allowing for gate length scaling. DC transfer measurements showed current density and transconductance of 150 mA/mm and 21.2 mS/mm, respectively. The f_T and f_{MAX} were measured to be 3.3 GHz and 12.9 GHz, respectively (Fig. 8). A maximum output power of 0.23 W/mm was achieved with a maximum PAE of 6.3%. It has been shown [130] that the Fe-related trap states in the UID buffer layer grown on Fe-doped semi-insulating Ga_2O_3 substrates negatively impact the transistor performance. Therefore, growth of thicker UID buffer layer leads to an improvement of DC-RF dispersion and enhancement of transport

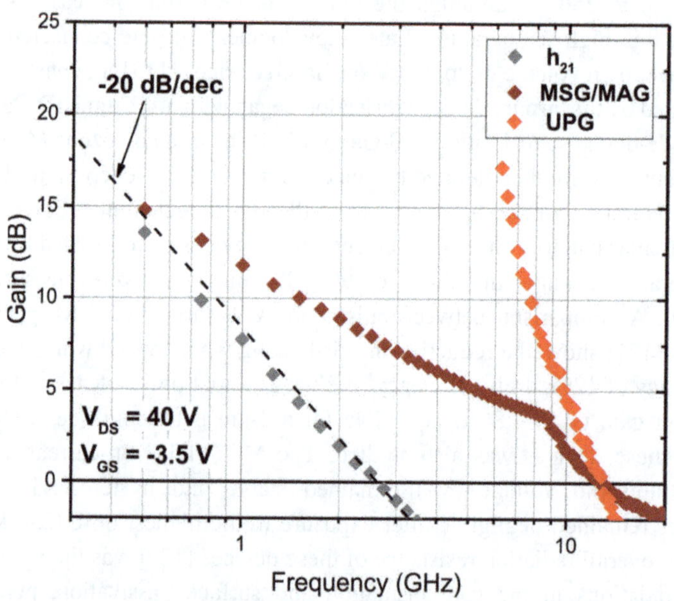

Fig. 8. Extrinsic small signal RF gain performance recorded at V_{GS} = -3.5V (peak gm) and V_{DS} = 40 V. A gain decay of -20 dB/dec is plotted with the dashed line [129]. Copyright 2017 IEEE.

properties. MOSFETs with ~2 kV breakdown voltage have been demonstrated using a 400-nm thick composite field plate oxide, with a combination of atomic layer deposited and plasma enhanced chemical vapor deposited SiO_2 layers [131]. An output current of 20 mA mm^{-1} and on-resistance of 520 $m\Omega \cdot cm^2$ s was achieved on these devices when highly-Si-doped Ga_2O_3 cap layer was used under the source and drain [132].

In power switching applications, E-mode transistors are typically more desirable over D-mode devices due to safety considerations and simplicity of gate-drive circuitry [21, 22, 133]. The E-mode Ga_2O_3 MOSFETs have been realized by depleting the channel via a wrap-gate fin-array field-effect transistor (finFET) structure [22], an unintentionally doped Ga_2O_3 with low carrier concentration as channel [134], and gate-recessed structure with ALD SiO_2 as the gate dielectric [133]. Chabak *et al.* [22] were the first to demonstrate E-mode MOSFETs in Sn-doped Ga_2O_3 wrap-gate finFETs on a native semi-insulating Mg-doped (100) β-Ga_2O_3 substrate. These finFETs demonstrated normally off operation with a threshold voltage between 0 and +1 V during high-voltage operation and an I_{ON}/I_{OFF} ratio of greater than 10^5 which was mainly limited by high on-resistance. The same group recently reported E-mode β-Ga_2O_3 transistors grown homoepitaxially by MBE, utilizing a recessed-gate process, which depleted the channel under the gate followed by deposition of ALD SiO_2 as the gate dielectric [21]. Wong *et al.* [134] also demonstrated E-mode MOSFETs with low series resistance using Si-ion implantation of the source/drain contacts and access region. In these devices, the channel was formed of an unintentionally doped Ga_2O_3 with low background carrier concentration. They achieved positive threshold voltage without additional constraints on the channel dimensions or device architecture. Their devices suffered from non-idealities associated with the Al_2O_3 gate dielectric, which caused huge hysteresis as shown in Fig. 9.

While breakdown voltage as high as 2 KV has been achieved on Ga_2O_3-based MOSFETs, the current densities reported so far are very low compared with their GaN counterparts which is mainly due to low channel mobility in this materials system. To address this issue, modulation-doped field effect transistor (MODFET) has been demonstrated [24, 25, 32, 83]. Ahmadi *et al.* [32] and Zhang *et al.* [24, 135] have separately reported $(Al_xGa_{1-x})_2O_3$-Ga_2O_3 modulation-doped FETs using Ge and Si as *n*-type dopants in the barrier, respectively. It has been shown that Ge incorporation in β-Ga_2O_3 films reduces as the substrate temperature increases [96]. On the other hand, higher substrate temperatures enable $(Al_xGa_{1-x})_2O_3$ films with larger Al content [102]. In contrast, Si incorporation does not depend on the substrate temperature, which allows high growth temperatures for $(Al_xGa_{1-x})_2O_3$ films [95]. Therefore, Ge-doped MODFET structures suffer from low Al content $(Al_xGa_{1-x})_2O_3$ barrier. It was expected that the introduction of β-$(Al_xGa_{1-x})_2O_3$ and formation of two-dimensional electron gas (2DEG) will lead to enhancement of electron mobility in these structures. However, the highest room-temperature electron mobility reported, so far, in β-$(Al_xGa_{1-x})_2O_3$-Ga_2O_3 double-heterostructures has been only 180 $cm^2/V \cdot s$ [24] limited by phonon scattering [136–138]. The electron mobility increased to 2790 $cm^2/V \cdot s$ at 50K. This high electron mobility allowed for observation of Shubnikov-de-Haas oscillations from which an electron

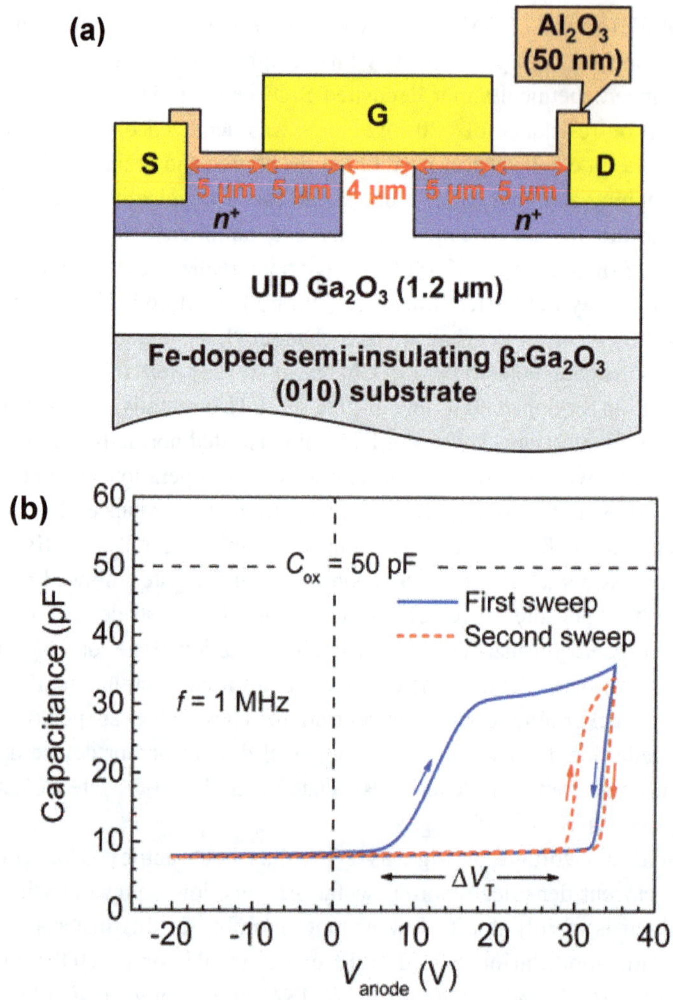

Fig. 9. (a) Cross section schematic of the enhancement-mode β-Ga_2O_3 (010) MOSFET with Si+-implanted source/drain contacts and access regions. (b) C–V characteristics of the Al_2O_3/Ga_2O_3 MOSCAP with V_{ANODE} swept from −20 to +35 V (forward) and back to −20 V (reverse). The MOSCAP showed hysteretic behavior, capacitance pinning below C_{OX}, and a large ΔV_T between the first and second forward sweeps [134]. Copyright 2017 Applied Physics Express.

effective mass of 0.33 m_e was extracted. A maximum drain current of I_{DS}=257 mA/mm, a peak g_m of 39 mS/mm and a pinch off voltage of -7V have been measured on MODFETs [25]. The three-terminal off-state breakdown measurement on the device with a gate-drain spacing (L_{GD}) of 1.55μm showed a breakdown voltage of 428V. Very recently, the same group demonstrated a MODFET with a high breakdown voltage of 1.37 kV for a gate-to-drain separation (L_{GD}) of 16 μm having a specific ON-resistance of 120.1 mΩ.cm^2 using SiNx as the passivation dielectric [139]. MODFETs with Sn-doped β-$(Al_{0.08}Ga_{0.92})_2O_3$ were shown very recently by Okumura *et al.* [140]. In this work, the epi-structure was grown by PAMBE. Ti/Au was deposited as Ohmic contact followed by an annealing in N_2 at 500 °C. The electron mobility on this device was measured to be only 8 cm^2/Vs for a sheet charge density of 2x10^{13} cm^{-2}, which is much lower than the MODFETs reported by other groups. These devices also suffered from a low maximum drain current density of 8 mA/mm.

For high voltage and high-power applications, vertical topologies are preferred to enhance packing density of devices and suppress the sensitivity of surface effects. Two types of vertical Ga_2O_3-based devices have been reported to date. Wong *et al.* [27] fabricated a current aperture vertical transistor (CAVET) with a Mg-implanted current blocking layer (Fig. 10(a)). Hu *et al.* [141], on the other hand, employed a deep-etch process to fabricate ~1 um thick vertical structures into a low-doped Ga_2O_3 substrate, as shown in Fig. 10(b). Very recently, normally-off Ga_2O_3 vertical FinFETs with a threshold voltage of 4 V, breakdown voltage of 1.6 kV, and a drain current density of 600 A/cm^2 were demonstrated [142]. The high threshold voltage was achieved by Fin-shaped channels with sub-micron widths.

Apart from conventional devices made with epi-layers, nano-membranes of Ga_2O_3 have been used to fabricate transistors [143–146]. Zhou *et al.* [145] demonstrated β-Ga_2O_3 D-mode and E-mode FETs by transferring nanomembranes of β-Ga_2O_3 to a SiO_2/Si substrate, followed by device fabrication. They showed that threshold voltage of the transistors can be varied by changing the nanomembrane thickness. They achieved E-mode operation using membranes thinner than 70 nm. Figure 11 shows a schematic of the device and an atomic force microscopy image of smooth β-Ga_2O_3 surface after the cleavage. Moreover, they demonstrated maximum I_D of 600/450 mA/mm for D/E-mode FETs and on/off ratio of 10^{10}. Ar bombardment was used to form oxygen vacancies prior to Ohmic metal deposition to reduce the contact resistance of source and drain. E-mode FET with source-drain separation of 0.9 μm showed a breakdown voltage of 185 V, corresponding to a field strength of 2 MV/cm. Higher $I_{D,MAX}$ of 1.5/1.0 A/mm for D/E-mode FETs was later demonstrated [147] by increasing the β-Ga_2O_3 doping concentration from 3.0x10^{18} to 8.0x10^{18} cm^{-3} and further scaling the channel length. Same group have also reported negative-capacitance FETs using similar device structure and using $HfZrO_2$ as the gate insulator [148]. They demonstrated minimum subthreshold slope value of 34.3 mV/dec at the reverse gate-voltage sweep and 53.1 mV/dec at the forward gate-voltage sweep at VDS = 0.5 V.

In conclusion, β-Ga_2O_3 is the most promising material to provide the next step in performance power electronics based on cost effective, large area, high quality substrates. We overviewed the progress on epitaxial growth and fabrication of electronic devices based on β-(Al,Ga)$_2$O$_3$ and its heterostructures. As discussed earlier in this review paper, HVPE, MOCVD, and MBE have been mainly employed for epitaxial growth of β-(Al,Ga)$_2$O$_3$. While HVPE is a high growth rate technique and more suitable for growth of thick films, for instance as drift layer in vertical devices, MBE and MOCVD can be utilized for growth of abrupt heterostructures. In contrast to MOCVD which results in relatively close growth rates on different crystal planes, the growth rate of films by MBE is much higher on non-cleavage planes (e.g. (010)) compared with that on cleavage planes (e.g. (001)). N-type conductivity of β-Ga_2O_3 has been demonstrated by the aforementioned growth techniques via Sn, Ge, or Si doping. So far, β-$(Al_xGa_{1-x})_2O_3$ films have been demonstrated by MBE, however, the Al content has been limited to lower than 26% and attempts to achieve high Al content has been unsuccessful. Both lateral and vertical β-Ga_2O_3-based FETs including MODFETs, MESFETs, MOSFETs, and FinFETs have been

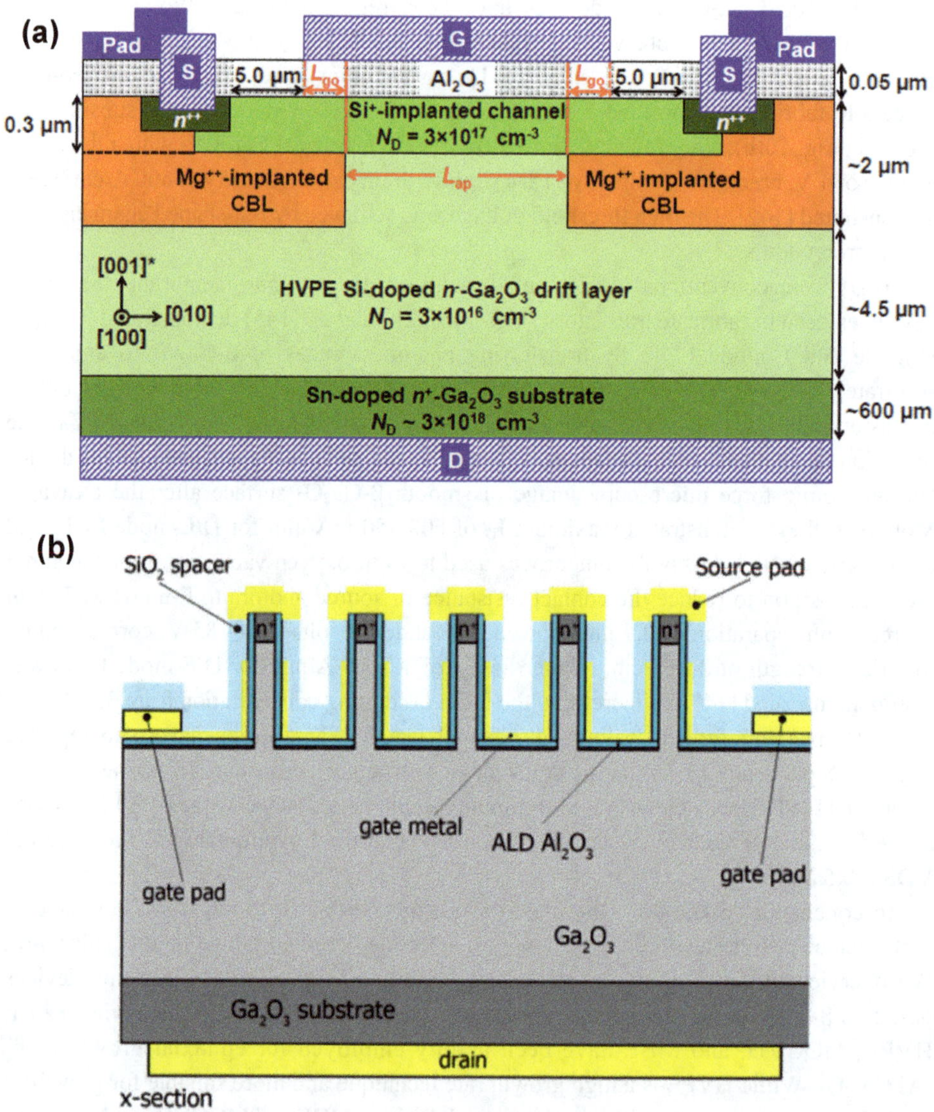

Fig. 10. Cross-section schematic of (a) Ga_2O_3 CAVET MOSFET with Lgo = 2.5 μm and Lap = 15 μm [27] and (b) Ga_2O_3 vertical Fin-FET power transistors [141]. Copyright 2017 IEEE.

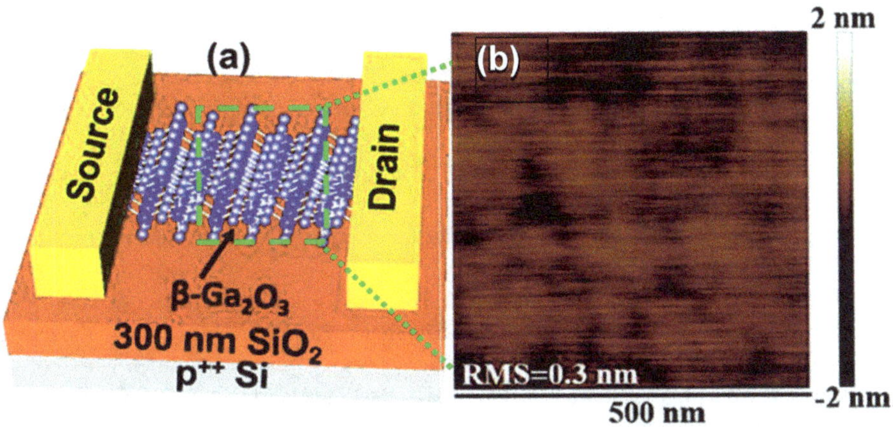

Fig. 11. (a) Schematic view of a GOOI FET with a 300 nm SiO_2 layer on Si substrate. (b) AFM image of β-Ga_2O_3 surface after cleavage [145]. Copyright 2017 IEEE.

demonstrated. Although breakdown voltages as high as 2 kV have been already achieved, these devices suffer from low current density compared to their GaN counterparts mainly due to very low electron mobility. Regardless of the significant progress that has been made in developing β-Ga_2O_3 technology, attempts for p-type doping of β-Ga_2O_3 have been unsuccessful and it is widely believed to be impossible (at least via conventional techniques). A relatively low electron mobility (limited by phonon scattering) and low thermal conductivity remain the two main obstacles that need to be overcome for β-Ga_2O_3 technology to succeed.

References

[1] Cheng Z, Hanke M, Vogt P, Bierwagen O and Trampert A 2017 Phase formation and strain relaxation of Ga2O3on c-plane and a-plane sapphire substrates as studied by synchrotron-based x-ray diffraction *Appl. Phys. Lett.* **111** 162104

[2] Baik K H, Irokawa Y, Ren F, Pearton S J, Park S S and Park Y J 2003 Design of junction termination structures for GaN Schottky power rectifiers *Solid. State. Electron.* **47** 975–9

[3] Marie P, Portier X and Cardin J 2008 Growth and characterization of gallium oxide thin films by radiofrequency magnetron sputtering *Phys. Status Solidi Appl. Mater. Sci.* **205** 1943–6

[4] Baik K H, Irokawa Y, Kim J, LaRoche J R, Ren F, Park S S, Park Y J and Pearton S J 2003 160-A bulk GaN Schottky diode array *Appl. Phys. Lett.* **83** 3192–4

[5] Zhang Y, Sun M, Liu Z, Piedra D, Hu J, Gao X and Palacios T 2017 Trench formation and corner rounding in vertical GaN power devices *Appl. Phys. Lett.* **110** 193506

[6] Cao Y, Chu R, Li R, Chen M, Chang R and Hughes B 2016 High-voltage vertical GaN Schottky diode enabled by low-carbon metal-organic chemical vapor deposition growth *Appl. Phys. Lett.* **108** 062103

[7] Otake H, Chikamatsu K, Yamaguchi A, Fujishima T and Ohta H 2008 Vertical GaN-based trench gate metal oxide semiconductor field-effect transistors on GaN bulk substrates *Appl. Phys. Express* **1** 011105

[8] Roy R, Hill V G and Osborn E F 1952 Polymorphism of Ga2O3and the System Ga2O3−H2O *J. Am. Chem. Soc.* **74** 719–22

[9] Higashiwaki M, Sasaki K, Kuramata A, Masui T and Yamakoshi S 2012 Gallium oxide (Ga 2O 3) metal-semiconductor field-effect transistors on single-crystal β-Ga 2O 3 (010) substrates *Appl. Phys. Lett.* **100** 1–4

[10] Tomm Y, Reiche P, Klimm D and Fukuda T 2000 Czochralski grown Ga2O3 crystals *J. Cryst. Growth* **220** 510–4

[11] Baliga B J 1989 Power semiconductor device figure of merit for high-frequency applications *IEEE Electron Device Lett.* **10** 455–7

[12] Higashiwaki M, Sasaki K, Murakami H, Kumagai Y, Koukitu A, Kuramata A, Masui T and Yamakoshi S 2016 Recent progress in Ga2O3power devices *Semicond. Sci. Technol.* **31** 34001

[13] Tomm Y, Ko J M, Yoshikawa A and Fukuda T 2001 Floating zone growth of-Ga O : A new window material for optoelectronic device applications *Sol. Energy Mater. Sol. Cells* **66** 369–74

[14] Aida H, Nishiguchi K, Takeda H, Aota N, Sunakawa K and Yaguchi Y 2008 Growth of β-Ga 2 O 3 Single Crystals by the Edge-Defined, Film Fed Growth Method *Jpn. J. Appl. Phys.* **47** 8506–9

[15] Sasaki K, Higashiwaki M, Kuramata A, Masui T and Yamakoshi S 2013 Ga2 O3 schottky barrier diodes fabricated by using single-crystal β-Ga2 O3 (010) substrates *IEEE Electron Device Lett.* **34** 493–5

[16] Higashiwaki M, Konishi K, Sasaki K, Goto K, Nomura K, Thieu Q T, Togashi R, Murakami H, Kumagai Y, Monemar B, Koukitu A, Kuramata A and Yamakoshi S 2016 Temperature-dependent capacitance-voltage and current-voltage characteristics of Pt/Ga2O3 (001) Schottky barrier diodes fabricated on n --Ga2O3 drift layers grown by halide vapor phase epitaxy *Appl. Phys. Lett.* **108** 1–6

[17] Oda M, Tokuda R, Kambara H, Tanikawa T, Sasaki T and Hitora T 2016 Schottky barrier diodes of corundum-structured gallium oxide showing on-resistance of 0.1 mΩ·cm ^{2} grown by MIST EPITAXY ® *Appl. Phys. Express* **9** 021101

[18] Nakagomi S, Sakai T, Kikuchi K and Kokubun Y 2019 β-Ga 2 O 3 /p-Type 4H-SiC Heterojunction Diodes and Applications to Deep-UV Photodiodes *Phys. Status Solidi Appl. Mater. Sci.* **1700796** 216

[19] Kokubun Y, Kubo S and Nakagomi S 2016 All-oxide p-n heterojunction diodes comprising p-type NiO and n-type β-Ga 2 O 3 *Appl. Phys. Express* **9** 9

[20] Mcglone J F, Xia Z, Zhang Y, Joishi C, Lodha S, Rajan S, Ringel S A and Arehart A R 2018 Trapping Effects in Si -Doped -Ga 2 O 3 MESFETs on an Fe-Doped -Ga 2 O 3 Substrate *IEEE Electron Device Lett.* **39** 1042–5

[21] Chabak K D, McCandless J P, Moser N A, Green A J, Mahalingam K, Crespo A, Hendricks N, Howe B M, Tetlak S E, Leedy K, Fitch R C, Wakimoto D, Sasaki K, Kuramata A and Jessen G H 2018 Recessed-Gate Enhancement-Mode β -Ga2O3 MOSFETs *IEEE Electron Device Lett.* **39** 67–70

[22] Chabak K D, Moser N, Green A J, Walker D E, Tetlak S E, Heller E, Crespo A, Fitch R, McCandless J P, Leedy K, Baldini M, Wagner G, Galazka Z, Li X and Jessen G 2016 Enhancement-mode Ga2O3 wrap-gate fin field-effect transistors on native (100) β-Ga2O3 substrate with high breakdown voltage *Appl. Phys. Lett.* **109**

[23] Johnson J, Brenner M, Arehart A R, Lodha S and Rajan S 2018 Modulation-doped β-(Al0.2Ga0.8)2O3/Ga2O3 field-effect transistor *Appl. Phys. Lett.* **023502** 3–7

[24] Zhang Y, Joishi C, Xia Z, Brenner M, Lodha S and Rajan S 2018 Demonstration of β-(Al x Ga 1-x) 2 O 3 /Ga 2 O 3 double heterostructure field effect transistors *Appl. Phys. Lett.* **112** 233503

[25] Zhang Y, Xia Z, Joishi C and Rajan S 2018 Design and Demonstration of (Al x Ga 1-x) 2 O 3 / Ga 2 O 3 Double Heterostructure Field Effect Transistor (DHFET) *2018 76th Device Res. Conf.* **0976** 1–2

[26] Wong M H, Goto K, Morikawa Y, Kuramata A, Yamakoshi S, Murakami H, Kumagai Y and Higashiwaki M 2018 All-ion-implanted planar-gate current aperture vertical Ga2O3 MOSFETs with Mg-doped blocking layer *Appl. Phys. Express* **11** 6

[27] Wong M H, Goto K, Kuramata A, Yamakoshi S, Murakami H, Kumagai Y and Higashiwaki M 2016 First Demonstration of Vertical Ga 2 O 3 MOSFET : Planar Structure with a Current Aperture *2017 75th Annu. Device Res. Conf.* **95** 4–5

[28] Okumura H, Kita M, Sasaki K, Kuramata A, Higashiwaki M and Speck J S 2014 Systematic investigation of the growth rate of β-Ga2O3(010) by plasma-assisted molecular beam epitaxy *Appl. Phys. Express* **3** 3–7

[29] Oshima Y, Ahmadi E, Kaun S and Wu F 2012 Growth and etching characteristics of (001) β -Ga 2 O 3 by plasma-assisted molecular beam epitaxy *Mater. Chem. Phys.* **133** 1–6

[30] Kaun S W, Wu F and Speck J S 2015 β-(Al$_x$Ga$_{1-x}$)$_2$O$_3$/Ga$_2$O$_3$ (010) heterostructures grown on β-Ga$_2$O$_3$ (010) substrates by plasma-assisted molecular beam epitaxy *J. Vac. Sci. Technol. A Vacuum, Surfaces, Film.* **33** 041508

[31] Ahmadi E, Koksaldi O S, Kaun S W, Oshima Y, Short D B, Mishra U K and Speck J S 2017 Ge doping of β-Ga 2 O 3 films grown by plasma-assisted molecular beam epitaxy *Appl. Phys. Express* **10** 041102

[32] Ahmadi E, Koksaldi O S, Zheng X, Mates T, Oshima Y, Mishra U K and Speck J S 2017 Demonstration of β-(Al $_x$ Ga $_{1-x}$) $_2$ O $_3$ /β-Ga $_2$ O $_3$ modulation doped field-effect transistors with Ge as dopant grown via plasma-assisted molecular beam epitaxy *Appl. Phys. Express* **10** 071101

[33] Han S-H, Mauze A, Ahmadi E, Mates T, Oshima Y and Speck J S 2018 n-type dopants in (001) β-Ga2O3 grown on (001) β-Ga2O3 substrates by plasma-assisted molecular beam epitaxy *Semicond. Sci. Technol.* **33** 045001

[34] Gogova D, Schmidbauer M and Kwasniewski A 2015 Homo- and heteroepitaxial growth of Sn-doped β-Ga2O3 layers by MOVPE *CrystEngComm* **17** 6744–52

[35] Wagner G, Baldini M, Gogova D, Schmidbauer M, Schewski R, Albrecht M, Galazka Z, Klimm D and Fornari R 2014 Homoepitaxial growth of β-Ga2O3 layers by metal-organic vapor phase epitaxy *Phys. Status Solidi Appl. Mater. Sci.* **211** 27–33

[36] Leedy K D, Chabak K D, Vasilyev V, Look D C, Boeckl J J, Brown J L, Tetlak S E, Green A J, Moser N A, Crespo A, Thomson D B, Fitch R C, Mccandless J P and Jessen G H 2018 Highly conductive homoepitaxial Si-doped Ga2O3 films on (010) β- Ga2O3 by pulsed laser deposition *Appl. Phys. Lett.* **012103** 2–6

[37] Kokubun Y, Miura K, Endo F and Nakagomi S 2007 Sol-gel prepared B-Ga2O3 thin films for ultraviolet photodetectors *Appl. Phys. Lett.* **90** 1–4

[38] Murakami H, Nomura K, Goto K, Sasaki K, Kawara K, Thieu Q T, Togashi R, Kumagai Y, Higashiwaki M, Kuramata A, Yamakoshi S, Monemar B and Koukitu A 2015 Homoepitaxial growth of β-Ga 2 O 3 layers by halide vapor phase epitaxy *Appl. Phys. Express* **8** 015503

[39] Thomas S R, Adamopoulos G, Lin Y H, Faber H, Sygellou L, Stratakis E, Pliatsikas N, Patsalas P A and Anthopoulos T D 2014 High electron mobility thin-film transistors based on Ga2O3grown by atmospheric ultrasonic spray pyrolysis at low temperatures *Appl. Phys. Lett.* **105**

[40] Lee S, Kaneko K and Fujita S 2016 Homoepitaxial growth of beta gallium oxide films by mist chemical vapor deposition *Jpn. J. Appl. Phys.* **55** 1202B8

[41] Zhang F B, Saito K, Tanaka T, Nishio M and Guo Q X 2014 Structural and optical properties of Ga2O3 films on sapphire substrates by pulsed laser deposition *J. Cryst. Growth* **387** 96–100

[42] Sasaki K, Higashiwaki M, Kuramata A, Masui T and Yamakoshi S 2013 MBE grown Ga2O3 and its power device applications *J. Cryst. Growth* **378** 591–5

[43] Sbrockey N M, Salagaj T, Coleman E, Tompa G S, Moon Y and Kim M S 2015 Large-Area MOCVD Growth of Ga2O3 in a Rotating Disc Reactor *J. Electron. Mater.* **44** 1357–60

[44] Wakabayashi R, Yoshimatsu K, Hattori M and Ohtomo A 2018 Epitaxial structure and electronic property of β -Ga2O3 films grown on MgO (100) substrates by pulsed-laser deposition *Appl. Phys. Lett.* **162101** 1–5

[45] Yu F-P, Ou S-L and Wuu D-S 2015 Pulsed laser deposition of gallium oxide films for high performance solar-blind photodetectors *Opt. Mater. Express* **5** 1240

[46] Garten L M, Zakutayev A, Perkins J D, Gorman B P, Ndione P F and Ginley D S 2016 Structure property relationships in gallium oxide thin films grown by pulsed laser deposition *MRS Commun.* **6** 348–53

[47] Zhang F, Arita M, Wang X, Chen Z, Saito K, Tanaka T, Nishio M, Motooka T and Guo Q 2016 Toward controlling the carrier density of Si doped Ga 2 O 3 films by pulsed laser deposition *Appl. Phys. Lett.* **109** 102105

[48] Wakabayashi R, Oshima T, Hattori M, Sasaki K, Masui T, Kuramata A, Yamakoshi S, Yoshimatsu K and Ohtomo A 2015 Oxygen-radical-assisted pulsed-laser deposition of β-Ga2O3 and β-(AlxGa1-x)2O3 films *J. Cryst. Growth* **424** 77–9

[49] Guo X C, Hao N H, Guo D Y, Wu Z P, An Y H, Chu X L, Li L H, Li P G, Lei M and Tang W H 2016 β-Ga2O3/p-Si heterojunction solar-blind ultraviolet photodetector with enhanced photoelectric responsivity *J. Alloys Compd.* **660** 136–40

[50] Feng Q, Li F, Dai B, Jia Z, Xie W, Xu T, Lu X, Tao X, Zhang J and Hao Y 2015 The properties of gallium oxide thin film grown by pulsed laser deposition *Appl. Surf. Sci.* **359** 847–52

[51] Meisenheimer P B, Kratofil T J and Heron J T 2017 Giant Enhancement of Exchange Coupling in Entropy-Stabilized Oxide Heterostructures *Sci. Rep.* **7** 13344

[52] Lee J, Safari A and Pfeffer R L 1992 Growth of epitaxial Pb(Zr,Ti)O3 films by pulsed laser deposition *Appl. Phys. Lett.* **61** 1643–5

[53] Sun X W, Xiao R F and Kwok H S 1998 Epitaxial growth of GaN thin film on sapphire with a thin ZnO buffer layer by liquid target pulsed laser deposition *J. Appl. Phys.* **84** 5776–9

[54] Kobayashi A, Kawano S, Kawaguchi Y, Ohta J and Fujioka H 2007 Room temperature epitaxial growth of m -plane GaN on lattice-matched ZnO substrates *Appl. Phys. Lett.* **90** 1–4

[55] Harbeke G, Krausbauer L, Steigmeier E F, Widmer A E, Kappert H F and Neugebauer G 1984 Growth and Physical Properties of LPCVD Polycrystalline Silicon Films *J. Electrochem. Soc.* **131** 675–82

[56] Joubert P 2006 The Effect of Low Pressure on the Structure of LPCVD Polycrystalline Silicon Films *J. Electrochem. Soc.* **134** 2541

[57] Meyerson B S and Olbricht W 1984 Phosphorus-Doped Polycrystalline Silicon via LPCVD: I . Process Characterization *J. Electrochem. Soc.* **131** 2361–5

[58] Joubert P, Loisel B, Chouan Y and Haji L 1987 The Effect of LPCVD Silicon Nitride Deposition on the Si-SiO2 Interface of Oxidized Silicon Wafers *J. Electrochem. Soc.* **134** 2541–5

[59] Zhang S L, Wang J T, Kaplan W and Östling M 1992 Silicon nitride films deposited from SiH2Cl2-NH3 by low pressure chemical vapor deposition: kinetics, thermodynamics, ... *Thin Solid Films* **213** 182–91

[60] Martin F, Bertin F, Sprey H and Granneman E 1991 LPCVD Si3N4growth retardation on silicon native oxide compared with in situ HF vapour-deglazed silicon substrates *Semicond. Sci. Technol.* **6** 1100–2

[61] Megiris C E and Glezer J H E 1992 Preparation of silicon dioxide films by low-pressure chemical vapor deposition on dense and porous alumina substrates *Chem. Eng. Sci.* **47** 3925–34

[62] Taylor R C 2006 Hexachlorodisilane as a Precursor in the LPCVD of Silicon Dioxide and Silicon Oxynitride Films *J. Electrochem. Soc.* **136** 2382

[63] Ohshita Y, Ogura A, Hoshino A, Hiiro S and Machida H 2001 HfO2 growth by low-pressure chemical vapor deposition using the Hf(N(C2H5)2)4/O2 gas system *J. Cryst. Growth* **233** 292–7

[64] Rafique S, Han L and Zhao H 2016 Synthesis of wide bandgap Ga2O3 (Eg~4.6–4.7 eV) thin films on sapphire by low pressure chemical vapor deposition *Phys. Status Solidi Appl. Mater. Sci.* **213** 1002–9

[65] Rafique S, Han L, Neal A T, Mou S, Boeckl J and Zhao H 2018 Towards High-Mobility Heteroepitaxial β-Ga2O3 on Sapphire Dependence on The Substrate Off-Axis Angle *Phys. Status Solidi Appl. Mater. Sci.* **215** 1700467

[66] Rafique S, Han L, Tadjer M J, Freitas J A, Mahadik N A and Zhao H 2016 Homoepitaxial growth of β-Ga2O3 thin films by low pressure chemical vapor deposition *Appl. Phys. Lett.* **108** 3–8

[67] Feng Z, Karim M R and Zhao H 2019 Low pressure chemical vapor deposition of β-Ga 2 O 3 thin films: Dependence on growth parameters *APL Mater.* **7**

[68] Rafique S, Han L, Neal A T, Mou S, Tadjer M J, French R H and Zhao H 2016 Heteroepitaxy of N-type β-Ga2O3 thin films on sapphire substrate by low pressure chemical vapor deposition *Appl. Phys. Lett.* **109**

[69] Rafique S, Karim M R, Johnson J M, Hwang J and Zhao H 2018 LPCVD homoepitaxy of Si doped β-Ga 2 O 3 thin films on (010) and (001) substrates *Appl. Phys. Lett.* **112**

[70] Rafique S, Han L, Mou S and Zhao H 2017 Temperature and doping concentration dependence of the energy band gap in β-Ga2O3 thin films grown on sapphire *Opt. Mater. Express* **7** 3561

[71] Joishi C, Zhang Y, Rafique S, Zhao H, Xia Z, Lodha S, Rajan S, Han L and Krishnamoorthy S 2018 Low-pressure CVD-grown β-Ga 2 O 3 bevel-field-plated Schottky barrier diodes *Appl. Phys. Express* **11** 031101

[72] Igawa, Takuto S F and K K 2018 *Ground breaking work on gallium oxide (Ga2O3) normally-off transistor*

[73] Xu Y, An Z, Zhang L, Feng Q, Zhang J, Zhang C and Hao Y 2018 Solar blind deep ultraviolet β-Ga 2 O 3 photodetectors grown on sapphire by the Mist-CVD method *Opt. Mater. Express* **8** 2941

[74] Tadjer M J, Mastro M A, Mahadik N A, Currie M, Wheeler V D, Freitas J A, Greenlee J D, Hite J K, Hobart K D, Eddy C R and Kub F J 2016 Structural, Optical, and Electrical Characterization of Monoclinic β-Ga2O3 Grown by MOVPE on Sapphire Substrates *J. Electron. Mater.* **45** 2031–7

[75] Alema F, Hertog B, Osinsky A, Mukhopadhyay P, Toporkov M and Schoenfeld W V 2017 Fast growth rate of epitaxial β – Ga 2 O 3 by close coupled showerhead MOCVD *J. Cryst. Growth* **475** 77–82

[76] Lv Y, Ma J, Mi W, Luan C, Zhu Z and Xiao H 2012 Characterization of β-Ga 2O 3 thin films on sapphire (0001) using metal-organic chemical vapor deposition technique *Vacuum* **86** 1850–4

[77] Du X, Mi W, Luan C, Li Z, Xia C and Ma J 2014 Characterization of homoepitaxial β -Ga 2 O 3 fi lms prepared by metal – organic chemical vapor deposition *J. Cryst. Growth* **404** 75–9

[78] Mi W, Ma J, Luan C, Lv Y, Xiao H and Li Z 2012 Characterization of β-Ga 2O 3 epitaxial films grown on MgO (111) substrates by metal-organic chemical vapor deposition *Mater. Lett.* **87** 109–12

[79] Nakagomi S and Kokubun Y 2017 Crystal orientation of monoclinic β -Ga 2 O 3 thin films formed on cubic MgO substrates with a c -Ga 2 O 3 interfacial layer *J. Cryst. Growth* **479** 67–74

[80] Chen Y, Liang H, Xia X, Shen R, Liu Y, Luo Y and Du G 2015 Effect of growth pressure on the characteristics of β-Ga 2 O 3 films grown on GaAs (100) substrates by MOCVD method *Appl. Surf. Sci.* **325** 258–61

[81] Mi W, Ma J, Luan C and Xiao H 2014 Structural and optical properties of β-Ga2O3 films deposited on MgAl2O4 (1 0 0) substrates by metal-organic chemical vapor deposition *J. Lumin.* **146** 1–5

[82] Baldini M, Albrecht M, Fiedler A, Irmscher K, Schewski R and Wagner G 2016 Editors' Choice—Si- and Sn-Doped Homoepitaxial β-Ga 2 O 3 Layers Grown by MOVPE on (010)-Oriented Substrates *ECS J. Solid State Sci. Technol.* **6** Q3040–4

[83] Zhang Y, Alema F, Mauze A, Koksaldi O S, Miller R, Osinsky A and Speck J S 2019 MOCVD grown epitaxial β -Ga 2 O 3 thin film with an electron mobility of 176 cm 2 /V s at room temperature *APL Mater.* **7**

[84] Feng Z, Anhar Uddin Bhuiyan A F M, Karim M R and Zhao H 2019 MOCVD homoepitaxy of Si-doped (010) β-Ga $_2$ O $_3$ thin films with superior transport properties *Appl. Phys. Lett.* **114** 250601

[85] Schewski R, Lion K, Fiedler A, Wouters C, Popp A, Levchenko S V., Schulz T, Schmidbauer M, Anooz S Bin, Grüneberg R, Galazka Z, Wagner G, Irmsche K, Scheffler M, Draxl C and Albrecht M 2019 Step-flow growth in homoepitaxy of β- Ga2O3 (100)— The influence of the miscut direction and faceting Step-flow growth in homoepitaxy of β -Ga 2 O 3 (100)— The influence of the miscut direction and faceting *APL Mater.* **7**

[86] Oshima T, Okuno T and Fujita S 2007 Ga2O3 thin film growth on c-plane sapphire substrates by molecular beam epitaxy for deep-ultraviolet photodetectors *Japanese J. Appl. Physics, Part 1 Regul. Pap. Short Notes Rev. Pap.* **46** 7217–20

[87] Ghose S, Rahman S, Hong L, Rojas-Ramirez J S, Jin H, Park K, Klie R and Droopad R 2017 Growth and characterization of β-Ga 2 O 3 thin films by molecular beam epitaxy for deep-UV photodetectors *J. Appl. Phys.* **122**

[88] Wiesinger M, Matich S, Aigner W and Stutzmann M 2019 Growth and characterization of β-Ga2O3 thin films on different substrates *Appl. Phys. Lett.* **105701**

[89] Sasaki K, Kuramata A, Masui T, Víllora E G, Shimamura K and Yamakoshi S 2012 Device-Quality β-Ga 2 O 3 Epitaxial Films Fabricated by Ozone Molecular Beam Epitaxy *Appl. Phys. Express* **035502** 5–8

[90] Sasaki K, Higashiwaki M, Kuramata A and Masui T 2014 Growth temperature dependences of structural and electrical properties of Ga2O3 epitaxial films grown on β-Ga2O3 (010) substrates by molecular beam epitaxy *J. Cryst. Growth* **392** 30–3

[91] Cheng Z, Hanke M and Galazka Z 2018 Growth mode evolution during (100) -oriented β - Ga 2 O 3 homoepitaxy *Nanotechnology* **4**

[92] Oshima T, Arai N, Suzuki N, Ohira S and Fujita S 2008 Surface morphology of homoepitaxial β -Ga 2 O 3 thin films grown by molecular beam epitaxy *Thin Solid Films* **516** 5768–71

[93] Tsai M-Y, Bierwagen O, White M E and Speck J S 2010 β-Ga2O3 growth by plasma-assisted molecular beam epitaxy *J. Vac. Sci. Technol. A Vacuum, Surfaces, Film.* **28** 354–9

[94] Vogt P and Bierwagen O 2016 Reaction kinetics and growth window for plasma-assisted molecular beam epitaxy of Ga2O3: Incorporation of Ga vs. Ga2O desorption *Appl. Phys. Lett.* **108** 1–5

[95] Krishnamoorthy S, Xia Z, Bajaj S, Brenner M and Rajan S 2017 Delta-doped β-gallium oxide field-effect transistor *Appl. Phys. Express Lett.* **051102** 1–4

[96] Ahmadi E, Koksaldi O S, Kaun S W, Oshima Y, Short D B, Mishra U K and Speck J S 2017 Ge doping of β-Ga2O3 films grown by plasma-assisted molecular beam epitaxy *Appl. Phys. Express* **10** 041102

[97] Feduniewicz A, Skierbiszewski C, Siekacz M, Wasilewski Z R, Sproule I, Grzanka S, Jakieła R, Borysiuk J, Kamler G, Litwin-Staszewska E, Czernecki R, Boćkowski M and Porowski S 2005 Control of Mg doping of GaN in RF-plasma molecular beam epitaxy *J. Cryst. Growth* **278** 443–8

[98] Farzana E, Ahmadi E, Speck J S, Arehart A R, Ringel S A and Ringel S A 2018 Deep level defects in Ge-doped (010) β- Ga2O3 layers grown by plasma-assisted molecular beam epitaxy *J. Appl. Phys.* **123** 161410

[99] Moser N, McCandless J, Crespo A, Leedy K, Green A, Neal A, Mou S, Ahmadi E, Speck J, Chabak K, Peixoto N and Jessen G 2017 Ge-doped β-Ga2O3 MOSFETs *IEEE Electron Device Lett.* **38** 775–8

[100] Neal A T, Mou S, Rafique S, Zhao H, Ahmadi E, Speck J S, Kevin T, Blevins J D, Thomson D B, Moser N, Chabak K D, Jessen G H, Neal A T, Mou S, Rafique S, Zhao H, Ahmadi E, Speck J S, Stevens K T, Blevins J D, Thomson D B, Moser N, Chabak K D and Jessen G H 2018 Donors and deep acceptors in β- Ga2O3 *Appl. Phys. Lett.* **062101** 1–6

[101] HILL V G, ROY R and OSBORN E F 1952 The System Alumina-Gallia-Water *J. Am. Ceram. Soc.* **35** 135–42

[102] Kaun S W, Wu F and Speck J S 2015 β-(AlxGa1−x)2O3/Ga2O3 (010) heterostructures grown on β-Ga2O3 (010) substrates by plasma-assisted molecular beam epitaxy *J. Vac. Sci. Technol. A Vacuum, Surfaces, Film.* **33** 041508

[103] Vogt P, Mauze A, Wu F, Bonef B and Speck J S 2018 Metal-oxide-catalyzed epitaxy: example of O plasma-assisted molecular beam epitaxy of β-(AlxGa1−x)2O3/β-Ga2O3 heterostructures *Appl. Phys. Express* **11** 11

[104] Vogt P, Brandt O, Riechert H, Lähnemann J and Bierwagen O 2017 Metal-Exchange Catalysis in the Growth of Sesquioxides: Towards Heterostructures of Transparent Oxide Semiconductors *Phys. Rev. Lett.* **119** 196001

[105] Krishnamoorthy S, Xia Z, Joishi C, Zhang Y, McGlone J, Johnson J, Brenner M, Arehart A R, Hwang J, Lodha S and Rajan S 2017 Modulation-doped β-(Al 0.2 Ga 0.8) 2 O 3 /Ga 2 O 3 field-effect transistor *Appl. Phys. Lett.* **111** 3–7

[106] Motoki K, Okahisa T, Matsumoto N, Matsushima M, Kimura H, Kasai H, Takemoto K, Uematsu K, Hirano T, Nakayama M, Nakahata S, Ueno M, Hara D, Kumagai Y, Koukitu A and Seki H 2001 Preparation of Large Freestanding GaN Substrates by Hydride Vapor Phase Epitaxy Using GaAs as a Starting Substrate *Jpn. J. Appl. Phys.* **40** L140–3

[107] Yoshida T, Oshima Y, Watanabe K, Tsuchiya T and Mishima T 2011 Ultrahigh-speed growth of GaN by hydride vapor phase epitaxy *Phys. status solidi c* **8** 2110–2

[108] Konishi K, Goto K, Togashi R, Murakami H, Higashiwaki M, Kuramata A, Yamakoshi S, Monemar B and Kumagai Y 2018 Comparison of O 2 and H 2 O as oxygen source for homoepitaxial growth of β-Ga 2 O 3 layers by halide vapor phase epitaxy *J. Cryst. Growth* **492** 39–44

[109] Murakami H, Nomura K, Goto K, Sasaki K, Kawara K, Thieu Q T, Togashi R, Kumagai Y, Higashiwaki M, Kuramata A, Yamakoshi S, Monemar B and Koukitu A 2015 Homoepitaxial growth of β-Ga $_2$ O $_3$ layers by halide vapor phase epitaxy *Appl. Phys. Express* **8** 015503

[110] Leach J H, Udwary K, Rumsey J, Dodson G, Splawn H and Evans K R 2019 Halide vapor phase epitaxial growth of β -Ga 2 O 3 and α -Ga 2 O 3 films *APL Mater.* **7** 1–8

[111] Nikolaev V I, Pechnikovab A I, StepanovbdI S I, Nikitinac P, A.N. S, Chikiryakac V, Sharofidinov S, V.E.Bougrovb, Romanov A E and Pechnikov A I 2008 Epitaxial growth of (-201) β-Ga2O3 on (0001) sapphire substrates by halide vapour phase epitaxy *Mater. Sci. Semicond. Process.* **47** 16–9

[112] Oshima Y and Shimamura K 2015 Quasi-heteroepitaxial growth of β-Ga2O3 on off-angled sapphire (0 0 0 1) substrates by halide vapor phase epitaxy *J. Cryst. Growth* **410** 53–8

[113] Xiu X, Zhang L, Li Y, Xiong Z, Zhang R and Zheng Y 2019 Application of halide vapor phase epitaxy for the growth of ultra-wide band gap Ga 2 O 3 *J. Semicond.* **40**

[114] Nomura K, Goto K, Togashi R, Murakami H, Kumagai Y, Kuramata A, Yamakoshi S and Koukitu A 2014 Thermodynamic study of β-Ga2O3growth by halide vapor phase epitaxy *J. Cryst. Growth* **405** 19–22

[115] Goto K, Konishi K, Murakami H, Kumagai Y, Monemar B, Higashiwaki M, Kuramata A and Yamakoshi S 2018 Halide vapor phase epitaxy of Si doped β-Ga 2 O 3 and its electrical properties *Thin Solid Films* **666** 182–4

[116] Higashiwaki M, Sasaki K, Goto K, Nomura K, Thieu Q T, Togashi R, Murakami H, Kumagai Y, Monemar B, Koukitu A, Kuramata A and Yamakoshi S 2015 Ga2O3Schottky barrier diodes with n--Ga2O3drift layers grown by HVPE *Device Research Conference - Conference Digest, DRC* vol 2015-Augus pp 29–30

[117] Higashiwaki M, Sasaki K, Wong M H, Kamimura T, Krishnamurthy D, Kuramata A, Masui T and Yamakoshi S 2013 Depletion-mode Ga2O3 MOSFETs on β-Ga 2O3 (010) substrates with Si-ion-implanted channel and contacts *Technical Digest - International Electron Devices Meeting, IEDM* vol 3 pp 707–10

[118] Higashiwaki M, Sasaki K, Kamimura T, Hoi Wong M, Krishnamurthy D, Kuramata A, Masui T and Yamakoshi S 2013 Depletion-mode Ga2O3 metal-oxide-semiconductor field-effect transistors on β-Ga2O3 (010) substrates and temperature dependence of their device characteristics *Appl. Phys. Lett.* **103** 1–5

[119] Wong M H, Sasaki K, Kuramata A and Yamakoshi S 2016 Field-Plated Ga 2 O 3 MOSFETs With a Breakdown Voltage of Over 750 V *IEEE Electron Device Lett.* **37** 212–5

[120] Wong M H, Sasaki K, Kuramata A, Yamakoshi S and Higashiwaki M 2016 Electron channel mobility in silicon-doped Ga2O3MOSFETs with a resistive buffer layer *Japanese Journal of Applied Physics* vol 55 pp 1202–11

[121] Singh M, Casbon M A, Uren M J, Pomeroy J W, Dalcanale S, Karboyan S, Tasker P J, Wong M H, Sasaki K, Kuramata A, Yamakoshi S, Higashiwaki M and Kuball M 2018 Pulsed Large Signal RF Performance of Field-Plated Ga2O3 MOSFETs *IEEE Electron Device Lett.* **39** 1572–5

[122] Wong M H, Takeyama A, Makino T, Ohshima T, Sasaki K, Kuramata A, Yamakoshi S and Higashiwaki M 2018 Radiation hardness of β -Ga2O3 metal-oxide-semiconductor field-effect transistors against gamma-ray irradiation *Appl. Phys. Lett.* **112**

[123] Zhang L, Verma A, Xing H (Grace) and Jena D 2017 Inductively-coupled-plasma reactive ion etching of single-crystal β-Ga 2 O 3 *Jpn. J. Appl. Phys.* **56** 030304

[124] Hogan J E, Kaun S W, Ahmadi E and Oshima Y 2016 Chlorine-based dry etching of β -Ga 2 O 3 *Semicond. Sci. Technol.* **31**

[125] Jian Z, Oshima Y, Wright S, Owen K and Ahmadi E 2019 Chlorine-based inductive coupled plasma etching of α-Ga 2 O 3 *Semicond. Sci. Technol.* **34** 035006

[126] Moser N A, McCandless J P, Crespo A, Leedy K D, Green A J, Heller E R, Chabak K D, Peixoto N and Jessen G H 2017 High pulsed current density β -Ga 2 O 3 MOSFETs verified by an analytical model corrected for interface charge *Appl. Phys. Lett.* **110** 143505

[127] Green A J, Chabak K D, Heller E R, Fitch R C, Baldini M, Fiedler A, Irmscher K, Wagner G, Galazka Z, Tetlak S E, Crespo A, Leedy K and Jessen G H 2016 3.8-MV/cm Breakdown Strength of MOVPE-Grown Sn-Doped β-Ga2O3 MOSFETs *IEEE Electron Device Lett.* **37** 902–5

[128] Chabak K D, Jr D E W, Green A J, Crespo A, Lindquist M, Leedy K, Tetlak S, Moser N A and Jessen G 2018 Sub-Micron Gallium Oxide Radio Frequency Field-Effect Transistors *2018 IEEE MTT-S Int. Microw. Work. Ser. Adv. Mater. Process. RF THz Appl.* 1–3

[129] Green A J, Chabak K D, Baldini M, Moser N, Gilbert R, Fitch R C, Wagner G, Galazka Z, McCandless J, Crespo A, Leedy K and Jessen G H 2017 β-Ga2O3 MOSFETs for radio frequency operation *IEEE Electron Device Lett.* **38** 790–3

[130] Joishi C, Xia Z, McGlone J, Zhang Y, Arehart A R, Ringel S, Lodha S and Rajan S 2018 Effect of buffer iron doping on delta-doped β-Ga2O3 metal semiconductor field effect transistors *Appl. Phys. Lett.* **113**

[131] Zeng K, Vaidya A and Singisetti U 2018 1.85 kV Breakdown Voltage in Lateral Field-Plated Ga2O3 MOSFETs *IEEE Electron Device Lett.* **39** 1385–8

[132] Zeng K, Vaidya A and Singisetti U 2019 A field-plated Ga2O3 MOSFET with near 2-kV breakdown voltage and 520 mΩ · cm2 on-resistance *Appl. Phys. express* **12** 8

[133] Chabak K, Green A, Moser N, Tetlak S, McCandless J, Leedy K, Fitch R, Crespo A and Jessen G 2017 Gate-recessed, laterally-scaled β-Ga2O3 MOSFETs with high-voltage enhancement-mode operation *2017 75th Annual Device Research Conference (DRC)* vol 38 (IEEE) pp 1–2

[134] Wong M H, Nakata Y, Kuramata A, Yamakoshi S and Higashiwaki M 2017 Enhancement-mode Ga2O3 MOSFETs with Si-ion-implanted source and drain *Appl. Phys. Express* **10** 4

[135] Zhang Y, Neal A, Xia Z, Joishi C, Johnson J M, Zheng Y, Bajaj S, Brenner M, Dorsey D, Chabak K, Jessen G, Hwang J, Mou S, Heremans J P and Rajan S 2018 Demonstration of high mobility and quantum transport in modulation-doped β-$(Al_xGa_{1-x})_2O_3/Ga_2O_3$ heterostructures *Appl. Phys. Lett.* **112** 173502

[136] Ghosh K and Singisetti U 2017 Ab initio velocity-field curves in monoclinic β-Ga2O3 *J. Appl. Phys.* **122** 035702

[137] Mengle K A, Shi G, Bayerl D and Kioupakis E 2016 First-principles calculations of the near-edge optical properties of β-Ga2O3 *Appl. Phys. Lett.* **109** 212104

[138] Ma N, Tanen N, Verma A, Guo Z, Luo T, Xing H (Grace) and Jena D 2016 Intrinsic electron mobility limits in β-Ga2O3 *Appl. Phys. Lett.* **109** 212101

[139] Joishi C, Zhang Y, Xia Z, Sun W, Arehart A R, Ringel S, Lodha S and Rajan S 2019 Breakdown characteristics of β-(Al0.22Ga0.78)2O3/ Ga2O3 field-plated modulation doped field effect transistors with SiNx passivation *IEEE Electron Device Lett.* **PP** 1–1

[140] Okumura H, Kato Y, Oshima T and Palacios T 2019 Demonstration of lateral field-effect transistors using Sn-doped β-(AlGa) 2 O 3 (010) *Jpn. J. Appl. Phys.* **58** SBBD12

[141] Hu Z, Nomoto K, Li W, Zhang L J, Shin J H, Tanen N, Nakamura T, Jena D and Xing H G 2017 Vertical fin Ga2O3 power field-effect transistors with on/off ratio >109 *Device Research Conference - Conference Digest, DRC* vol 034001 (IEEE) pp 1–2

[142] Hu Z, Nomoto K, Li W, Jinno R, Nakamura T, Jena D and Xing H G 2019 1.6 kV Vertical Ga2O3 FinFETs With Source- Connected Field Plates and Normally-off Operation *Ispsd19* **3** 483–6

[143] Kim J, Kim J, Oh S and Mastro M A 2016 Exfoliated β-Ga2O3 nano-belt field-effect transistors for air-stable high power and high temperature electronics *Phys. Chem. Chem. Phys.* **18** 15760–4

[144] Ahn S, Ren F, Kim J, Oh S, Kim J, Mastro M A and Pearton S J 2016 Effect of front and back gates on β-Ga2O3 nano-belt field-effect transistors *Appl. Phys. Lett.* **109** 062102

[145] Zhou H, Si M, Alghamdi S, Qiu G, Yang L and Ye P D 2017 High-Performance Depletion/ Enhancement-mode β-Ga2O3 on Insulator (GOOI) Field-Effect Transistors With Record Drain Currents of 600/450 mA/mm *IEEE Electron Device Lett.* **38** 103–6

[146] Hwang W S, Verma A, Protasenko V, Rouvimov S, Xing H G, Seabaugh A, Haensch W, Van De Walle C, Galazka Z, Albrecht M, Fornnari R and Jena D 2013 Nanomembrane β-Ga 2 O 3 high-voltage field effect transistors *Device Research Conference - Conference Digest, DRC* (IEEE) pp 207–8

[147] Zhou H, Maize K, Qiu G, Shakouri A and Ye P D 2017 β-Ga2O3 on insulator field-effect transistors with drain currents exceeding 1.5 A/mm and their self-heating effect *Appl. Phys. Lett.* **111** 092102

[148] Si M, Yang L, Zhou H and Ye P D 2017 β-Ga2O3 Nanomembrane Negative Capacitance Field-Effect Transistors with Steep Subthreshold Slope for Wide Band Gap Logic Applications *ACS Omega* **2** 7136–40

Opportunities and Challenges in MOCVD of β-Ga_2O_3 for Power Electronic Devices

M. A. Mastro*, J. K. Hite, C. R. Eddy, Jr., and M. J. Tadjer

U.S. Naval Research Laboratory, Washington, DC 20375, USA
**michael.mastro@nrl.navy.mil*

S. J. Pearton

Department of Materials Science and Engineering, University of Florida, Gainesville, FL 32611, USA

F. Ren

Department of Chemical Engineering, University of Florida, Gainesville, FL 32611, USA

J. Kim

Department of Chemical and Biological Engineering, Korea University, Seoul 02841, Korea

Recent breakthroughs in bulk crystal growth of β-Ga_2O_3 by the edge-defined film-fed technique has led to the commercialization of large-area β-Ga_2O_3 substrates. Standard epitaxy approaches are being utilized to develop various thin-film β-Ga_2O_3 based devices including lateral transistors. This article will discuss the challenges for metal organic chemical vapor deposition (MOCVD) of β-Ga_2O_3 and the design criteria for use of this material system in power electronic device structures.

Keywords: Gallium oxide; MOCVD; power electronic device.

1. Properties

β-Ga_2O_3-based transistors and diodes possess fundamental electronic properties that make them ideal candidates for high power devices (Table 1). A number of these properties derive directly from the wide band-gap of β-Ga_2O_3 (E_g = 4.85 eV) including an exceptionally high electric breakdown field (approximately 8 MV/cm). This high breakdown field allows β-Ga_2O_3-based devices to be biased at a high drain voltage ($V_{break\text{-}down} \gg 10$ kV) while maintaining a large dynamic range. Furthermore, the wide band-gap of β-Ga_2O_3 allows device operation at elevated temperature without degradation. Additionally, Ga_2O_3 has a high saturation electron velocity ($v_{sat} = 2 \times 10^7$ cm/s), which is partially accountable for the high current density, I_{max} (where $I_{max} \approx qnv_{sat}$, q is the elementary charge, and n is the charge density) in devices [1].

Table 1. Properties of relevant semiconductor materials and normalized unipolar power-device figures of merit (FOM). The Johnson FOM describes the power-frequency capability, the Baliga FOM gives the specific on-resistance in the drift region, the combined FOM combines the power, frequency, voltage metrics, the Baliga high-frequency FOM provides a measure of switching losses, the Keyes FOM describes the thermal capability to handle high power density at high frequency. The Johnson and Baliga FOMs are remarkably high for Ga_2O_3 [2].

Properties	Si	GaAs	Diamond	4H-SiC	GaN	Ga_2O_3
Bandgap E_g [eV]	1.12	1.42	5.5	3.25	3.4	4.85
Dielectric Constant, ε	11.8	12.9	5.7	9.7	9	10
Breakdown Field, E_c [MV/cm]	0.3	0.4	20	2.5	3.3	8
Electron Mobility, μ [cm^2/V·s]	1500	8500	4500	1000	1250	250
Maximum Velocity, v_s [10^7 cm/s]	1	1	2.5	2	3	2
Thermal Conductivity, λ, [W/cm·K]	1.5	0.5	24	4.9	2.3	0.23
Figure of Merits / relative to Si						
Johnson = $E_c^2 \cdot V_s^2/4\pi^2$	1	1.8	27777	277	1089	2844
Baliga = $\varepsilon \cdot \mu \cdot E_c^3$	1	14.7	429378	317	846	3214
Combined = $\lambda \cdot \varepsilon \cdot \mu \cdot V_s \cdot E_c^2$	1	3.7	257627	248	353	37
Baliga High Frequency = $\mu \cdot E_c^2$	1	10.1	13333	46	100	142
Keyes = $\lambda \cdot [(c \cdot Vs)/(4\pi \cdot \varepsilon)]^{1/2}$	1	0.3	23.0	3.6	1.8	0.2

Power semiconductor devices, used in three-terminal switches or two-terminal rectifiers, when forward biased should have minimal resistance in the on-state, $R_{on\text{-}sp}$, and support a large blocking voltage, V_B, in the off-state [3]. In a standard device design, increasing the thickness, L_N, or decreasing the doping, N_d, of an n^- drift region increases the on-resistance as described by

$$R_{on-sp} = \frac{L_N}{q\mu_n N_d} \tag{1}$$

Avalanche breakdown occurs when the electric field in the deletion region exceeds the material dependent critical value, E_c [4]. For an abrupt junction, the depletion layer extends almost entirely in the lightly doped side as described by

$$W_D = \sqrt{\frac{\varepsilon_S V_B}{q N_d}}, \tag{2}$$

where ε_s is the permittivity [5]. The linearly decreasing field across the depletion layer has a maximum at the junction. For a drift layer thickness sufficient to support this depletion width, $L_N \approx W_D$, the maximum breakdown is set by critical electric field, $V_B = \frac{E_C W_D}{2}$.

Relating these equations gives the maximum blocking voltage inversely related to the doping density in drift layer by

$$V_B = \frac{\varepsilon_S E_c^2}{2qN_d} \tag{3}$$

Again combining these equations shows the inherent tradeoff between on-resistance and blocking voltage

$$R_{on-sp} = \frac{4V_B^2}{\varepsilon_S \mu_n E_c^3}, \tag{4}$$

where the denominator of the equation is the Baliga figure of merit [6]. The simplest method to break this design tradeoff is to move to a semiconductor material with a higher critical electric field (Fig. 1).

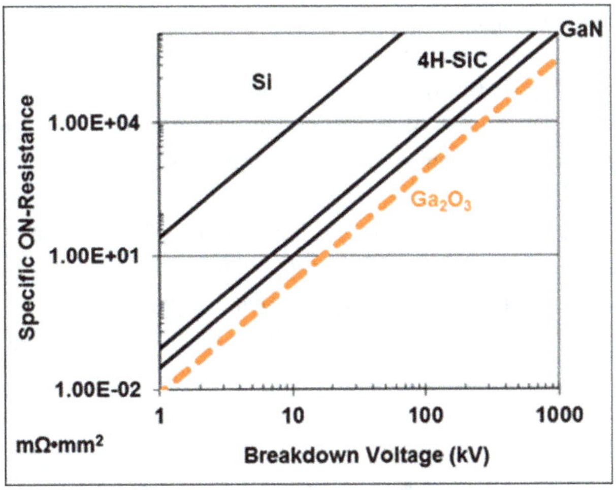

Fig. 1. Each material faces an inherent tradeoff of specific on-resistance to breakdown field as $R_{on-sp} \sim V_B^2$. The high critical electric field of Ga_2O_3 provides an inherent advantage compared to GaN and SiC given that $R_{on-sp} \sim E_c^{-3}$ [7].

Examination of the bandgap and breakdown field for various semiconductors reveals a simple relationship given by $\varepsilon_c = a(E_g)^n$ where a and n are fitting parameters. Specifically, the parameters for indirect semiconductors are a = 2.38x10^5 and n = 1.995, direct semiconductors are a = 1.73x10^5 and n = 2.506, and all semiconductors are a = 1.75x10^5 and n = 2.359 [8]. The generally accepted value for breakdown field of β-type of Ga_2O_3 is 8 MV/cm although this exceptionally high value has not been experimentally confirmed [9].

Early demonstrations of high-breakdown β-Ga_2O_3 electronic devices are promising, e.g., a critical field strength of 3.8 MV/cm and a 1kV vertical Schottky diode, yet fall short of the predicted levels. Several fundamental material issues are limiting the capability of β-Ga_2O_3 for power electronic devices. Experimental measurements of mobility are less than half of the theoretical predictions. Key scattering mechanisms are still unclear including the role of point defects and defect complexes as well as structural stacking faults. Further studies are needed to understand the limitations in saturation electron velocity and breakdown field [9].

2. Crystal Structure of β-Ga_2O_3

Examining the structure of the β-Ga_2O_3 crystal helps to frame a number of issues in the growth and behavior of Ga_2O_3. The valence band maximum in β-Ga_2O_3 forms from weakly interacting O 2p orbital states with contribution of Ga 3d and 4s orbitals while the conduction band minimum forms from Ga 4s states [10]. Closer examination of Fig. 2 shows that the Ga^{3+} cations (with a green coloring) have two distinct bonding coordinations. The Ga (I) cation has a distorted tetrahedral coordination with four bonds and the Ga (II) cations has an octahedral coordination with six bonds. Among the common n-type dopants for β-Ga_2O_3, Si, and Ge donors prefer the tetrahedral coordination of the Ga(I) site while Sn donors prefer the octahedral coordination of Ga(II) Site.

Dramatic progress has been made over the past few years in the bulk crystal growth of β-Ga_2O_3. The edge-defined film-fed growth (EFG) technique involves pulling the boule along the [010] direction at a rate of approximately 15mm/hr. The EFG process has led to the production of up to 4-inch diameter Ga_2O_3 substrates oriented in ($\bar{2}01$) or (001) planes. The float zone technique is able to produce (100), (010), and (001) oriented boules with a diameter of 1 inch at a growth rate approaching 5mm/hr. The Czochralski technique has produced, at a pull rate of 2 mm/hr, (100) oriented boules of 2-inch diameter with potential for larger diameter boules in the future [11]. The dislocation density of current bulk wafers is of the order 10^3 cm^{-2}, a key result for making large area power devices [9].

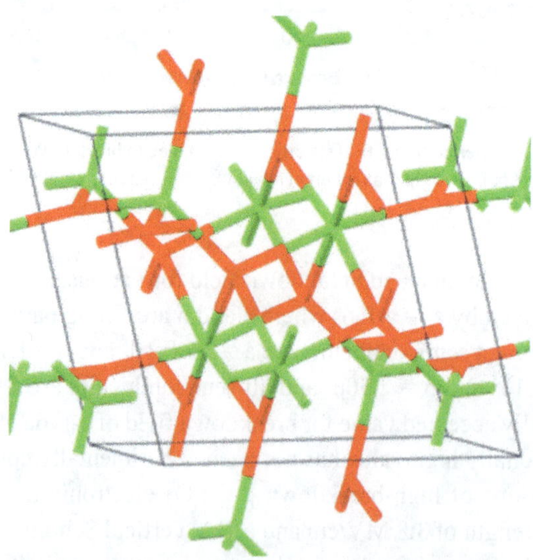

Fig. 2. Bonding structure of monoclinic (C2/M group symmetry) β-Ga_2O_3 phase.

3. Homoepitaxy on (100) Plane

The availability from the Czochralski process of (100) orientated wafers provided a basis for a series of homoepitaxial studies [12, 13] as described in [14]. Schewski *et al.* found that double positioning (180° rotation) of the monoclinic crystal growth (100) plane leads to twin lamellae formation and stacking mismatch boundaries. This double positioning creates a number of possible monoclinic Ga_2O_3 stacking faults including a half unit cell twin layer, a twin layer at surface, and diagonal stacking fault that serves to restore lattice stacking in the direction of growth [15].

Schewski *et al.* reported that (100) substrates offcut towards the $[00\bar{1}]$ direction provided steps to align the crystal. The growth proceeds in a step-flow manner where impinging adatoms diffuse to a terrace edge. It was observed that an optimal miscut of 6° leads to a density of twins of approximately zero [15]. Conceptually it follows that too small of an offcut can lead to isolated nucleation of islands on large terraces with double positioning and resultant formation of twin lamellae and stacking mismatch boundaries.

In order to describe this process of adatom diffusion to step edges vs. formation of (potentially twinned) islands, Schewski *et al.* extended a model of Bales and Zangwill [16]. This model required the coupled solution to a set of s ordinary differential equations ODEs for the density of adatoms, $\langle n_1 \rangle$,

$$\frac{1}{F}\frac{d\langle n_1\rangle}{dt} = \gamma - \frac{D}{F}\xi^{-2}\langle n_1\rangle - \frac{D}{F}\chi\langle n_1\rangle - k_1\langle n_1\rangle - \sum_{s=1}^{\infty} k_s\langle n_s\rangle, \tag{5}$$

and density of island of size s, $\langle n_s \rangle$,

$$\frac{1}{F}\frac{d\langle n_s\rangle}{dt} = \frac{D}{F}\sigma_{s-1}\langle n_1\rangle\langle n_{s-1}\rangle - \frac{D}{F}\sigma_s\langle n_1\rangle\langle n_s\rangle + k_{s-1}n_{s-1} - k_s\langle n_s\rangle \quad s = 2, 3, \ldots, \tag{6}$$

with equations for flux of incoming adatoms, F, adatom attachment, γ, diffusion to Island, ξ, diffusion to step edge, χ, and deposition on existing island detailed in [15]. The solution to these equations is displayed in Fig. 3 at four growth rates that define the adatom flux to the surface.

This calculation provides some guidance on how to control the stacking fault formation for homoepitaxy on (100) plane. Examination of Fig. 3 indicates that a layer with minimal stacking faults requires a flux of atoms to the surface in less than the time for these atoms to diffuse to a step edge. The diffusion constant of 7×10^{-9} cm^2s^{-1} employed in this calculation is based on a growth temperature of 850°C. Increasing the growth temperature is a clear lever to increase diffusion rate; however, decomposition of the surface is expected at higher temperatures. Nevertheless, reports of Si doping as well as In doping or $(In_xGa_{1-x})_2O_3$ alloy formation have shown a surfactant-mechanism that can increase the effective adatom diffusion rate [17].

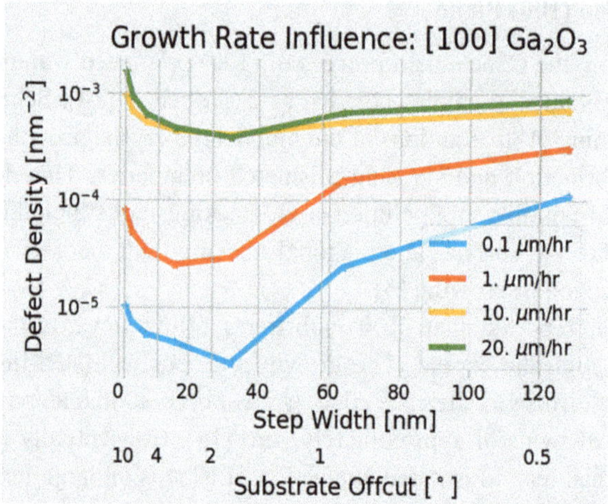

Fig. 3. Calculation of stacking fault density as a function of terrace step width for various growth rates. At low offcut, the adatom surface diffusion length is less than the step terrace width, which leads to island growth with twin and related stacking fault formation. In contrast, a step-bunching mechanism will occur at high offcut [15].

4. Facet Stability

A follow-on study by Schewski *et al.* found that the substrate monoclinic offcut direction of $[00\bar{1}]$ and [001] were not equivalent [18]. A Ga_2O_3 substrate offcut towards $[00\bar{1}]$ resulted in steps that reconstruct as the $(\bar{2}01)$ facet. Subsequent deposition proceeded in a 2D step-flow manner and the resultant film had high mobility. In contrast, a Ga_2O_3 substrate offcut towards [001] reconstructed as a twin $(\bar{2}01)$ defect nucleated at a (001)-B step. The film deposited on this substrate possessed a high density of stacking faults (that behaved as acceptor-like electron trap states) and displayed low electron mobility.

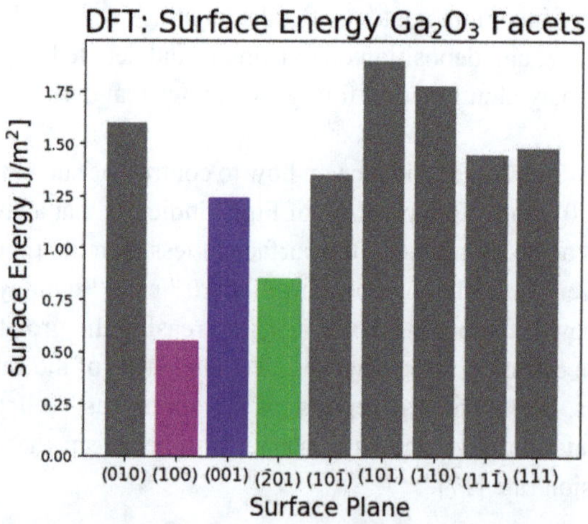

Fig. 4. Density functional theory calculation of surface energy of relevant Ga_2O_3 surfaces [18].

An earlier ab-initio studies by Bermudez reported a low surface energy of β-Ga_2O_3 on the (100) plane [19]. The general understanding of facet energetics was based on this report by Bermudez that only examined that the (100), (010), (001) and $(10\bar{1})$ faces of β-Ga_2O_3. The energy of these crystal planes is shown in Fig. 4 with the inclusion of the $(\bar{2}01)$ plane [18]. The reconstruction of the {001} step edge for the <001> offcut is apparent given that the surface energy of the $(\bar{2}01)$ plane is lower than the (001) plane. As discussed by Schewski *et al.*, the asymmetry of the monoclinic (100) crystal results in an asymmetry of the reconstruction. Specifically the (001)-B step cannot form the $(\bar{2}01)$ plane without a stacking fault [18].

Although the preceding discussion was based on homoepitaxy on (100) plane, it is clear that the crystal facet energetics determine the preferred orientations in bulk crystal growth (as discussed above) and the defect formation mechanisms of thin-film growth. Rafique *et al.* found that hetero-epitaxy on (0001) c-plane sapphire produced $(\bar{2}01)$ oriented Ga_2O_3 with in-plane rotational domains. The use of (0001) sapphire offcut towards $(11\bar{2}0)$ favored the formation of one domain with the highest mobility found for the film grown on (0001) sapphire with an offcut of 6° [20]. Another effect of the crystal facet energetics was seen in the surface morphology of homoepitaxy (010) Ga_2O_3, which formed with a striped surface morphology along [001] [21].

5. Carrier Concentration / Compensation

There has also been progress in the development in the epitaxy of doped β-Ga_2O_3 by a number of techniques, including MOCVD, HVPE, and MBE with reports of n-type doping over the range 10^{15} to 10^{19} cm^{-3} using Sn or Si shallow donors [9]. It is critical to understand the influence of precursor (trimethylgallium (TMGa) vs. triethylgallium (TEGa)), dopant type (Sn vs. Si) and carrier gas (Ar vs. N_2). This interplay in the growth environment can be expected given that in β-Ga_2O_3 the top of the valence and the bottom of the conduction band, respectively, are made up of the anionic (O 2p states with contributions from Ga 3d and 4s orbitals) and cationic states (Ga 4s states). Similarly, the carrier behavior has been shown to dramatically change under annealing, e.g., N_2 annealing creates deep acceptor states in n-type β-Ga_2O_3 [22, 23].

Figure 5 provides a comparison of conductivity for each particular set of precursor and dopant. Baldini *et al.* found in MOCVD efficient activation of the Si dopant to produce free carriers in the range of $1\text{x}10^{17}$ to $8\text{x}10^{19}$ cm^{-3} in β-Ga_2O_3 films on (010) β-Ga_2O_3 substrates [21]. In contrast, incorporation of the Sn dopant was hampered above a concentration of $1\text{x}10^{19}$ cm^{-3}. This behavior is framed by the earlier discussion that Si (and Ge) prefer the tetrahedral coordination of Ga(I) site and Sn prefers octahedral coordination of Ga(II) site. Lastly, Baldini *et al.* reported that a memory effect of Sn in the reactor produced films with an unintentional Sn concentration of approximately $4\text{x}10^{17}$ cm^{-3} [21].

It is commonly understood for MOCVD of compound semiconductor films that the TMGa reaction pathway of $Ga(CH_3)_3 \rightarrow Ga\text{-}CH_2(surface) + CH_4(g)$ can leave a high level of carbon in the films. This is in contrast to the TEGa sequential β-elimination pathway of $Ga(C_2H_5)_3 \rightarrow (C_2H_5)_2GaH + C_2H_4(g)$ that should yield a semiconductor film with relatively

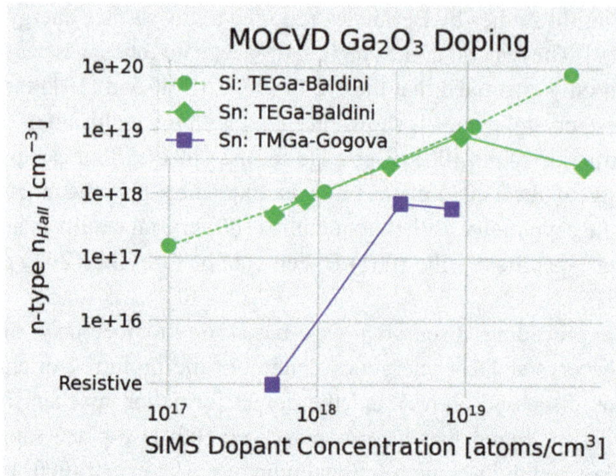

Fig. 5. Comparison of Si and Sn doping during growth of β-Ga_2O_3 with TMGa and TEGa precursors (adapted from [21] and [24]). With a TEGa source, moderate levels of Sn and Si dopant demonstrate linear n-type incorporation in β-Ga_2O_3.

less carbon. Similarly it is generally understood in an MOCVD growth environment with a gallium metalorganic precursor that a decrease in growth temperature will increase the relative level of carbon, which may act as a deep acceptor. This may be especially pertinent given the deposition temperature of β-Ga_2O_3 is approximately 300 degrees lower than the typical MOCVD temperature for GaN. Additionally, an increase in growth rate will generally increase the relative level of carbon.

It has been generally observed that MOCVD films produced via a Ga source of TMGa are resistive except at high n-type doping and with H_2O as the oxygen source. Using TMGa and H_2O, Gogova *et al.* doped β-Ga_2O_3 with Sn on (0001) sapphire and (100) β-Ga_2O_3 substrates (Fig. 5). Raman spectroscopy of the films found C-H-related bands. Their analysis stated that Ga vacancy-related defects and the carbon-related complexes act as acceptors compensating for the Sn donors [24].

Tuomisto *et al.* employed positron annihilation spectroscopy to relate the concentration of negative and neutral vacancies to the conductivity of doped and undoped β-Ga_2O_3 thin films [25]. These results as depicted in Fig. 6 show that MOCVD with a Ga source of TEGa resulted in a low concentration of gallium vacancies while MOCVD with a TMGa precursor resulted in a high concentration of gallium vacancies. Not shown in the plot is that all films with a vacancy concentration equal to or greater than $1x10^{17}$ cm^{-3} were insulating. An interesting conclusion from Tuomisto *et al.* [25] is that growth kinetics and chemical reactions at the MOCVD growth surface dictate the Ga vacancy formation, i.e., not the Fermi level potential of the crystal in a thermodynamic equilibrium condition.

An underlying mechanism for unintentional conductivity in β-Ga_2O_3 was suggested to be based on hydrogen [26, 27]. Interstitial hydrogen (H_i) or hydrogen trapped at oxygen vacancies (H_O) is predicted to act as a shallow donors [9]. Qin *et al.* employed infrared

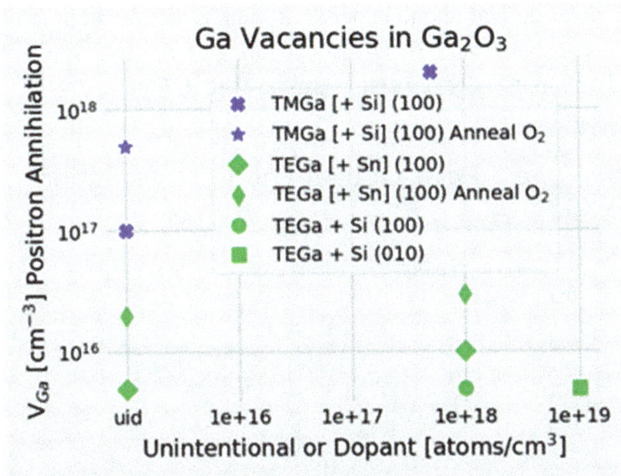

Fig. 6. Comparison of gallium vacancy, V_{Ga}, concentration as measured by positron annihilation spectroscopy as a function of dopant concentration for comparing TEGa and TMGa precursors as well as with and without an oxygen anneal (as adapted from [25]). All films deposited with a TMGa precursor displayed a vacancy concentration equal to or greater than $1\text{x}10^{17}$ cm^{-3} and were electrically insulating (not shown).

spectroscopy to study β-Ga_2O_3 annealed in a H_2 or D_2 ambient. This study found a hidden reservoir of hydrogen that is composed of various hydrogen centers coupled to a gallium vacancy as well as a variety of other species. Vibrational spectroscopy assigned the dominant hydrogen center to a neutral complex composed of two equivalent hydrogen at a relaxed gallium vacancy, $V_{Ga(I)}$-2H [28].

6. Mobility Limits

The reported theoretical room temperature mobility of β-Ga_2O_3 is in the range of 200 to 300 cm^2/V·s; however, typical experimental mobilities are reported in 50 to 150 cm^2/V·s range [9]. Referring to Table 1, the mobility of β-Ga_2O_3 is low relative to other wide bandgap semiconductors. It is constructive to understand what limits the electron mobility in the β-Ga_2O_3 crystal. As discussed by Ma *et al.*, β-Ga_2O_3 has an effective mass, m_c^*, of approximately $0.25m_0$, which is comparable to the effective mass of GaN of $0.21m_0$ although the mobility of GaN is approximately 1200 cm^2/V·s [29].

It is known that the bonding of Ga and O has a large difference in electronegativity, X_{AB}, or similarly a large Pauling ionicity given by $f_P = 1 - exp(\frac{-X_{AB}^2}{4})$. A useful value for predicting the influence of electron to polar optical phonon interaction in polar semiconductors is the Fröhlich coupling constant,

$$\alpha_F = \frac{q^2}{8\pi\varepsilon_0}\sqrt{\frac{2m_c^*}{\hbar\omega_0}}\left(\frac{1}{\varepsilon_\infty} - \frac{1}{\varepsilon_s}\right), \tag{7}$$

where $\hbar\omega_0$ is the polar optical phonon energy, ε_∞ and ε_s are the high- and low-frequency relative dielectric constants, and ε_0 is the vacuum permitvitty. The Fröhlich coupling constant is quite large in β-Ga_2O_3 crystal as can be seen in Fig. 7.

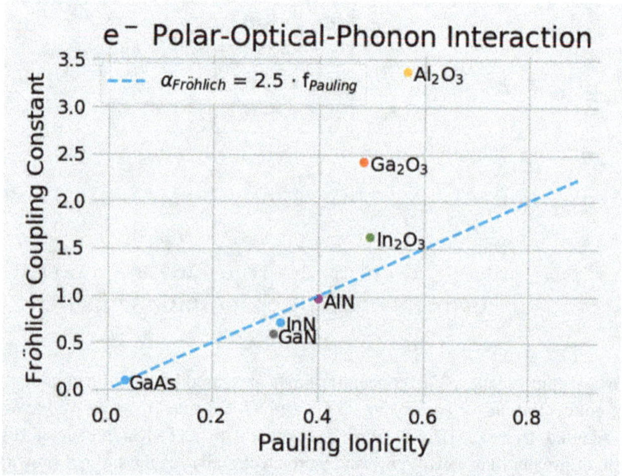

Fig. 7. The exceptionally strong electron to polar optical phonon interaction as measured by the dimensionless Fröhlich coupling constant in β-Ga_2O_3 greatly exceeds what would be expected from a naive linear prediction given the strong ionic character (as measured by the Pauling ionicity) in the Ga-O bond [29].

The mobility in semiconductors is controlled by several scattering mechanisms with one mechanism usually dominant for a given set of dopant levels and temperature. It is common in polar semiconductors at room temperature for phonon scattering to limit the mobility at low donor levels while impurity scattering limits the mobility at high impurity levels. The impact of the predicted large Fröhlich coupling constant is apparent in Fig. 8a, which displays the mobility of β-Ga_2O_3 as a function of doping. It is clear in Fig. 8a that the electron scattering by polar optical phonons limits the mobility to approximately 200 $cm^2/V{\cdot}s$ for β-Ga_2O_3 with donor densities less approximately $5x10^{18}$ cm^{-3}.

Examining the mobility at a low donor level in Fig. 8b reveals that electron mobility is controlled by polar optical phonon scattering at temperatures above 200K. The general understanding of semiconductor transport is that phonon scattering is inherent to the particular crystal and only marginal improvement may be possible through strain engineering [30].

In contrast, the mobility at a donor level of $1x10^{20}$ cm^{-3} is depicted in Fig. 8c. As mentioned above, neutral and ionized impurity scattering dominates at above approximately $5x10^{18}$ cm^{-3}. In Fig. 8c, the high donor level leaves a large density of neutral impurities that are the dominant scattering source in the temperature range depicted.

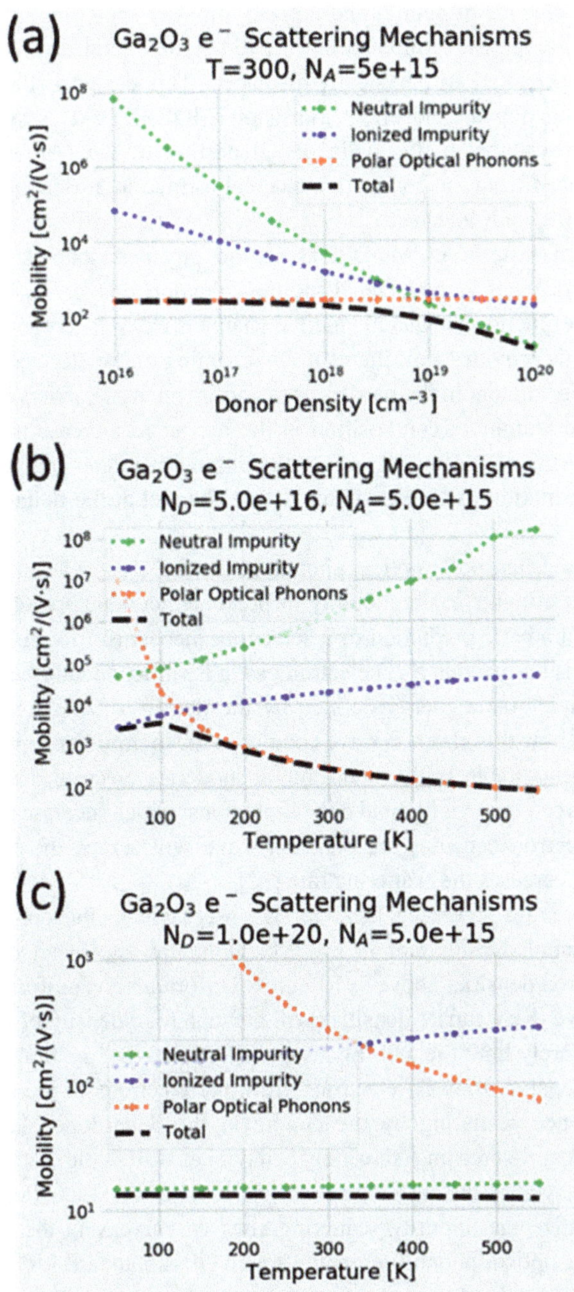

Fig. 8. Calculated mobility and relevant scattering mechanisms in β-Ga_2O_3 (a) at 300K as a function of donor density, (b) as a function of temperature at a low donor level ($5x10^{16}$ cm^{-3}) where polar optical phonon scattering is the dominant mechanism limiting the mobility above 200K, and (c) as a function of temperature at a high donor level ($1x10^{20}$ cm^{-3}) where impurity scattering controls the mobility.

The strong influence of neutral and ionized impurity scattering at high doping levels motivated the development of modulation doped heterostructures where the donor source and the transport channel are physically displaced. This design is the basis of the well-known modulation doped field effect transistor (MODFET). This structure is depicted in Fig. 9a, where donors in the delta doped portion of the $(Al_xGa_{1-x})_2O_3$ barrier are physically displaced from the conductive channel formed in a triangular potential well at the $(Al_xGa_{1-x})_2O_3$ / Ga_2O_3 interface.

A key design criteria in MODFETs is the physical spacing between the two-dimensional electron gas in the channel and the delta doped layer. As is visible in Fig. 9a, the electron wavefunction for the conductive channels extends to the donors in the delta-doped barrier. This provides a mechanism for screening of the electrons in the channel by neutral and ionized donors in the barrier. This confinement of carriers is further improved by increasing the aluminum composition in the barrier to increase the conduction band offset between $(Al_xGa_{1-x})_2O_3$ and Ga_2O_3. An additional bandgap engineering constraint is to prevent the formation of a second conductive channel at the delta-doped layer in the barrier.

As discussed earlier, polar optical phonon scattering creates a fundamental limitation that is difficult to mitigate in the mobility of polar semiconductors. Ghosh and Singisetti recently made a dramatic prediction of a screening mechanism for polar optical phonons in Ga_2O_3 [32]. It is known that a 2D electron gas in a semiconductor behaves at a plasmon wave with a characteristic energy defined by the density of electrons [31]. Ghosh and Singisetti studied how this plasmon wave couples to the longitudinal optical phonon modes in Ga_2O_3. This remarkable study found that at moderate electron densities the plasmon wave will anti-screen the longitudinal optical phonons, which increases the scattering rate; while at high electron densities the plasmon wave will screen the longitudinal optical phonons, which decreases the scattering rate [32].

This anti-screening/screening behavior is observable in the dot-dashed line in the mobility vs. channel density plot in Fig. 9b. Dramatic improvements in mobility are predicted at electron densities above 5×10^{18} cm^{-3}. Unfortunately, uniformly doping a single epilayer to achieve these carrier densities will create a high density of impurity scattering sites that will severely limit the mobility as was discussed for Fig. 8. Again the solution is to separate the dopants from 2D electrons. Still, the electrons in the triangular potential well can experience scattering by the donors in the delta-doped layer. The mobility accounting for the anti-screening/screening behavior as well as the scattering by the remote impurities is displayed by the dashed line in Fig. 9b. As can be seen, there is a clear design tradeoff to minimize the impurity scattering sites vs. increasing the channel density to increase the polar optical phonon screening - as well as standard MODFET design rules [4, 33, 34].

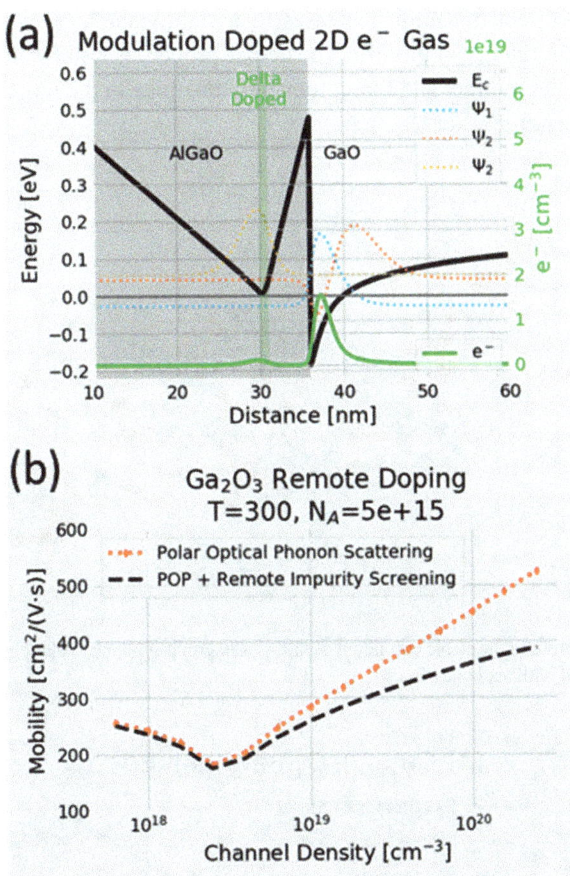

Fig. 9. β-$(Al_xGa_{1-x})_2O_3$ / Ga_2O_3 modulation doped structure with a conductive channel formed in a triangular potential well. (a) Calculation of conduction band energy, electron concentration, and electron wavefunction for a structure composed of 35nm delta-doped $(Al_xGa_{1x})_2O_3$ on Ga_2O_3. A small portion of the electron wavefunction in the channel extends to the delta-doped donors in the $(Al_xGa_{1-x})_2O_3$ layer. (b) The dot-dashed line depicts a mobility model that accounts for polar optical phonon anti-screening at moderate electron densities and screening at high electron densities [32]. The dashed line depicts the mobility including screening by the remote impurities in the delta-doped layer.

7. Alloy Formation

From the previous section it is clear that research is needed in epitaxy of $(Al_xGa_{1-x})_2O_3$ / Ga_2O_3 heterostructures. The $(Al_xGa_{1-x})_2O_3$ alloy is challenging as α-Al_2O_3 is stable as the corundum phase (Fig. 10) while β-Ga_2O_3 is stable as the monoclinic phase.

The phase of $(Al_xGa_{1-x})_2O_3$ that forms not only has a different bandgap energy but presents a different transition state. It is known that the β-Ga_2O_3 is theoretically an indirect semiconductor but this direct gap transition is so similar in energy that β-Ga_2O_3 effectively behaves as a direct gap semiconductor. Examination of Fig. 11, which is based on the model in Peelaers, *et al.* [35], shows that the energy difference of the indirect transition compared to the direct transition increases to a significant level with increasing alloy composition. The influence of this effect on device properties is not well studied. Similarly, the separation of the indirect and direct transition energies in the $(Al_xGa_{1-x})_2O_3$ corundum phase is also shown in Fig. 11.

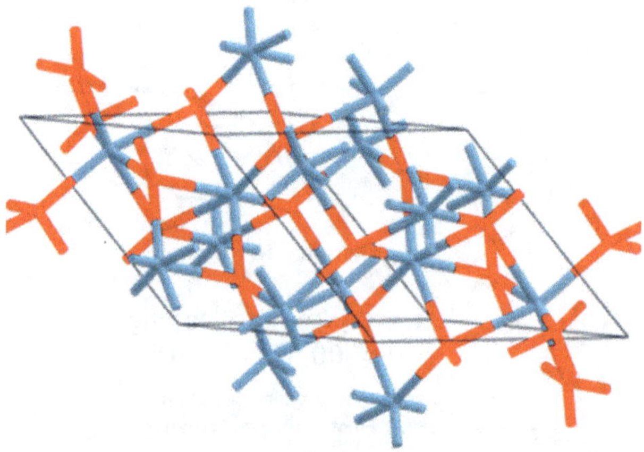

Fig. 10. Bonding structure of the stable corundum α-Al_2O_3 (R3c group symmetry) phase where the Al cations are octahedral coordinated with six bonds.

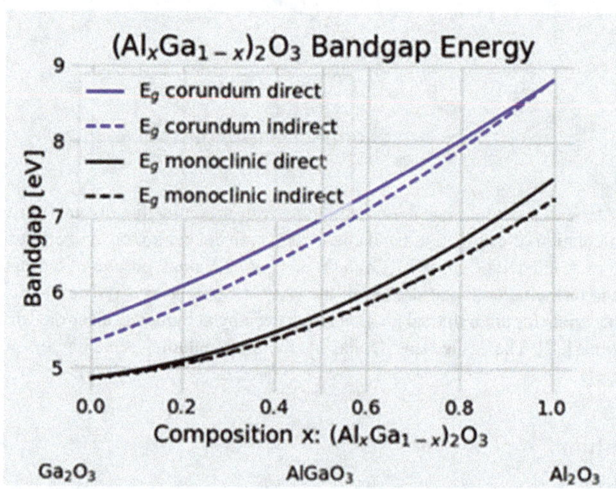

Fig. 11. Direct and indirect bandgap energies of the monoclinic and corundum phases of the $(Al_xGa_{1-x})_2O_3$ alloy [35].

It is known that forming alloys of $(Al_xGa_{1-x})_2O_3$ is difficult. Examination of Fig. 12 based on the enthalpy of formation model of Peelaers, *et al.* [35] provides some insight. The calculation suggests that the monoclinic phase is stable for $x < 0.71$. The stability local minimum in the formation energy for $AlGaO_3$ ($x = 0.5$) is logical given the crystal structure of the binary compounds. The bonding structure of the corundum α-Al_2O_3 phase is composed of Al cations only at octahedral coordinated sites with six bonds each (Fig. 10). In contrast, Fig. 2 shows that the monoclinic β-Ga_2O_3 phase is composed of equal parts of Ga (I) cations in a distorted tetrahedral coordination with four bonds and Ga (II) cations in an octahedral coordination with six bonds. At the stability local minima at x = 0.5, the Ga (I) cations are at the tetrahedral coordinate site while, critically, the Al cations are only present at the octahedral coordination site.

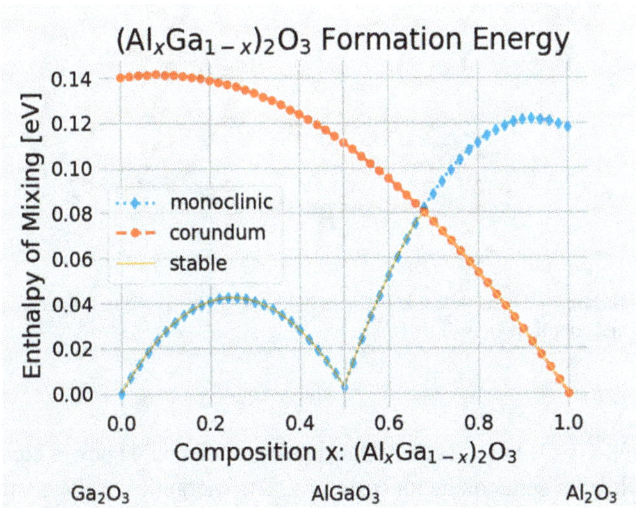

Fig. 12. Theoretical formation energy of $(Al_xGa_{1-x})_2O_3$ based on the model of Peelaers, *et al.* [35]. The monoclinic phase is stable for x < 0.71 and the corundum phase is stable at higher mole fractions.

Early work by Hill *et al.* into the equilibrium diagram of Al_2O_3 in β-Ga_2O_3 found the presence of a stable phase of $AlGaO_3$ [36]. This work reported that this phase required a temperature of 800 °C, which may preclude lower temperature growth techniques such as MBE. The exploration of $AlGaO_3$ growth conditions and the role of strain is an important area of future research.

The addition of $(In_xGa_{1-x})_2O_3$ to heterostructures would also be beneficial for the development of electronic devices. Again, the In_2O_3 crystal possesses a different stable phase (as seen in Fig. 13), which contributes to the difficulty in forming a $(In_xGa_{1-x})_2O_3$ stable alloy. Single crystal monoclinic structures were only reported at low indium content (x < 0.15). At high indium content (x > 0.8) the cubic bixbyite phase is formed while at intermediate values an additional rhombohedral $InGaO_3$(II) crystallographic phase formed [37]. Regardless, as discussed above, reports indicate that indium behaves as a surfactant so its key role may be in improving the diffusivity of the gallium atom during growth.

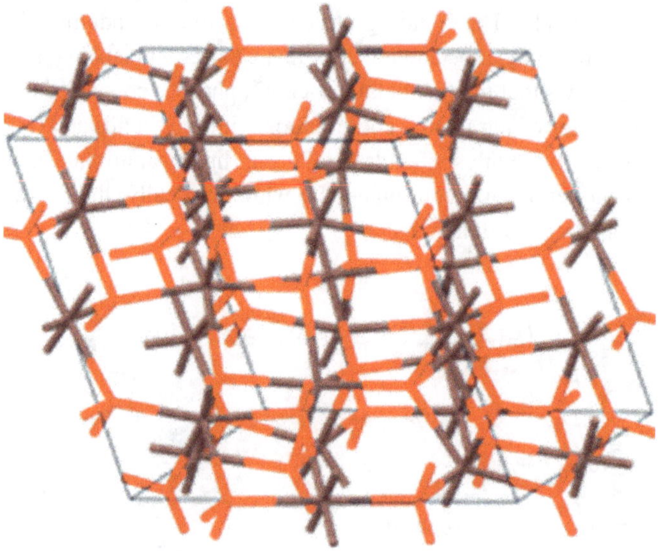

Fig. 13. Bonding structure of the stable cubic bixbyite In_2O_3 (Ia=3 [206]) group symmetry phase with the In cations in an octahedral coordination with six bonds.

8. Conclusion

Development of β-Ga_2O_3-based MOCVD technology will enable power electronic devices not possible with other semiconductor materials. The energetics of the crystal facets has a strong influence on growth morphology and defect formation for a given substrate orientation and offcut, growth temperature, and growth rate. It is clear that the growth environment, particularly metalorganic precursor selection and the presence of hydrogen, has a strong impact on the formation of gallium vacancies and the resulting compensation of intentional dopants. These compensating centers may be native defects or complexes from either sublattice. The understanding of the role of hydrogen including in the formation the vacancy complex is still evolving. The mobility in β-Ga_2O_3 is limited at low donor densities by polar optical phonons. The possibility to screen the longitudinal optical phonon modes by the electron gas plasmon further motivates development of modulation doped $(Al_xGa_{1-x})_2O_3$ / β-Ga_2O_3 heterostructures. To achieve efficient devices based on this structure requires additional development into the growth of $(Al_xGa_{1-x})_2O_3$ particularly at high alloy mole fraction.

Acknowledgments

The work at NRL was partially supported by DTRA Grant No. HDTRA1-17-1-0011 (Jacob Calkins, monitor) and the Office of Naval Research. The work at UF is partially supported by HDTRA1-17-1-0011. The project or effort depicted is sponsored by the Department of the Defense, Defense Threat Reduction Agency. The content of the information does not necessarily reflect the position or the policy of the federal government, and no official endorsement should be inferred. The work at Korea University was supported by the Korea Institute of Energy Technology Evaluation and Planning (KETEP) and the Ministry of Trade, Industry and Energy (MOTIE) of Korea (Grant No. 20172010104830) and Space Core Technology Development Program (2017M1A3A3A02015033) through the National Research Foundation of Korea funded by the Ministry of Science, ICT and Future Planning of Korea.

References

1. M.A. Mastro, A. Kuramata, J. Calkins, J. Kim, F. Ren, S.J. Pearton, ECS J. Solid State Sci. Technol. 6, P356 (2017)
2. B.J. Baliga, IEEE Electron Device Letters, 10, 455 (1989)
3. S. Dimitrijev, J. Han, D. Haasmann, H.A. Moghadam, and A. Aminbeidokhti, MRS Bulletin 40(05), 43-46 (2014)
4. S.S. Li, Semiconductor Physical Electronics, Plenum (1993)
5. S. Dimitrijev, Principles of Semiconductor Devices, Second Edition, New York, Oxford: Oxford University Press (2012)
6. B.J. Baliga, J. Appl. Phys. 53, 1759 (1982)
7. M.A. Mastro, Power MOSFETs and Diodes. In S. Pearton, F. Ren, M. Mastro (Eds.) Gallium Oxide Technology, Devices and Applications, p. 401-418, Elsevier (2019)
8. M.A. Mastro, Fundamentals and future of semiconductor device technology in III-V Compound Semiconductors: Integration with Silicon-Based Microelectronics, CRC/Taylor & Francis (2011)
9. S.J. Pearton, J. Yang, P.H. Cary, F. Ren, J. Kim, M.J. Tadjer, M.A. Mastro, Appl. Phys. Rev. 5, 011301 (2018)
10. H. Peelaers, C.G. Van de Walle, Phys. Rev. B 94, 195203 (2016)
11. Z. Galazka, R. Uecker, D. Klimm, K. Irmscher, M. Naumann, M. Pietsch, A. Kwasniewski, R. Bertram, S. Ganschow, M. Bickermann, ECS J. Solid State Sci. and Tech., 6(2), q3007 (2017)
12. A. Fiedler, R. Schewski, M. Baldini, Z. Galazka, G. Wagner, M. Albrecht, K. Irmscher, Applied Physics 122, 165701 (2017)
13. R. Schewski, G. Wagner, M. Baldini, D. Gogova, Z. Galazka, T. Schulz, T. Remmele, T. Markurt, Holger von Wenckstern, M. Grundmann, O. Bierwagen, P. Vogt, M. Albrecht, Appl. Phys. Express 8, 011101 (2015)
14. R. Fornari, Progress in MOVPE growth of Ga_2O_3. In S. Pearton, F. Ren, M. Mastro (Eds.) Gallium Oxide Technology, Devices and Applications, p. 3-30, Elsevier (2019)
15. R. Schewski, M. Baldini, K. Irmscher, A. Fiedler, T. Markurt, B. Neuschulz, T. Remmele, T. Schulz, G. Wagner, Z. Galazka, M. Albrecht, J.Applied Physics, 120(22), 225308 (2016)
16. G. Bales, A. Zangwill, Phys. Rev. B, 41, 5500 (1989)
17. M. Baldini, M. Albrecht, D. Gogova, R. Schewski, G. Wagner, Semi. Sci. and Tech. 30 (2), 024013 (2015)
18. R. Schewski, K. Lion, A. Fiedler, C. Wouters, A. Popp, S. V. Levchenko, T. Schulz, M. Schmidbauer, S. Bin Anooz, R. Grüneberg, Z. Galazka, G. Wagner, K. Irmscher, M. Scheffler, C. Draxl, M. Albrecht, APL Materials, 7(2), 022515 (2019)
19. V.M. Bermudez, Chem. Phys. 323, 193 (2006)
20. S. Rafique, L. Han, A. T. Neal, S. Mou, J. Boeckl, H. Zhao, Phys. Status Solidi A 215, 1700467 (2017)

21. M. Baldini, M. Albrecht, A. Fiedler, K. Irmscher, R. Schewski, G. Wagner, ECS J. Solid State Sci. Technol. 6, Q3040 (2017)
22. A. Kuramata, K. Koshi, S. Watanabe, Y. Yamaoka, T. Masui, S. Yamakoshi, Jpn. J. Appl. Phys. 55, 1202A2 (2016)
23. O. Ueda, N. Ikenaga, K. Koshi, K. Iizuka, A. Kuramata, K. Hanada, T. Moribayashi, S. Yamakoshi, and M. Kasu, Jpn. J. Appl. Phys. 55, 1202BD (2016)
24. D. Gogova, M. Schmidbauer, A. Kwasniewski, CrystEngComm 17-35, (2015)
25. F. Tuomisto, A. Karjalainen, V. Prozheeva, I. Makkonen, G. Wagner, M. Baldini, Proc. SPIE 10919, 1091910-1 (2019)
26. M.R. Lorenz, J.F. Woods, R.J. Gambino, J. Phys. Chem. Solids, 28, 403 (1967).
27. J.B. Varley, J.R. Weber, A. Janotti, C.G. Van de Walle, Appl. Phys. Lett., 97, 142106 (2010)
28. Y. Qin, M. Stavola, W. Beall Fowler, P. Weiser, S.J. Pearton, ECS J. Solid State Sci. Technol. 8-7, Q3103 (2019)
29. N. Ma, N. Tanen, A. Verma, Z. Guo, T. Luo, H. (Grace) Xing, D. Jena, Appl. Phys. Lett., 109, 212101 (2016)
30. K.A. Khair, S.S. Ahmed, 2017 IEEE 17th International Conference on Nanotechnology (IEEE-NANO) (2017)
31. M. A. Mastro, ECS Journal of Solid State Science and Technology, 6 (11) S3044 (2017)
32. K. Ghosh, U. Singisetti, J. Mater. Res. 32-4142 (2017)
33. M.A. Mastro, J.R. LaRoche, N.D. Bassim, C.R. Eddy, Jr., Microelectronics Journal, 36, 705 (2005)
34. M.A. Mastro, D.V. Tsvetkov, V.A. Soukhoveev, A. Usikov, V. Dmitriev, B. Luo, F. Ren, K.H. Baik, S.J. Pearton, Solid State Electronics, 48, 179 (2004)
35. H. Peelaers, J.B. Varley, J.S. Speck, C.G. Van de Walle, Appl. Phys. Lett. 112, 242101 (2018)
36. V.G. Hill, R. Roy, and E.F. Osborn, J. Am. Ceram. Soc. 35, 135 (1952)
37. H. von Wenckstern, D. Splith, M. Purfürst, Z. Zhang, Ch. Kranert, S. Müller, M. Lorenz, M. Grundmann, Semicond. Sci. Technol. 30, 024005 (2015)

Theory of High Field Transport in β-Ga_2O_3

Krishnendu Ghosh[1,2,*] and Uttam Singisetti[2,†]

[1]*TCAD Department, Intel Corporation, Hillsboro OR 97124, USA*

[2]*Electrical Engineering Department, University at Buffalo, Buffalo NY 14260, USA*

[†]*uttamsin@buffalo.edu*

We present a comprehensive review of high-field transport properties in an emerging and trending ultra-widebandgap semiconductor β-Ga_2O_3. The focus is on the theoretical understanding of the microscopic mechanisms that control the dynamics of far-from-equilibrium electrons. A manifold of density functional calculations and Boltzmann theory based transport formalism unravels the behavior of the electron distribution under a varied range of external electric fields. The key high-field transport properties that govern electronic device performance, like velocity and ionization co-efficients, are enlightened in detail with physical insights. Anisotropies in the above transport co-efficients are probed from the microscopic investigation of bandstructure, electron-phonon interactions, and electron-electron interactions.

Keywords: Ga_2O_3; velocity; impact ionization; high field transport.

1. Introduction

With the increasing demand for energy, as the conventional silicon based power devices are hitting the limits of efficiency, widebandgap (WBG) materials have brought the promise to push the limits forward. WBG semiconductors are promising for higher temperature operation, higher voltage withstanding capability, higher power density and more accessibility to the visible spectrum of light compared to narrow bandgap semiconductors. The progress of crystal growth and device fabrication technologies is further making WBG materials a suitable alternative for silicon. For power electronic applications a low resistance and a high breakdown voltage are preferred. GaN (bandgap, $E_g = 3.4$ eV) and 4H-SiC ($E_g = 3.3$ eV) have been the two most popular WBG materials since the last two decades. The progress of GaN technology in the past decade is a remarkably significant advancement in power electronics. However, to keep up the progress wider bandgap materials need to be

*The previously published computational studies reported in this review work were performed while K.G. was at the University at Buffalo.

studied. A recently emerged ultra-widebandgap semiconductor, β-Ga_2O_3 [1–3] has experimentally shown record performance as a candidate for next-generation power electronics.

1.1. β-Ga_2O_3 — potential and applications

An important yardstick to evaluate the performance of materials for power devices is the Baliga's figure-of-merit (BFoM) [4]. Mathematically, it is defined as $BFoM = \mu\epsilon E_{br}^3$, where μ, ϵ, and E_{br} are the mobility, dielectric constant, and breakdown field for the material respectively. Table 1 shows [1] the comparison of BFoM of different WBG materials with respect to the BFoM of silicon. It is clear that apart from diamond, β-Ga_2O_3 outperforms all other competitors significantly. The other component shown on Table 1 is thermal conductivity (κ_c). A high κ_c is crucial for power electronic applications in order to avoid localization of Joule heating and remains a bottleneck for β-Ga_2O_3, although careful phonon engineering can improve that as well. However, in the current review we only focus on the high-field electron transport properties in β-Ga_2O_3. Record breakdown voltage of more than 1 kV [5] has already been experimentally reported in β-Ga_2O_3 power MOSFETs. Mature crystal growth techniques [6–12] and applicability of conventional device processing methods [13–16] make the feasibility of realizing commercial β-Ga_2O_3 devices more prominent. Ease of achieving controllable n-type doping and difficulty of p-type doping make electrons the primary charge carriers in this material. Other than reports on power MOSFETs [17–26] and Schottky diodes [27–35], there are several reports on deep ultra violet photo-detectors [36–45] reflecting the promise of the material in optoelectronics. Figure 1(a) shows the relative position of β-Ga_2O_3 among other WBG materials in terms of bandgap and projected breakdown field. On the other hand, from a device perspective, one can improve the breakdown voltage by using a higher drain access length but that causes an increase in on-resistance. The trade-off between on-resistance and breakdown voltage for several semiconductors is shown on Fig. 1(b). The further right this trade-off line for a material is, the better it is for power electronic applications. β-Ga_2O_3 performs better than most other WBG materials. Very recently, usage of field-plates has

Table 1. Comparison of the properties of different WBG materials with respect to Si for power electronic applications [1].

	Si	GaAs	4H-SiC	GaN	β-Ga_2O_3	Diamond
E_g (eV)	1.1	1.4	3.3	3.4	4.5-4.9	5.5
μ ($cm^2/V.s$)	1400	8000	1000	1200	300	2000
ϵ	11.8	12.9	9.7	9.0	10.0	5.5
$BFoM$	1	15	340	870	3444	24664
κ_c ($(W(cm.K)^{-1}$)	1.5	0.55	2.7	2.1	0.27[010] 0.11[100]	10

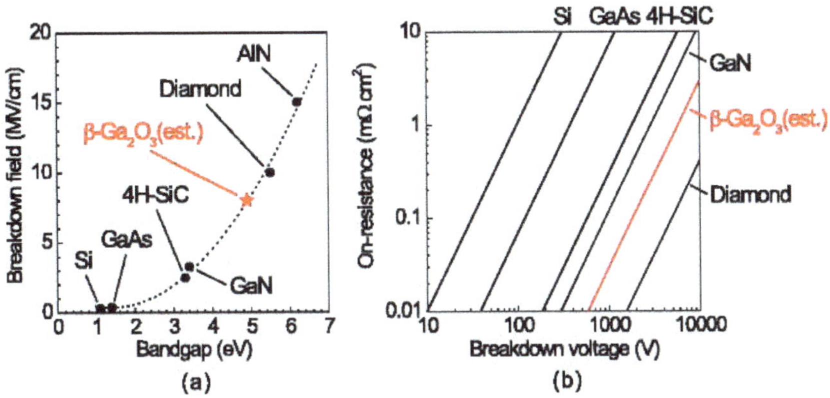

Fig. 1. (a) A phenomenological trend of breakdown field with respect to bandgap. β-Ga_2O_3 is expected to show a superior breakdown field compared to other existing power semiconductors like GaN and SiC. (b) Trade-offs between on resistance in a transistor and the breakdown voltage for different semiconductors. β-Ga_2O_3 is positioned better than most WBGs. (a-b) is reproduced from [17] with permission from AIP.

been shown to boost the breakdown voltage drastically [46]. Electronic structure calculations [47–51], optical properties [52–58], dielectric and thermal properties [59–64] in β-Ga_2O_3 have been reported several times. An isotropic effective mass of 0.28 times the free electron mass is computed theoretically [50]. Gallium and oxygen vacancies produce paramagnetic centers as computed and observed from electron paramagnetic resonance measurements [65, 66]. The bandgap can range from 4.5eV-4.8eV [57], where the variation occurs since the different polarization of the incident light promotes electrons from different valence bands owing to selection rules. Investigating the dynamics of electrons under different regimes of externally applied electric field is an important direction of research. A series of theoretical studies on electron transport in β-Ga_2O_3 has been reported by the current authors [67–71] along with other recent theoretical [72–74] and experimental [75] papers from colleagues.

1.2. *Motivation for transport studies in β-Ga_2O_3*

As β-Ga_2O_3 has started gaining a rising impact in power electronics and optoelectronics, knowledge on electron transport properties is becoming increasingly important. Achieving a high electron mobility is important to reduce the on-state resistance of power transistors and thus improving the on current. Device engineering, like making hetero-structures to form two-dimensional electron gas (2DEG), requires a solid understanding of the electronic transport behavior to optimize performance. On the other hand, a high breakdown voltage is mandatory for power transistors. Understanding the physics of breakdown mechanisms, including impact ionization, requires a deep understanding of high-field electron transport. Currently, there are several experimental reports that show Hall mobility measurements [75], trends of mobility with temperature and electron density, and anisotropies [76] in mobility as well. Also, there are a handful of theoretical reports on low-field electron transport that attempts to point out dominant mobility limiting mechanisms [67, 72, 73]. On the other hand, high-field transport in β-Ga_2O_3 is premature with only a very few work [68, 70] being reported by the current authors. So it is of high importance to motivate and identify future directions of both theoretical and experimental high-field transport work in β-Ga_2O_3 based on the existing limited literature.

This review paper tries to address these important points standing on the recent reports of electron transport in β-Ga_2O_3. On Sec. 2 we briefly discuss on the different microscopic interactions, like electron phonon interactions and electron-electron interactions, that control in the electron transport phenomena. On Sec. 3 we focus on the high-field transport with the details of velocity-field curves, impact ionization coefficients, their respective anisotropies, and implication of ionization on breakdown performance of devices before concluding with a brief summary on Sec. 4.

2. Microscopic Interactions

Electron-phonon interaction (EPI) and electron-electron interaction (EEI) play a very important role in the diffusive regime of electron transport. The workflow to compute the interaction elements begins with the computation of electronic wave-functions and single-particle energies which are usually computed using density functional theory (DFT). Next one computes the phonon energies, their displacement patterns, and the perturbation to the self-consistent potential due to the lattice vibrations. Then the matrix elements of the perturbation in potential with respect to the electronic wave-function produce the EPI elements. The EEI elements are obtained as direct and exchange Coulomb interactions between an electron in the conduction band and another electron in the valence band. Before discussing on the the behavior of these interactions in β-Ga_2O_3, we briefly highlight on the main characteristics of the electronic bands and lattice vibration modes.

2.1. *Electronic structure and lattice vibrations*

Figure 2(a) shows a conventional unit cell of monoclinic β-Ga_2O_3 [77] while Fig. 2(b) shows the Brillouin zone (BZ) along with the reciprocal vectors. The Cartesian axes convention followed in this paper is also shown on Fig. 2(a). The electronic bandstructure under plane-wave pseudopotential based DFT using hybrid functionals [50, 58] have been reported several times in the literature. Excited states calculations are also performed under GW [53, 54]. The optical bandgap, strictly speaking is indirect [50], but does not differ much from the direct gap in both energy and position in reciprocal space. The minima of the conduction band follows

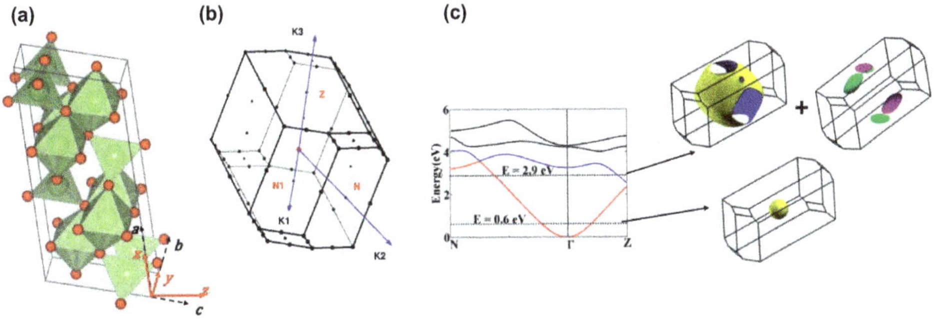

Fig. 2. (a) Crystal structure of β-Ga_2O_3 and the Cartesian direction convention followed in this work. Red atoms denote oxygen, while the green ones are gallium. (b) The Brillouin zone of β-Ga_2O_3 primitive cell with the three reciprocal vectors. (c) Conduction bands of β-Ga_2O_3, the isoenergy surfaces are shown for two different energy levels (see text for details). (a) and (c) are reproduced from [68] and (b) is reproduced from [67] with permission from AIP.

a well-defined parabolic behavior with an effective mass of about 0.28 times the free electron mass and shows a very small amount of anisotropy. Figure 2(c) shows the first few conduction bands. It is to to be noted that the first remote valley at Z occurs at a significantly higher energy (2.5 eV) from the conduction band minima compared to that in other conventional semiconductors like GaAs [78] and GaN [79]. This has a very crucial role in high-field transport properties as will be discussed later in Sec. 3. Figure 2(c) also shows the equi-energy surfaces on the BZ. The surface is spherically symmetric at a low electron energy making the the spherical symmetry assumptions amenable to the low-field transport calculations in this material. However, at higher energies a significant amount of anisotropy in the bands and also non-parabolicity emerges in.

2.2. *Electron-phonon interactions*

Electron-phonon interaction calculations have seen a significant development [80–84] in the recent years primarily due to the advent of density functional methods. EPI in polar materials can be categorized into two types, namely polar (long-range) EPI and non-polar (short-range) EPI. The first type originates from the coupling of the electrons with the net dipole oscillation arising from partially ionic bonds, while the second one is a result of the coupling of the electrons with the perturbation in the self-consistent crystal potential arising purely from the vibration of the lattice (without considering the dipole oscillation). While density functional perturbation theory (DFPT) computed perturbations in the potential (ΔV_{scf}) account for both long-range and short-range effects, it is important to isolate their effects primarily due to two reasons. (I) Electron transport in different regimes of the external electric field get effected by the two types of interactions differently, like, the long-range polar interaction controls the low-field transport properties predominantly while the opposite is true for high-field transport. Hence it is important to probe the two different types of interactions separately. (II) DFPT calculations are computationally very intensive and ΔV_{scf}, in practice, is computed on a very coarse-grid of phonon wave-vectors ($\mathbf{q}$ points). However, convergence in transport properties require a fine resolution of the EPI elements. The common practice is to interpolate the EPI elements using maximally localized Wannier functions (MLWF) [85–92] that utilize the periodicity of the crystal. However, the long-range component of the EPI is not amenable to such interpolation schemes. The way to circumvent this inconsistency is to subtract the long-range part of the EPI on the coarse-grid and then interpolate only the short-range part using MLWF. The long-range EPI is explicitly computed on the fine grid, as it is much cheaper to compute just the long-range part. In the following we highlight on the key features of both types of EPI in β-Ga_2O_3.

2.2.1. *Polar EPI*

DFPT [93] perturbatively computes the dynamical matrix of a system whose eigen values and eigen vectors provide the phonon energies and displacement patterns respectively. There are a total of 30 phonon modes in β-Ga_2O_3 arising from the 10 atom primitive unit cell. It is to be noted that like other polar materials β-Ga_2O_3 also has partially ionic bonds that give rise to net dipole moments. This brings a splitting of the phonon energies near the Γ point giving rise to longitudinal optical (LO) and transverse optical (TO) modes. The magnitude of the splitting between LO and TO modes account for the strength of the net dipole moment projected along a given wave-vector. For β-Ga_2O_3 this splitting is highly dependent upon the direction [59, 67] of the phonon wave-vector. The modes that possess a net dipole moment are IR active and β-Ga_2O_3 contains 12 such IR active modes, 8 of which, known as the B_u modes, are polarized on the xz plane, while 4, known as the A_u modes, are polarized along the y direction. Table 2 shows the LO-TO splittings for different directions of the phonon wave-vector. This highly anisotropic splitting plays a pivotal role in determining the physics of the polar optical phonon

Table 2. The LO-TO splitting of the different polar modes [67] for the wave-vector along the three Cartesian directions.

x		y		z	
TO(meV)	LO(meV)	TO(meV)	LO(meV)	TO(meV)	LO(meV)
21.30	21.68	13.02	13.15	21.30	28.49
29.30	32.68			29.30	29.43
34.30	34.98	36.30	41.63	34.30	34.23
43.00	44.64			43.00	43.13
52.00	59.18	55.90	67.61	52.00	57.84
70.20	77.13			70.20	79.99
83.70	88.92	82.00	93.14	83.70	84.87
90.10	92.76			90.10	96.15

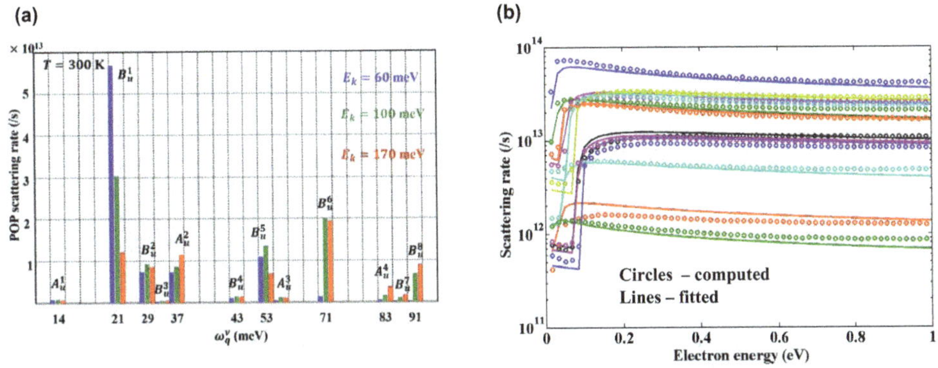

Fig. 3. (a) The relative contribution of different polar phonon modes in the total scattering rate. (b) The scattering rates with respect to electron energy. Circles show spherically averaged *ab initio* scattering rates while the solid lines show the fitted rates for the modes with significantly high contribution. (a) is reproduced from [67] with permissions from AIP.

scattering. The LO-TO splitting shown on Table 2 is consistent with ellipsometry observations [59].

The anisotropy of the LO-TO splitting in β-Ga_2O_3 gets reflected in the polar EPI as well. Among the B_u modes, B_u^1 has the highest contribution along the z direction whose LO value is 28.49 meV, and there are other modes with significantly high coupling strengths like B_u^5, B_u^6, and B_u^8. Among the 4 A_u modes, A_u^2 and A_u^3 are the dominant modes. Overall, the polar coupling is the strongest along the z direction that has important consequences on electron transport. The fact that B_u^1 has the strongest coupling strength has also been reported in a recent publication [74]. Figure 3(a) shows the polar EPI mediated scattering rates (split modewise) computed using the Fermi Golden rule. It is not obvious to identify any particular mode as the sole dominating mode. For lower electron energies (typically < 100 meV), B_u^1 provides the highest scattering rate. With increasing electron energy the dominant phonon shows a red-shift, like for an electron energy of 170 meV the dominant phonon mode is B_u^6. Under low electric fields (below 10 kV/cm), the average electron energy lies below 100 meV and hence it is prudent to take the B_u^1 mode as the dominant mobility limiting mode. The scattering rates with respect to electron energy is shown on Fig. 3(b). On a recent work [71], the scattering rates are also fitted on an analytical model for treating multi-phonon based polar scatterings. Figure 3(b) shows the fitted scattering rates as well and the fitting parameters can be obtained from Ref. 71. The idea behind this fitting is to enable modeling of the multi-phonon scattering rates using analytical equations [94].

2.2.2. *Non-polar EPI*

High-field transport properties are controlled by both polar and non-polar EPI. Figure 4(a) shows a comparative plot of polar and non-polar EPI in β-Ga_2O_3 with

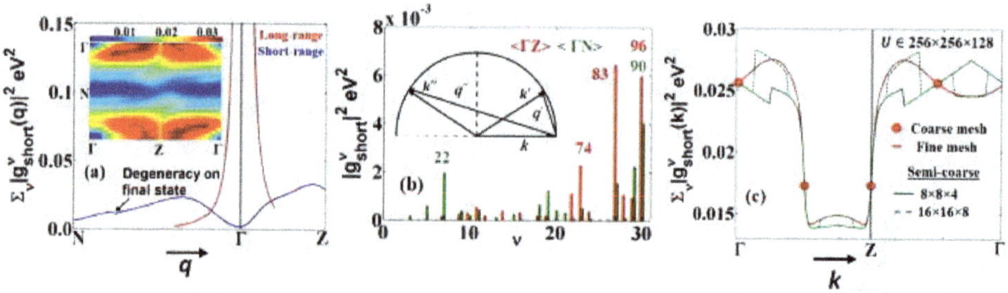

Fig. 4. (a) Polar and non-polar EPI elements along two reciprocal vector directions. (Inset) A contour plot of non-polar EPI. (b) Relative contribution of different phonon modes to the non-polar EPI. (Inset) Relation between electron and phonon wave-vectors for high momentum relaxation (see text for details). (c) Non-polar EPI with respect to electron wave-vector. (a-c) is reproduced from [68] with permissions from AIP.

respect to the phonon wavevector **q**. The polar EPI drops rapidly with increasing **q** due to the long-range nature of Coulomb interaction. On the other hand, the non-polar EPI shows a non-monotonic nature. The important aspect to note here is that, while computing scattering rates from the EPI elements one needs to implement energy conservation. Now, for β-Ga_2O_3 the bottom of the conduction band is isolated from any other valleys. Under a very low electric field, only $q \rightarrow 0$ takes part in the scattering process and the polar EPI is several orders of magnitude higher in that limit making non-polar scattering negligible. Under a moderately high electric field, **q** vectors beyond a certain threshold do not take part in the scattering process and one needs to consider both polar and non-polar EPI, although polar EPI is still the stronger one. However, under even higher electric fields, when the electrons are far from the Γ point, short-range (large **q**) scatterings become important. Although, polar scattering still occurs frequently, they are mostly long-ranged providing a weaker momentum relaxation. Hence, non-polar scatterings are primarily responsible for strong momentum relaxation due to significantly high short range contributions. This results to effects like velocity overshoot as will be discussed in Sec. 3.1. The inset on Fig. 4(a) shows a contour plot of the non-polar EPI (summed over all phonon modes) along a 2D plane on the BZ formed by two of the reciprocal lattice vectors.

Figure 4(b) shows a mode-wise comparison of the non-polar EPI elements along two different directions in the BZ. The high energy phonon modes show a higher strength of coupling. However, such mode-wise picture is highly dependent upon the wave-vector of the electron and that of the phonon. For the case shown on Fig. 4(b), **k** and **q** on the green bar plots are $0.125 \times \mathbf{\Gamma N}$ and $0.25 \times \mathbf{\Gamma N}$, while that for the red bar plots are $0.25 \times \mathbf{\Gamma Z}$ and $0.5 \times \mathbf{\Gamma Z}$ respectively. The choice of such q vectors represent a case where high momentum relaxation can occur. The alignment of **k** and **q** ensures the high momentum relaxation as depicted on the inset on Fig. 4(b) [68]. Figure 4(c) shows the dependence of the non-polar EPI on the electronic wave-vector computed via MLWF scheme [68].

2.3. *Electron-electron interactions*

EEI plays the central role in the impact ionization process. A high energy electron in the conduction band interacts with an electron in the valence band to promote (ionize) the latter to the conduction band thus creating an electron-hole pair. This process is schematically shown on Fig. 5(a) where, a conduction band electron at a higher energy state loses energy and drops to the bottom, while a valence band electron gains that energy and rises to the conduction band. Energy and momentum have to be conserved in the process as shown on the Feynman diagram. The electron at state $|m\mathbf{k}\rangle$, where m is the band index, in the conduction band interacts with the electron $|n'\mathbf{k}'\rangle$ in the valence band. This results to the latter moving to the state $|m'\mathbf{k}' - \mathbf{q}\rangle$ and the former transits to $|n\mathbf{k} + \mathbf{q}\rangle$.

The ionization rate for an electron at state $|m\mathbf{k}\rangle$ is computed summing over all the internal degrees of freedom, which includes integrations over $\mathbf{k}'$ and $\mathbf{q}$ and summations over m', n', and n. In case of β-Ga_2O_3, for realistic values of the external electric field, it is imperative to set $n = 1$ due to the separation of the first conduction band from the higher bands near the Γ point. This is because the high energy electron at $|m\mathbf{k}\rangle$ has to lose an energy higher than the bandgap (4.9 eV) in order to cause ionization. The remote valley at the Z point on the BZ occurs at an energy 2.5 eV above the first conduction band valley. So for n to have a value higher than, the electron at $|m\mathbf{k}\rangle$ needs to have an energy higher than 7.4 eV which is very unlikely at realistic values of electric field in a power transistor. Also, ionization is more likely to be caused by electrons near the Γ point for the same reason. The long-range nature of the Coulomb interaction prohibits a zone-edge electron to relax to near the Γ point valley. This explains the nature of the ionization rate along the high symmetry lines shown Fig. 5(b). Moreover, note that since the primary electron needs to have at least an energy of 4.9 eV, ionization can be only caused by electrons at $m > 4$. In principle, there are states on $m = 4$ that have an energy higher than 4.9 eV, but those states usually lie far

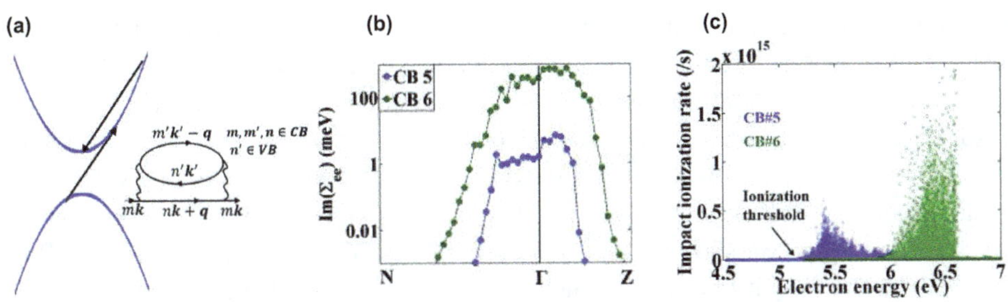

Fig. 5. (a) A schematic to show the impact ionization process. The Feynman diagram shows all the internal degrees of freedom involved in the process. The bubble represents the electron-hole pair. (b) The imaginary part of the electron-electron self-energy (ionization rate) along two reciprocal vector directions in the BZ. (c) The ionization rates with respect to energy for different $\mathbf{k}$-points. (a-c) was reproduced from [70] with permission from AIP.

from the Γ point, making their ionization rates negligibly low. Now on the valence band side, the energy of the electron at $|n'\mathbf{k}'\rangle$ can not be below a certain threshold, as otherwise it will again require the primary electron at $|m\mathbf{k}\rangle$ to have an extremely high energy to cause ionization. Typically considering the top four valence bands keeps that primary electron energy below 6.5 eV. So this provides a gauge on the range of the degrees of freedom m', n', and n.

Figure 5(c) shows the ionization rates with respect to energy, however without averaging over the $\mathbf{k}$ points. The anisotropy on the ionization rate is apparent from its wide range of variation at a given energy. The ionization onsets at an energy slightly higher than 5 eV. Near the threshold the ionization rates approximately follow a power law and shows a soft ionization behavior.

3. High-Field Transport

High-field transport in β-Ga_2O_3 is very little explored except a few reports by the current authors [68, 70]. However, there are several experimental reports on breakdown field measurements [5, 22, 23, 46] that entails high-field transport phenomena. Under moderately high electric fields (< 0.5 MV/cm), the essential physics are the negative differential conductivity and velocity saturation, while at higher electric fields impact ionization becomes important. In the following we attempt to summarize some key findings of fullband Monte Carlo study of high-field transport in β-Ga_2O_3. The relevant scattering rates are computed from the first-principles as described in Sec. 2.

3.1. *Velocity-field curves*

During the FBMC simulation, an ensemble of electrons is subjected to an external electric field and their trajectories on the reciprocal space is recorded for a time until a steady state in their distribution is reached. The trajectories evolve with time through an interplay between drift along the direction of electric field and randomization of momentum due to scattering. At a given moment of time, the drift velocity is extracted as an ensemble average of the group velocity of all the electrons. As steady state is reached, the drift velocity becomes fixed over time. Figure 6(a) shows the transient response of the ensemble of electrons in β-Ga_2O_3 for different electric fields. At lower fields, the drift velocity increases with time initially before reaching a steady-state. However, as the electric field is increased, the transient response shows a non-monotonic behavior. The initial phase (before the peak in the velocity occurs) of such transport is dominated by polar scattering when the momentum relaxation rate is much lower than the scattering rate. However as the electrons start drifting to higher energy states, non-polar scattering becomes significant and provides strong momentum relaxation leading to a drop in drift velocity. To utilize this velocity overshoot phenomenon in transistors to improve RF performance, one needs to make the channel length low enough such that the

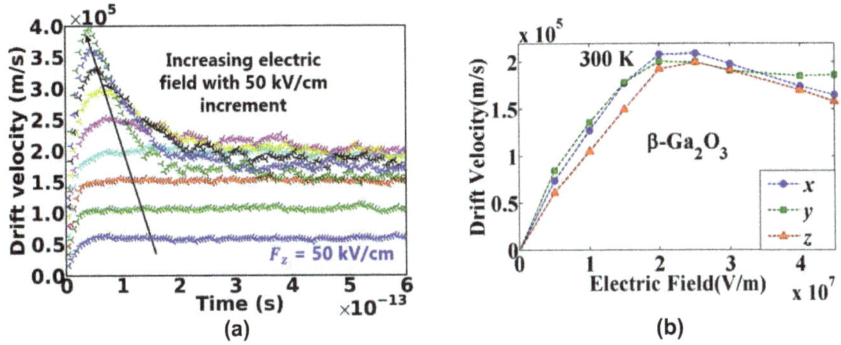

Fig. 6. (a) The transient response of electron velocity in β-Ga_2O_3 probed by FBMC simulation. (b) Velocity-field curves in β-Ga_2O_3 along the three Cartesian directions. (a-b) are reproduced from [68] with permission from the AIP.

electrons reach from source to drain prior to the velocity starts dropping off. From the transient electron dynamics in β-Ga_2O_3, the required channel length, at an electric field of 450 kV/cm, becomes smaller than 20 nm which is challenging but feasible with the state-of-the-art fabrication techniques.

Figure 6(b) shows the velocity-field curves in β-Ga_2O_3 along the three Cartesian directions. The peak velocity is about 2×10^7 cm/s at an electric field of 200 kV/cm. The peak velocity is slightly lower than that reported in GaN [79, 95]. The anisotropy in the curves at low electric fields is due to polar scattering as explained in Sec. 2.2.1. The extracted mobility is about are 112 cm^2 /V.s (140 cm^2 /V.s) along z (x and y) direction [68] which is in good agreement with the low-field iterative calculations [67]. At higher electric fields the anisotropy is a result of that in the bandstructure. Along the y direction the non-parabolicity in the bands begin at a higher value of the group velocity making the drift velocity in that direction higher. Finally, to comment on the negative differential conductivity (NDC) in β-Ga_2O_3, we note that as the remote valleys occur at a much higher energies, the occurrence of intervalley scattering is less likely at moderate values of the electric field unlike GaAs [78]. Hence, NDC in β-Ga_2O_3 is a joint result of band non-parabolicity and short-range intra-valley scatterings [68].

3.2. *Impact ionization co-efficients*

As the electric field is increased further, electrons gain high enough energy to excite valence band electrons to the conduction bands. Figure 7(a) shows a fractional population of electrons in different bands as the electric field is increased up to 8 MV/cm. It can be noted that the population in bands 5 and 6 are minuscule. On the other hand, we discussed in Sec. 2.3 that electrons up to band 4 do not contribute to ionization. But since impact ionization is a cumulative process, it is important to account for that even though the number of electrons capable of

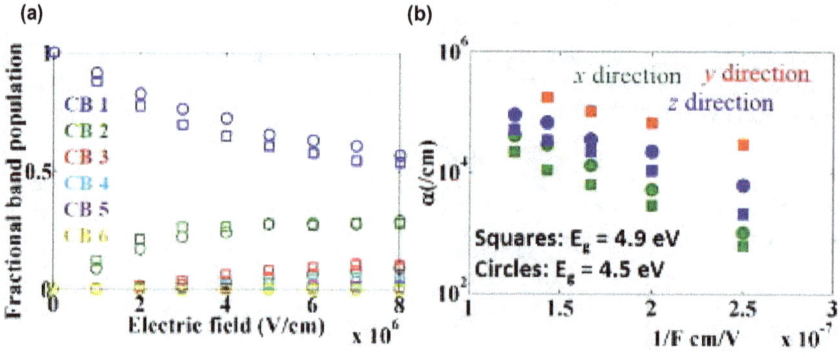

Fig. 7. (a) Electron population of different conduction bands with increasing electric field. (b) The electron impact ionization co-efficients along three different Cartesian directions. (a-b) are reproduced from [70] with permission from the AIP.

causing ionization is low. Also, since beyond the ionization threshold, the ionization rate rises sharply and surpasses the phonon mediated scattering rates quickly, it is very likely that an electron beyond the threshold is going to cause ionization.

Impact ionization coefficients (IIC) represent the inverse of the distance traversed by an electron between two successive ionization events. Figure 7(b) shows the IIC in β-Ga_2O_3 for three different Cartesian directions. The IIC magnitudes are lower than that in GaN [96] as expected due to a larger bandgap. The anistropy in IIC is significantly high due to that in the ionization rates. The x direction IIC is the lowest as the primary electrons are less likely to go to high energy states in that direction. The work reported in [70] did not consider interband tunneling of the electrons which is also likely to happen at such high electric fields. The only possible way of interband transition considered in [70] is via non-polar phonon scatterings. Also, as the calculation reported in [70] uses a mean-field computed bandstructure, the conduction bands are rigid shifted to match experimentally reported bandgap. Figure 7(b) shows the IIC for two extreme values of the bandgap. The x and y directions provide slightly higher mobility than that in the z direction [69] and here we see that the IIC is lowest along the x direction. So we can conclude that transport in x direction is expected to provide an overall superior power electronic performance compared to that in the other two directions.

3.3. *Avalanche phenomenon*

Avalanche breakdown in devices is a cumulative process where the impact ionization becomes self-sustainable, meaning that the generated secondary carriers cause further ionization to generate more carriers. Mathematically, if electron and hole initiated IIC are comparable, avalanche breakdown can happen [97] when $\int_0^L \alpha(x)dx > 1$, where α is the IIC and L is the distance traversed by electrons (typically the length of the space charge region in a p-n junction). The integral,

Table 3. The Chynoweth parameters, ionization integrals, an critical breakdown field estimates in β-Ga$_2$O$_3$ [70].

	a (/cm)	b (V/cm)	I ($\alpha_n \sim \alpha_p$)	I($\alpha_n \sim \alpha_p$)	$E_c(\alpha_n \sim \alpha_p)$
x	0.79×10^6	2.92×10^7	0.38	0.32	10.2
y	2.16×10^6	1.77×10^7	Breakdown	Breakdown	4.8
z	0.70×10^6	2.10×10^7	Breakdown	0.70	7.6

known as the ionization integral, is the key quantity to predict breakdown field of materials. To make the computation of the integral more tractable, the IIC is expressed as an analytical function of the electric field so that if the spatial dependence of the electric field is known computation of the ionization integral becomes straightforward. The analytical relation connecting the electric field to the IIC is known as the Chynoweth relation [98] and has the form $\alpha(F) = a.e^{-\frac{b}{F}}$, where a and b are the Chynoweth parameters. The avalanche breakdown is a joint result of electron and hole initiated impact ionization events, but the hole IIC is still an unknown in β-Ga$_2$O$_3$ and is going to be a very important direction of work. Table 3 shows [70] the estimated electronic Chynoweth parameters, computed ionization integrals for two different limits (α_n and α_p denote electron and hole IIC respectively) and also predicts the breakdown field. The breakdown field in this case is estimated considering a triangular electric field profile which closely resembles the case is a p-n junction. A very recent work [99, 100] reports a TCAD simulation of field-plated β-Ga$_2$O$_3$ MOSFET using the impact ionization parameters reported here. The corresponding breakdown voltage was 1.78 kV which is in excellent agreement with another experimentally observed breakdown voltage report [46].

4. Summary

We presented a bottom-up review of some recent theoretical advances in the high-field electron transport phenomena in β-Ga$_2$O$_3$. Estimates of transport coefficients including velocity and impact ionization coefficients are studied with quantitative understanding and physical insights. The microscopic interaction that dominates transport up to moderately high electric field is the polar coupling between electrons and ${B_u}^1$ phonons. At higher fields significant non-polar electron-phonon scattering and electron-electron scattering become important. A peak velocity of about 2×10^7 cm/s at an electric field of 200 kV/cm, and an estimated breakdown field ranging between 4.8 MV/cm - 10.2 MV/cm based on direction of transport summarizes the essential high-field transport parameters. x direction is identified as the most suitable direction of transport for power electronic applications.

However, given that β-Ga$_2$O$_3$ is still at a very early stage of exploration, in the coming years we expect to see a lot more development on intrinsic transport properties and also on the influence of external factors like dielectric environment, defects, stress, and others on the electronic transport properties. For example, (a) the development of high electron mobility transistors (HEMTs) [101] requires the investigation of high-field transport properties of 2DEG using Monte Carlo techniques, (b) for the accurate prediction of breakdown fields, it is essential to predict the impact ionization co-efficients of holes. Complete atomistic description of the microscopic interactions will become too expensive to be tractable for certain cases, like 2DEG. Mesoscale models with parameters fitted using intrinsic bulk interactions might provide with a suitable path to surmount these challenges. Also, careful device level simulations incorporating geometry related effects are required to predict realistic device performance.

References

1. M. Higashiwaki, K. Sasaki, H. Murakami, Y. Kumagai, A. Koukitu, A. Kuramata, T. Masui and S. Yamakoshi, *Semiconductor Science and Technology* **31**, 034001 (2016).
2. S. J. Pearton, J. Yang, P. H. Cary, F. Ren, J. Kim, M. J. Tadjer and M. A. Mastro, *Applied Physics Reviews* **5**, 011301 (2018).
3. M. Higashiwaki, H. Murakami, Y. Kumagai and A. Kuramata, *Japanese Journal of Applied Physics* **55**, 1202A1 (2016).
4. B. J. Baliga, *Semiconductor Science and Technology* **28**, 074011 (2013).
5. Z. Hu, K. Nomoto, W. Li, N. Tanen, K. Sasaki, A. Kuramata, T. Nakamura, D. Jena and H. G. Xing, *IEEE Electron Device Letters* **39**, 869 (2018).
6. Y. Tomm, P. Reiche, D. Klimm and T. Fukuda, *Journal of crystal growth* **220**, 510 (2000).
7. E. G. Víllora, K. Shimamura, Y. Yoshikawa, K. Aoki and N. Ichinose, *Journal of Crystal Growth* **270**, 420 (2004).
8. H. Aida, K. Nishiguchi, H. Takeda, N. Aota, K. Sunakawa and Y. Yaguchi, *Japanese Journal of Applied Physics* **47**, 8506 (2008).
9. Z. Galazka, R. Uecker, K. Irmscher, M. Albrecht, D. Klimm, M. Pietsch, M. Brützam, R. Bertram, S. Ganschow and R. Fornari, *Crystal Research and Technology* **45**, 1229 (2010).
10. K. Sasaki, M. Higashiwaki, A. Kuramata, T. Masui and S. Yamakoshi, *Journal of Crystal Growth* **378**, 591 (2013).
11. Z. Galazka, K. Irmscher, R. Uecker, R. Bertram, M. Pietsch, A. Kwasniewski, M. Naumann, T. Schulz, R. Schewski, D. Klimm *et al.*, *Journal of Crystal Growth* **404**, 184 (2014).
12. A. Kuramata, K. Koshi, S. Watanabe, Y. Yamaoka, T. Masui and S. Yamakoshi, *Japanese Journal of Applied Physics* **55**, 1202A2 (2016).
13. K. Sasaki, M. Higashiwaki, A. Kuramata, T. Masui and S. Yamakoshi, *Applied Physics Express* **6**, 086502 (2013).
14. F. Ren, S. J. Pearton, J. Yang, P. Carey, S. Ahn, R. Khanna, K. Bevlin, D. Geerpuram and A. Kuramata, Exploration of process techniques for Ga_2O_3 based electronics, in *Meeting Abstracts*, (23) (2018), pp. 1419–1419.
15. K. Sasaki, Q. T. Thieu, D. Wakimoto, Y. Koishikawa, A. Kuramata and S. Yamakoshi, *Applied Physics Express* **10**, 124201 (2017).
16. S. Cui, Z. Mei, Y. Zhang, H. Liang and X. Du, *Advanced Optical Materials* **5**, 1700454 (2017).
17. M. Higashiwaki, K. Sasaki, A. Kuramata, T. Masui and S. Yamakoshi, *Applied Physics Letters* **100**, 013504 (2012).

18. M. J. Tadjer, N. A. Mahadik, V. D. Wheeler, E. R. Glaser, L. Ruppalt, A. D. Koehler, K. D. Hobart, C. R. Eddy and F. J. Kub, *ECS Journal of Solid State Science and Technology* **5**, P468 (2016).
19. A. J. Green, K. D. Chabak, M. Baldini, N. Moser, R. Gilbert, R. C. Fitch, G. Wagner, Z. Galazka, J. Mccandless, A. Crespo *et al.*, *IEEE Electron Device Letters* **38**, 790 (2017).
20. N. Moser, J. McCandless, A. Crespo, K. Leedy, A. Green, A. Neal, S. Mou, E. Ahmadi, J. Speck, K. Chabak *et al.*, *IEEE Electron Device Letters* **38**, 775 (2017).
21. K. Zeng, J. S. Wallace, C. Heimburger, K. Sasaki, A. Kuramata, T. Masui, J. A. Gardella and U. Singisetti, *IEEE Electron Device Letters* **38**, 513 (2017).
22. M. H. Wong, K. Sasaki, A. Kuramata, S. Yamakoshi and M. Higashiwaki, *IEEE Electron Device Letters* **37**, 212 (2016).
23. A. J. Green, K. D. Chabak, E. R. Heller, R. C. Fitch, M. Baldini, A. Fiedler, K. Irmscher, G. Wagner, Z. Galazka, S. E. Tetlak *et al.*, *IEEE Electron Device Letters* **37**, 902 (2016).
24. M. Higashiwaki, K. Sasaki, T. Kamimura, M. Hoi Wong, D. Krishnamurthy, A. Kuramata, T. Masui and S. Yamakoshi, *Applied Physics Letters* **103**, 123511 (2013).
25. M. H. Wong, Y. Nakata, A. Kuramata, S. Yamakoshi and M. Higashiwaki, *Applied Physics Express* **10**, 041101 (2017).
26. H. Zhou, M. Si, S. Alghamdi, G. Qiu, L. Yang and P. D. Ye, *IEEE Electron Device Lett* **38**, 103 (2017).
27. T. Oishi, Y. Koga, K. Harada and M. Kasu, *Applied Physics Express* **8**, 031101 (2016), DOI: 10.1142/S012915641940008110.7567/apex.8.031101.
28. K. Sasaki, M. Higashiwaki, A. Kuramata, T. Masui and S. Yamakoshi, *IEEE electron device letters* **34**, 493 (2013).
29. M. Higashiwaki, K. Sasaki, K. Goto, K. Nomura, Q. T. Thieu, R. Togashi, H. Murakami, Y. Kumagai, B. Monemar, A. Koukitu *et al.*, β-Ga_2O_3 schottky barrier diodes with n^--β-Ga_2O_3 drift layers grown by hvpe, in *2015 73rd Annual Device Research Conference (DRC)*, (2015), pp. 29–30.
30. J. Yang, S. Ahn, F. Ren, S. Pearton, S. Jang and A. Kuramata, *IEEE Electron Device Letters* **38**, 906 (2017).
31. K. Sasaki, D. Wakimoto, Q. T. Thieu, Y. Koishikawa, A. Kuramata, M. Higashiwaki and S. Yamakoshi, *IEEE Electron Device Letters* **38**, 783 (2017).
32. Y. Yao, R. Gangireddy, J. Kim, K. K. Das, R. F. Davis and L. M. Porter, *Journal of Vacuum Science & Technology B, Nanotechnology and Microelectronics: Materials, Processing, Measurement, and Phenomena* **35**, 03D113 (2017).
33. J. Yang, S. Ahn, F. Ren, R. Khanna, K. Bevlin, D. Geerpuram, S. Pearton and A. Kuramata, *Applied Physics Letters* **110**, 142101 (2017).
34. W. Li, K. Nomoto, Z. Hu, D. Jena and H. G. Xing, *Applied Physics Express* **12**, 061007 (2019).
35. A. S. Pratiyush, Z. Xia, S. Kumar, Y. Zhang, C. Joishi, R. Muralidharan, S. Rajan and D. N. Nath, *IEEE Photonics Technology Letters* **30**, 2025 (2018).
36. T. Oshima, T. Okuno, N. Arai, N. Suzuki, S. Ohira and S. Fujita, *Applied physics express* **1**, 011202 (2008).
37. Y. Li, T. Tokizono, M. Liao, M. Zhong, Y. Koide, I. Yamada and J.-J. Delaunay, *Advanced Functional Materials* **20**, 3972 (2010).
38. D. Guo, Z. Wu, P. Li, Y. An, H. Liu, X. Guo, H. Yan, G. Wang, C. Sun, L. Li *et al.*, *Optical Materials Express* **4**, 1067 (2014).
39. B. Zhao, F. Wang, H. Chen, Y. Wang, M. Jiang, X. Fang and D. Zhao, *Nano letters* **15**, 3988 (2015).
40. D. Guo, Z. Wu, Y. An, X. Guo, X. Chu, C. Sun, L. Li, P. Li and W. Tang, *Applied Physics Letters* **105**, 023507 (2014).

41. Q. Feng, L. Huang, G. Han, F. Li, X. Li, L. Fang, X. Xing, J. Zhang, W. Mu, Z. Jia *et al.*, *IEEE Transactions on Electron Devices* **63**, 3578 (2016).
42. W.-Y. Kong, G.-A. Wu, K.-Y. Wang, T.-F. Zhang, Y.-F. Zou, D.-D. Wang and L.-B. Luo, *Advanced Materials* **28**, 10725 (2016).
43. R. Zou, Z. Zhang, Q. Liu, J. Hu, L. Sang, M. Liao and W. Zhang, *Small* **10**, 1848 (2014).
44. B. Zhao, F. Wang, H. Chen, L. Zheng, L. Su, D. Zhao and X. Fang, *Advanced Functional Materials* **27**, 1700264 (2017).
45. G. Hu, C. Shan, N. Zhang, M. Jiang, S. Wang and D. Shen, *Optics express* **23**, 13554 (2015).
46. K. Zeng, A. Vaidya and U. Singisetti, *IEEE Electron Device Letters* **39**, 1385 (2018).
47. K. Yamaguchi, *Solid state communications* **131**, 739 (2004).
48. Y. Zhang, J. Yan, G. Zhao and W. Xie, *Physica B: Condensed Matter* **405**, 3899 (2010).
49. H. He, R. Orlando, M. A. Blanco, R. Pandey, E. Amzallag, I. Baraille and M. Rérat, *Physical Review B* **74**, 195123 (2006).
50. H. Peelaers and C. G. Van de Walle, *physica status solidi (b)* **252**, 828 (2015).
51. T. Lovejoy, R. Chen, E. Yitamben, V. Shutthanadan, S. Heald, E. Villora, K. Shimamura, S. Zheng, S. Dunham, F. Ohuchi *et al.*, *Journal of Applied Physics* **111**, 123716 (2012).
52. M. Rebien, W. Henrion, M. Hong, J. Mannaerts and M. Fleischer, *Applied physics letters* **81**, 250 (2002).
53. K. A. Mengle, G. Shi, D. Bayerl and E. Kioupakis, *Applied Physics Letters* **109**, 212104 (2016).
54. A. Schleife, Optical properties of Ga_2O_3 from first principles: excitons, free-carriers, and dielectric screening (conference presentation), in *Oxide-based Materials and Devices IX*, (2018), p. 1053305.
55. X. Ma, Y. Zhang, L. Dong and R. Jia, *Results in physics* **7**, 1582 (2017).
56. Y. Jinliang and Q. Chong, *Journal of Semiconductors* **37**, 042002 (2016).
57. F. Ricci, F. Boschi, A. Baraldi, A. Filippetti, M. Higashiwaki, A. Kuramata, V. Fiorentini and R. Fornari, *Journal of Physics: Condensed Matter* **28**, 224005 (2016).
58. H. Peelaers and C. G. Van de Walle, *Applied Physics Letters* **111**, 182104 (2017).
59. M. Schubert, R. Korlacki, S. Knight, T. Hofmann, S. Schöche, V. Darakchieva, E. Janzén, B. Monemar, D. Gogova, Q.-T. Thieu *et al.*, *Physical Review B* **93**, 125209 (2016).
60. M. D. Santia, N. Tandon and J. Albrecht, *MRS Advances* **1**, 109 (2016).
61. M. D. Santia, N. Tandon and J. Albrecht, *Applied Physics Letters* **107**, 041907 (2015).
62. M. Handwerg, R. Mitdank, Z. Galazka and S. F. Fischer, *Semiconductor Science and Technology* **30**, 024006 (2015).
63. Z. Guo, A. Verma, X. Wu, F. Sun, A. Hickman, T. Masui, A. Kuramata, M. Higashiwaki, D. Jena and T. Luo, *Applied Physics Letters* **106**, 111909 (2015).
64. C. J. Szwejkowski, N. C. Creange, K. Sun, A. Giri, B. F. Donovan, C. Constantin and P. E. Hopkins, *Journal of Applied Physics* **117**, 084308 (2015).
65. B. Kananen, L. Halliburton, E. Scherrer, K. Stevens, G. Foundos, K. Chang and N. Giles, *Applied Physics Letters* **111**, 072102 (2017).
66. W. R. Lambrecht, D. Skachkov, A. Ratnaparkhe, H. J. von Bardeleben, U. Gerstmann, Q. D. Ho and P. Déak, Computational studies of beta-ga 2 o 3 band structure and the electron paramagnetic resonance spectra of the ga-vacancy defects (conference presentation), in *Oxide-based Materials and Devices IX*, (2018), p. 1053304.
67. K. Ghosh and U. Singisetti, *Applied Physics Letters* **109**, 072102 (2016), DOI: 10.1142/S0129156419400081 10.1063/1.4961308.
68. K. Ghosh and U. Singisetti, *Journal of Applied Physics* **122**, 035702 (2017).
69. K. Ghosh and U. Singisetti, *Journal of Materials Research* **32**, 4142 (2017).
70. K. Ghosh and U. Singisetti, *Journal of Applied Physics* **124**, 085707 (2018).
71. A. Parisini, K. Ghosh, U. Singisetti and R. Fornari, *Semiconductor Science and Technology* **33**, 105008 (2018).
72. Y. Kang, K. Krishnaswamy, H. Peelaers and C. G. Van de Walle, *Journal of Physics: Condensed Matter* **29**, 234001 (2017).
73. N. Ma, N. Tanen, A. Verma, Z. Guo, T. Luo, H. Xing and D. Jena, *Applied Physics Letters* **109**, 212101 (2016).

74. K. Mengle and E. Kioupakis, *AIP Advances* **9**, 015313 (2019).
75. A. Parisini and R. Fornari, *Semiconductor Science and Technology* **31**, 035023 (2016), DOI: 10.1142/S0129156419400081l0.1088/0268-1242/31/3/035023.
76. M. H. Wong, K. Sasaki, A. Kuramata, S. Yamakoshi and M. Higashiwaki, *Japanese Journal of Applied Physics* **55**, 1202B9 (2016).
77. S. Geller, *The Journal of Chemical Physics* **33**, 676 (1960).
78. S. Bosi and C. Jacoboni, *Journal of Physics C: Solid State Physics* **9**, 315 (1976).
79. B. Gelmont, K. Kim and M. Shur, *Journal of applied physics* **74**, 1818 (1993).
80. K. Carva, M. Battiato and P. M. Oppeneer, *Physical review letters* **107**, 207201 (2011).
81. B.-L. Huang and M. Kaviany, *Physical Review B* **77**, 125209 (2008).
82. J. Sjakste, N. Vast and V. Tyuterev, *Physical review letters* **99**, 236405 (2007).
83. N. Sergueev, D. Roubtsov and H. Guo, *Physical review letters* **95**, 146803 (2005).
84. B. Liao, J. Zhou, B. Qiu, M. S. Dresselhaus and G. Chen, *Physical Review B* **91**, 235419 (2015).
85. I. Souza, N. Marzari and D. Vanderbilt, *Physical Review B* **65**, 035109 (2001).
86. N. Marzari and D. Vanderbilt, *Physical review B* **56**, 12847 (1997).
87. A. Calzolari, N. Marzari, I. Souza and M. B. Nardelli, *Physical Review B* **69**, 035108 (2004).
88. A. A. Mostofi, J. R. Yates, Y.-S. Lee, I. Souza, D. Vanderbilt and N. Marzari, *Computer physics communications* **178**, 685 (2008).
89. F. Giustino, M. L. Cohen and S. G. Louie, *Physical Review B* **76**, 165108 (2007).
90. S. Poncé, E. R. Margine, C. Verdi and F. Giustino, *Computer Physics Communications* **209**, 116 (2016).
91. C. Verdi and F. Giustino, *Physical review letters* **115**, 176401 (2015).
92. F. Giustino, *Reviews of Modern Physics* **89**, 015003 (2017).
93. A. D. C. S. Baroni, S. De. Gironcoli, *Reviews of Modern Physics* **73**, 515 (2001).
94. Y.-C. Chang and R. James, *Physical Review B* **53**, 14200 (1996).
95. F. Bertazzi, M. Moresco and E. Bellotti, *Journal of Applied Physics* **106**, 063718 (2009).
96. M. Moresco, F. Bertazzi and E. Bellotti, *Journal of Applied Physics* **106**, 063719 (2009).
97. S. M. Sze and K. K. Ng, *Physics of semiconductor devices* (John wiley & sons, 2006).
98. A. Chynoweth, *physical review* **109**, 1537 (1958).
99. I. Lee, A. Kumar, K. Zeng, U. Singisetti and X. Yao, Modeling and power loss evaluation of ultra wide bandgap Ga_2O_3 device for high power applications, in *Energy Conversion Congress and Exposition (ECCE), 2017 IEEE*, (2017), pp. 4377–4382.
100. I. Lee, A. Kumar, K. Zeng, U. Singisetti and X. Yao, Mixed-mode circuit simulation to characterize Ga_2O_3 MOSFETs in different device structures, in *Wide Bandgap Power Devices and Applications (WiPDA), 2017 IEEE 5th Workshop on*, (2017), pp. 185–189.
101. Y. Zhang, C. Joishi, Z. Xia, M. Brenner, S. Lodha and S. Rajan, *Applied Physics Letters* **112**, 233503 (2018).

Ultra-Wide Bandgap $Al_xGa_{1-x}N$ Channel Transistors

Towhidur Razzak* and Siddharth Rajan

Electrical and Computer Engineering,
The Ohio State University, Columbus OH 43210, USA
**razzak.1@osu.edu*

Andrew Armstrong

Sandia National Laboratories,
Albuquerque, New Mexico 87185, USA

High Al-composition $Al_xGa_{1-x}N$, an emerging class of materials, is gaining significant traction due to its high critical breakdown electric field exceeding that of GaN and high electron saturation velocity that is comparable to GaN. High Al-composition $Al_xGa_{1-x}N$ holds promise for applications such as highly scaled next generation RF devices and power devices. However, significant strides remain to be made before $Al_xGa_{1-x}N$ can take a share of the limelight. Encouraging progress has been made in recent years, including multiple reports of RF operation of $Al_xGa_{1-x}N$ channel transistors with reported unity current gain cutoff frequency as high as 40 GHz, high current density devices with the maximum reported current density exceeding 600 mA/mm and reports of breakdown fields in lateral transistors that are almost 3× the reported breakdown field for GaN channel devices of similar dimensions. This paper focuses on the lateral devices demonstrated thus far and discusses breakthrough performance results achieved for $Al_xGa_{1-x}N$ channel transistors together with the key challenges that are yet to be properly addressed. The paper starts with a discussion of contact formation to $Al_xGa_{1-x}N$ films – the key difficulty of fabricating $Al_xGa_{1-x}N$ transistors. This is followed by a discussion of the types of devices that are typically used. This is followed up with a discussion of the approach required to make high quality devices with $Al_xGa_{1-x}N$ as a template together with some recent result highlights.

Keywords: High Al-composition AlGaN channel; transistors; high power RF.

1. Introduction

Ultra-wide bandgap (UWBG) semiconductors have become an area of interest due to the high breakdown fields in these materials [1–7]. $Al_xGa_{1-x}N$, a ternary alloy of GaN and AlN, display a broad spectrum of bandgaps from 3.4 eV (GaN) to 6.2 eV (AlN) and have various device applications ranging from optoelectronics to RF amplification and power switching applications [8]. These materials are expected to have electron saturation velocity (v_{sat}) comparable to GaN [9, 10] and therefore should show similar levels of saturation current density. The primary advantage of UWBG AlGaN compared to GaN is the former's higher breakdown characteristics [11–13]. In fact, UWBG AlGaN is expected to have high critical breakdown field, F_{BR}, exceeding 11 MV/cm for x > 0.7 [12]. The high F_{BR} can be attributed to the ultra-wide band gap in these materials, where F_{BR} is expected to scale as the 2.5-power of the bandgap [14]. Because of its higher F_{BR} and comparably high v_{sat} for Al-rich $Al_xGa_{1-x}N$ relative to GaN, UWBG AlGaN is expected to have Johnson figure of merit (JFOM)

$$JFOM = v_{sat} \cdot \frac{F_{BR}}{2\pi} \quad (1)$$

much higher than GaN. Thus, $Al_xGa_{1-x}N$-based devices offer many properties which are beneficial for the fabrication of high power, high voltage devices and are expected to play an important role in advancing semiconductor device technology.

However, despite the potential benefits of $Al_xGa_{1-x}N$-based transistors, they are still quite a distance away from wide industry adoption. One of the key reasons is the difficulty of forming ohmic contacts to these materials, which arises from their low electron affinity. Reports of multiple methods, such as direct metal-semiconductor contact and hetero-structure based contacts, exist in the literature, but none have been able to provide a comprehensive solution to this problem [1–3, 5, 11, 15, 16]. Nonetheless, there have been many reports where these devices have displayed glimpses of their potential, achieving certain metrics that cannot be achieved with current GaN technology. Critical breakdown fields exceeding 3 MV/cm have become commonplace with lateral Al-rich AlGaN devices [1, 5, 12] compared to ~1 MV/cm that is typical for GaN devices with similar dimensions. Vertical devices have demonstrated much higher F_{BR} = 8.1 MV/cm reported for PIN diodes [17]. In terms of current densities, there have been reports of device operation with current densities exceeding 600 mA/mm with the highest reported current density being 635 mA/mm [12, 18]. There have also been a multitude of reports on small-signal RF performance with the highest reported f_T being 40 GHz [15, 16]. These are only initial reports, and, as with any emerging field, these numbers are expected to improve.

In this paper we focus on lateral $Al_xGa_{1-x}N$ channel transistors. We first discuss the current status of all the different methods that have been employed to achieve low resistance contacts to high Al-composition $Al_xGa_{1-x}N$ and the current status of low resistance contact formation. This is followed by some of the latest research results on lateral $Al_xGa_{1-x}N$ channel transistors.

2. Ohmic Contacts to High Al-Composition $Al_xGa_{1-x}N$ Films

An ohmic contact is a non-rectifying electrical junction which ideally passes the required current without dropping any voltage across it. This is typically achieved in a transistor by forming a metal-semiconductor contact to the source and drain regions with the metal work-function being ideally as close as possible to the semiconductor electron affinity for n-type contacts [19]. In most cases, however, there will be a positive metal-semiconductor energy barrier at the metal-semiconductor interface, but with a high enough doping in the semiconductor (typically greater than 10^{19} cm^{-3}) the width of the depletion region at the metal-semiconductor interface can be made very thin, on the order of a few nanometers, such that carriers can readily tunnel across such barrier [20]. The heavy doping can be achieved either by annealing the contact, thus causing a reaction between the metal and the semiconductor, and/or heavily doping the semiconductor contact regions [7, 21].

2.1. *Contact formation to GaN*

The most common ohmic contact formation strategy to n-GaN has been to use metals such and Ti and/or Al with a work function close to the electron affinity of GaN. These metals form low Schottky barriers to n-GaN. Such contact schemes have primarily employed a metallization scheme of Ti/Al/(Ni, Ti, Mo, Pt, or Nb)/Au [22–28]. The two main mechanisms for this low resistance ohmic contact formation for AlGaN/GaN high electron mobility transistors (HEMTs) have been postulated to be either low Schottky barrier contacts and/or tunneling contacts [29]. High temperature rapid thermal annealing (RTA), typically at a temperature above 800° C, has been used extensively with great success for the formation of such contacts. During this process, interfacial reaction between the metals and the semiconductor results in spike-like structure formation driven by the dislocations in the AlGaN barrier layer, which directly touches the two-dimensional electron gas (2DEG) [30]. Besides this, during the annealing process Ti reacts with GaN, leading to the formation of TiN, which in turn produces N vacancies (V_N) that act as donors and heavily dopes the metal-semiconductor interface [30]. This results in ultra-thin depletion regions which are virtually transparent to electron conduction [31]. The low Schottky barrier coupled with heavy doping and additional interface doping by annealing, have been successful in demonstrating contact resistance as low as 0.1 Ω.mm [32].

While this approach has worked well with GaN technology, the same success has not been reproducible with high Al-composition $Al_xGa_{1-x}N$. The primary reason for this is the low electron affinity in these set of materials [33]. Due to reduced electron affinity, the barrier heights of the metal semiconductor contact are higher for $Al_xGa_{1-x}N$ channel devices than GaN devices, which results in the formation of rectifying contacts instead of ohmic contacts [20]. Moreover, this problem is exacerbated with an increase of the Al-content in the channel [7]. This has essentially been the main impediment to the wide adoption of

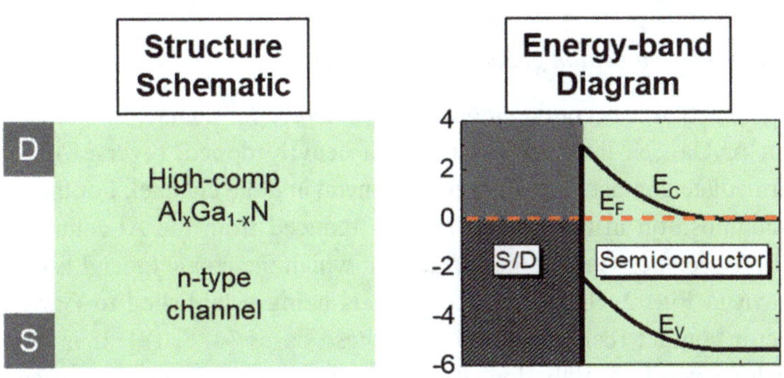

Fig. 1. (a) Schematic diagram of a metal semiconductor contact to n-type high Al-composition $Al_xGa_{1-x}N$ channel transistor (b) Energy-band diagram under the source/drain access region of the metal-semiconductor contact layer simulated via 1D Schrodinger-Poisson simulator.

$Al_xGa_{1-x}N$ channel devices that otherwise provide excellent material properties well suited for high performance power and RF transistors. The schematic of metal-semiconductor contacts to a typical high Al-composition $Al_xGa_{1-x}N$ film, together with the energy-band diagram under the source/drain access region, has been shown in Fig. 1.

2.2. *Contact formation to AlGaN*

To date, the efforts in achieving low resistance ohmic contacts to high Al-content $Al_xGa_{1-x}N$ can be generally categorized into two approaches:

1. Direct metal-semiconductor contact formation to UWBG $Al_xGa_{1-x}N$
2. Using a reverse Al-composition graded contact layer such that metal-semiconductor contact is established to GaN

2.2.1. *Metal-semiconductor based approaches*

Approaches based on metal-semiconductor contact formation to $Al_xGa_{1-x}N$ films have tried to improve the contact resistance by utilizing various metallization schemes such as Ti/Al/Ni/Au, Zr/Al/Mo/Au, V/Al/V/Au, Nb/Ti/Al/Mo/Au etc. [22–28]. Although there has not been any explicit investigation to understand the physical mechanism which could aid the formation of low resistance contacts with these metallization schemes, the aim of these studies were to either explore ways to reduce the metal-semiconductor barrier heights and/or more effectively dope the metal-semiconductor interface. Efforts have also focused on actively introducing [Si^+] as an n-type dopant to reduce the depletion region thickness and increase the tunneling probability by approaches such as Si ion implantation [5, 6], introduction of SiN_x interlayer between the metal and semiconductor etc. [25, 34]. Such efforts have proven to be useful for cases with $x < 0.5$, where contact resistance as low as 1.65 Ω.mm has been achieved for $Al_xGa_{1-x}N$ channel metal-oxide semiconductor high mobility field effect transistors (MOSHFET) for $x = 0.40$ [18]. However, for Al-composition above 0.5, these methods have not been as successful, with the best reported result being 24.6 Ω.mm [1].

2.2.2. *Reverse Al-composition graded contact layer*

Another approach that is becoming more widely adopted for contact formation to high Al-composition $Al_xGa_{1-x}N$, involves the use of a heavily doped, reverse Al-composition graded intermediate contact layer between the metal and the channel. For this scheme, the $Al_xGa_{1-x}N$ composition in the contact layer is reduced from the Al-composition in the channel until it becomes GaN at the surface, on which the contact metal is deposited [4]. This is shown in Fig. 2. Since metal contact is being established to GaN, the metal-semiconductor contact resistance is low in these cases [4, 5, 12]. It is of paramount importance, however, to heavily dope the contact layer with n-type dopants such as [Si^+]. This serves two purposes: 1) keep the contact region highly conductive to minimize any additional resistance being introduced by the contact layer and 2) compensate for any

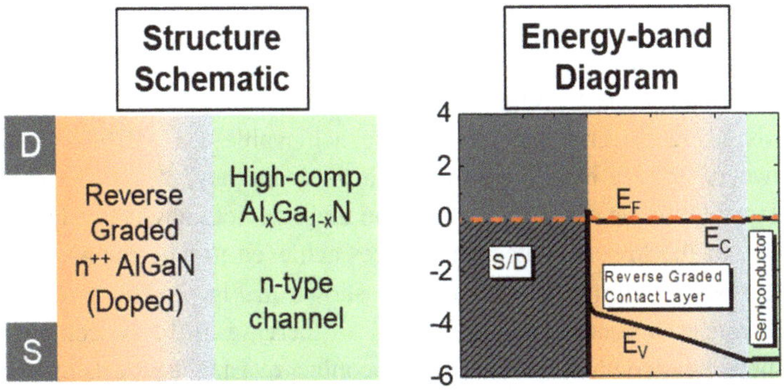

Fig. 2. (a) Schematic diagram of a reverse Al-composition graded contact layer for high Al-composition $Al_xGa_{1-x}N$ channel transistor (b) Energy-band diagram under the source/drain access region of the metal-semiconductor contact layer simulated via 1D Schrodinger-Poisson simulator.

negative polarization charge introduced because of the reverse Al-composition grading of the $Al_xGa_{1-x}N$ in the contact region [12]. Due to the strong polarization effect displayed by AlGaN material system, the reverse Al-composition grading produces the negative polarization charge [35] for the typical metal-face growth orientation.

The first report employing such reverse graded contacts came out in 2015, where the authors deployed the contact layer for a metal-insulator semiconductor field effect transistor (MISFET) [4]. In this report, the authors employed molecular beam epitaxy (MBE) to grow the transistor channel and the graded contact on a single growth. However, even though contact resistances as low as 0.3 Ω.mm were achieved, the carrier mobility in the channel was low and limited the current density to 60 mA/mm [4]. There are currently no reports of high mobility high Al-composition $Al_xGa_{1-x}N$ channel transistors grown by MBE reported in the literature.

In contrast, metalorganic chemical vapor deposition (MOCVD) based growth of high Al-composition $Al_xGa_{1-x}N$ channel transistors have routinely achieved high electron mobility with some reports claiming mobilities exceeding 200 $cm^2/V\cdot s$ [1, 15, 16]. However, uniform heavy n-type doping via [Si^+] of the Al-composition graded contact layers, one of the key requirements of growing low resistance reverse Al-composition graded contact layers, has proven difficult for MOCVD grown contact layers [12]. Multiple experiments performed to gauge the feasibility of such all-MOCVD grown devices have shown this to be true. While initial experiments yielded results with resistivity above 10^{-4} $\Omega.cm^2$, recent experiments which carefully considered the design of the contact layers have been able to bring down the contact resistance to low 10^{-5} $\Omega.cm^2$, which is still well above 10^{-6} $\Omega.cm^2$ demonstrated with MBE contacts [5, 12, 36]. The likely reason for this difference is that in $Al_xGa_{1-x}N$ films grown via MOCVD, dopant incorporation is a strong function of the Al-composition of the film, and high dopant incorporation becomes more challenging for higher Al-content $Al_xGa_{1-x}N$. As a result, the growth of reverse-graded contact layers is challenging [12]. This approach is still currently an active area of research and will likely improve over time with growth and design optimization.

To circumvent this problem, there have been multiple reports which demonstrated an MBE regrown reverse Al-composition graded contact layer on an MOCVD grown channel. Such a hybrid design aims to utilize the high mobility MOCVD grown channel while simultaneously using lower contact resistance achievable via MBE regrown contacts [11, 15]. However, such efforts are currently at their infancy. Working transistors have been demonstrated, but the contact resistance on these devices have been far from ideal.

A summary of the lowest contact resistances achieved to date for $Al_xGa_{1-x}N$ channel devices for various Al-compositions has been shown in Fig. 3. The general trend is an exponential increase in contact resistance with an increase in the Al-composition in the channel for metal-semiconductor contacts. The contact resistance reports using the reverse Al-composition graded devices break away from that trend, as expected, since metal-semiconductor contact is established with GaN.

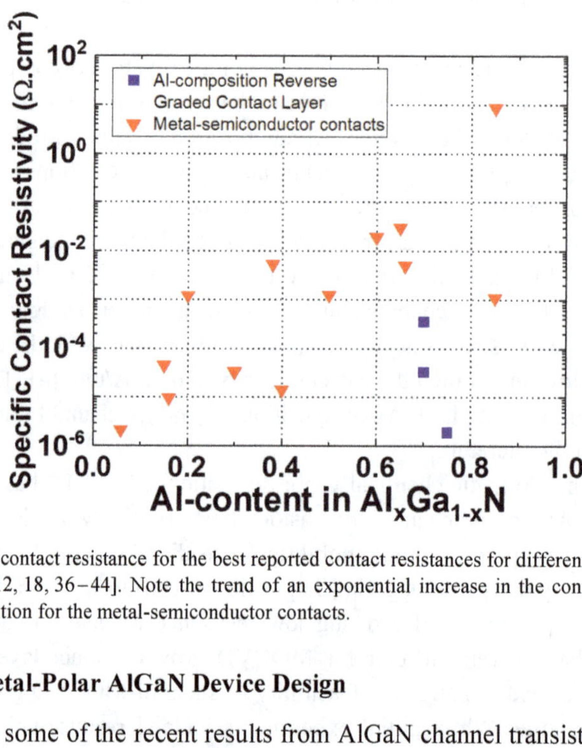

Fig. 3. Comparison of contact resistance for the best reported contact resistances for different $Al_xGa_{1-x}N$ channel devices [1, 2, 4, 5, 7, 12, 18, 36–44]. Note the trend of an exponential increase in the contact resistance with increasing Al-composition for the metal-semiconductor contacts.

3. Advanced Metal-Polar AlGaN Device Design

Before discussing some of the recent results from AlGaN channel transistors, it is worthwhile to discuss some of the device structures typically employed for $Al_xGa_{1-x}N$-based devices. $Al_xGa_{1-x}N$ channel lateral transistors are typically field-effect transistors (FETs) and can be widely categorized into two types:

1. Transistors with charge originating from doping of the material
2. Transistors with charge originating from polarization effect of $Al_xGa_{1-x}N$

In the first type of devices, the channel conductivity is controlled by the addition of dopants during the epitaxial growth of the channel and are commonly called metal semiconductor field effect transistors (MESFETs) or metal insulator semiconductor field effect transistors (MISFETs). These structures are akin to junction field-effect transistors (JFETs) with a Schottky (metal-semiconductor) or metal-insulator-semiconductor (MIS) junction for channel charge modulation. Typical metals used for gate contacts are Ni, Pt, W etc. with the barrier heights ranging from 1.5 to 2.5 eV [33, 35].

The second type of devices uses the polarization effect of $Al_xGa_{1-x}N$ to introduce charge in the channel. $Al_xGa_{1-x}N$ material system, owing to the large ionicity of the Al(Ga)-N bonds, possesses a large piezoelectric polarization component. On top of that, the uniaxial nature of the wurtzite crystal structure of $Al_xGa_{1-x}N$ with a non-ideal c/a ratio also displays a large spontaneous polarization. One such transistor, called a high electron mobility transistor (HEMT), uses a two-dimensional electron gas (2DEG) for channel charge conduction. In these devices, a thin layer of metal-face $Al_xGa_{1-x}N$ film (barrier) is epitaxially grown on a lower Al-composition metal-face $Al_xGa_{1-x}N$ film (channel), which causes a 2DEG to be formed at the heterojunction. There is no need for modulation doping for these HEMTs, as is required for AlGaAs/GaAs heterostructures HEMTs, as the charge is fully generated by the polarization effect. The 2DEG charge density is dependent on the thickness of the barrier layer and the Al-composition difference between the barrier and the channel [35].

Using the same polarization effect, it is also possible to fabricate another type of transistor called a polarization-doped field effect transistor (PolFET), whereby the 2DEG is spread into a three-dimensional electron slab (3DES) instead of charge confinement in a quasi-2D plane. In these transistors, instead of abruptly growing a higher Al-composition $Al_xGa_{1-x}N$ barrier on the channel, the Al-composition is gradually increased to induce a 3DES. PolFETs behave similarly to a MESFET but displays higher carrier mobility due to the absence of ionized impurity ions. These devices also show higher linearity of the transconductance-gate bias profile compared to HEMTs because they do not suffer from velocity degradation behavior of HEMTs [35, 45].

4. $Al_xGa_{1-x}N$ Channel Device Demonstrations

Despite obvious challenges, number of reports on $Al_xGa_{1-x}N$ channel devices have been recently on the rise. As with any new technology, these initial results have not been able to challenge incumbent technologies in all fronts. However, the importance of the exploration of $Al_xGa_{1-x}N$ as a possible template for next generation RF and power electronics transistors has been unequivocally proven by these demonstrations. One key challenge of fabricating high-performance RF devices using materials such as β-Gallium Oxide (β-Ga_2O_3) and high Al-composition $Al_xGa_{1-x}N$ is the low mobility in these materials. In the case of $Al_xGa_{1-x}N$, the main reason for the low mobility is the presence of high levels of alloy scattering, especially for $Al_xGa_{1-x}N$ with Al-composition, x, roughly between 0.2 to 0.8. As a result, the theoretically predicted mobility achievable in this range of $Al_xGa_{1-x}N$ is of the order 200-300 $cm^2/V{\cdot}s$. Correspondingly, the sheet resistance, R_{SH}, is high in these materials with

$$R_{SH} = \frac{1}{en_s\mu} \tag{2}$$

where e is the elementary electronic charge, n_s is the sheet charge density and μ is the carrier mobility. For RF devices, two important parameters of interest are the unity current gain cutoff frequency, f_T, and unity power gain cutoff frequency, f_{MAX}. Both these parameters are strong functions of the source and drain resistance, R_S and R_D, which are dependent on the sheet resistance of the channel.

$$f_T = \frac{g_m}{2\pi(C_{gs} + C_{gd})\left(1 + \frac{(R_s + R_d)}{R_0}\right) + g_m C_{gd}(R_s + R_d)}$$

$$f_{max} \approx \frac{f_T}{2\sqrt{\frac{R_i + R_s + R_g}{R_0} + (2\pi f_T)R_g C_{gd}}} \tag{3}$$

While $Al_xGa_{1-x}N$ displays high F_{BR}, the drawbacks of having high contact resistance and high sheet resistance make realizing high performance RF devices a challenge. To combat the high sheet resistance, device scaling should be utilized to reduce the source and drain resistance and correspondingly boost the high frequency performance. Further it should be noted that the high critical breakdown field in these materials enable more aggressive scaling than GaN because they can sustain much higher breakdown voltages than GaN for devices of similar dimensions.

The feasibility of $Al_xGa_{1-x}N$ channel transistors for RF performance up to 100 GHz has been demonstrated in the literature via simulation [13]. The schematic and the electric field profile of the device are shown in Fig. 4. In this study, using the test case of $Al_{0.75}Ga_{0.25}N$ channel and assuming an electron mobility of 100 $cm^2/V{\cdot}s$ and a contact resistance of

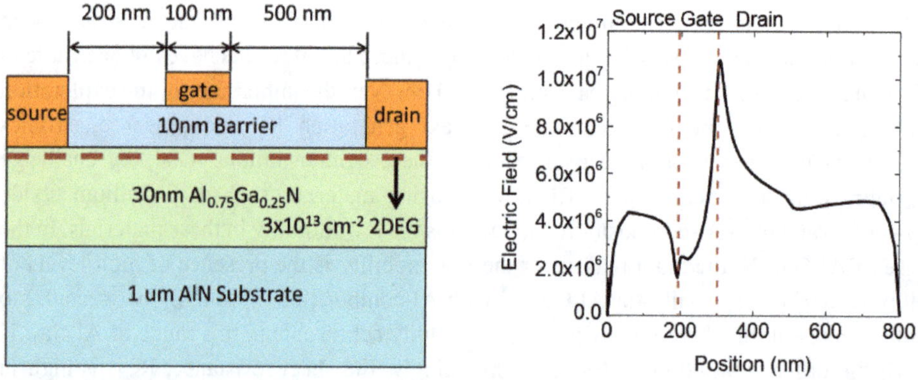

Fig. 4. (a) Schematic of a 100 nm $Al_{0.75}Ga_{0.25}N$ channel HEMT with a 10 nm barrier and (b) Electric field profile of the device at a drain bias of 100 V.

0.1 Ω.mm, it has been demonstrated that for a scaled transistor with a gate length of 100 nm and a source-drain spacing of 800 nm, f_T can potentially exceed 100 GHz with a breakdown voltage of 100 V and a current density of 2.5 A/mm. The electron saturation velocity and sheet charge density were assumed to be 1×10^7 cm/s and 3×10^{13} cm^{-2} respectively. It should be noted that this study did not account for gate-leakage related breakdown, which might be a limiting factor. However, the use of a high-k gate dielectric and field plates should help to mitigate such issues [46].

The channel resistance is an important parameter for power devices as well, as the Baliga Figure of Merit (BFOM) is inversely proportional to on-resistance, as can be seen below

$$BFOM = \epsilon \cdot \mu \cdot E_C^3 = \frac{4 \cdot V_{BR}^2}{R_{on}} \tag{4}$$

where V_{BR} is the breakdown voltage of the device and R_{on} is the on-resistance. Hence, scaling is also critical for fabricating power devices with $Al_xGa_{1-x}N$ channel transistors. Again, this can be enabled by the high critical breakdown fields in these materials which compensates for the low mobility. Another benefit of scaling is the diminishing size of the integrated circuit (IC) chips, which results in increased number of dies per wafer hence reduced cost of production.

Recently, there have been a multitude of reports of high Al-composition $Al_xGa_{1-x}N$ channel devices which demonstrated both high power and high frequency potential of these devices [1, 3, 5, 11–13, 15, 16, 43, 47]. In terms of RF performance, the highest reported f_T of 40 GHz has been demonstrated for an $Al_{0.60}Ga_{0.40}N$ channel HEMT [15]. The reported epitaxial layer structure grown on an AlN on sapphire substrate, consisted of a 0.25 μm $Al_{0.60}Ga_{0.40}N$ channel and a 30 nm n-doped ($[Si^+] = 2\times10^{18}$ cm^{-3}) $Al_{0.75}Ga_{0.25}N$ barrier. The sheet charge density, n_s, of the reported devices were 8.5×10^{12} cm^{-2} with electron mobility of 175 $cm^2/V\cdot s$. The maximum current density was found to be 460 mA/mm for devices with gate length of 130 nm. The contact to the channel was achieved by utilizing the hybrid MBE contact regrowth scheme on MOCVD channel, discussed earlier. Another recent report of RF operation on an $Al_{0.70}Ga_{0.30}N$ channel HEMT demonstrated an f_T of 28.4 GHz with a maximum current density of 80 mA/mm for devices with gate length of 80 nm [16]. Breakdown fields up to 2.5 MV/cm, which is more than 2× higher than demonstrated with GaN channel devices with similar dimensions, were measured for devices with gate to drain spacing of 2 microns. The device structure was grown via MOCVD and consisted of a $Al_{0.70}Ga_{0.30}N$ buffer and channel layer with a thickness of unintentionally (UID) doped 400 nm followed by a 30 nm n-doped ($[Si^+] = 3.5\times10^{18}$ cm^{-3}) $Al_{0.85}Ga_{0.15}N$ barrier layer. The 2DEG sheet resistance resistivity was found to be approximately 2200 Ω/□ with a pinch-off voltage of approximately -6.0 V. The mobility and sheet charge were estimated to be 390 $cm^2\,V^{-1}s^{-1}$ and 7.2×10^{12} cm^{-2}.

Although decent RF performance was demonstrated by the above-mentioned devices, the contact resistances were high in the above-mentioned devices with highly non-linear source to drain current-voltage relationship. However, a more recent report on all-MOCVD grown MESFETs with a reverse Al-composition graded AlGaN contact layer, demonstrated contact resistances as low as 5 Ω.mm with linear source to drain current-voltage relationship. The device structure consisted of an AlN on sapphire template with a 40 nm of $Al_{0.70}Ga_{0.30}N$ channel with a doping density of 4.5×10^{18} cm^{-3}. The electron mobility was found to be 56 cm^2/V·s with a sheet charge of 1.8×10^{13} cm^{-2}. Devices with gate-length of 600 nm displayed a maximum current density of 635 mA/mm [12]. Breakdown performance measured on devices with a gate to drain spacing of 770 nm showed no breakdown up to 200 V which corresponds to an average breakdown field of 2.86 MV/cm, which is around 3× higher than demonstrated with GaN channel devices with similar dimensions.

The highest reported average breakdown field of an AlGaN channel device is 3.6 MV/cm, demonstrated for an all-MOCVD grown $Al_{0.70}Ga_{0.30}N$ channel MISFET with reverse Al-composition graded contact layer [5]. The channel sheet charge was reported to be 1×10^{13} cm^{-2} with a mobility of 90 cm^2/V·s. Similar to the all-MOCVD MESFETs [12], these all-MOCVD also displayed linear source to drain current-voltage relationship, albeit with a high contact resistance of 14 Ω.mm. The maximum current density was found to be 350 mA/mm for devices with gate length of 1 μm. However, the authors demonstrated that the current density was contact resistance limited and estimated a current density greater than 500 mA/mm achievable in these structures by using wide-contact geometry I-shaped devices which reduces the fractional contribution of the contact resistance to the overall resistance.

Although the above-mentioned results are promising, they are far from the simulated values reported earlier. The reason for the lower than expected performance is the significantly lower injection velocity in these devices compared to what is achievable with further device growth and design optimization [15]. Further, performance is expected to improve with improvement in the contact resistance and device scaling.

5. Future Directions

Despite significant progress, $Al_xGa_{1-x}N$ channel devices have remained significantly more challenging than GaN or GaAs technology in terms of making lower resistance contacts. While low mobility is a fundamental limitation for these devices, it is not a deal-breaker because this limitation can be circumvented by careful device design. Another issue that we have not touched upon is the need for high quality dielectric material for these UWBG semiconductors. As the technology matures, it will be important to find good quality gate dielectric materials that have comparable critical electric fields and maintain large band offsets relative to high Al-composition $Al_xGa_{1-x}N$. At this important juncture for $Al_xGa_{1-x}N$ channel transistors, the three key challenges that must be solved in the coming years are thus:

1. Significant reduction in contact resistance
2. Ultra-scaled devices
3. Good quality gate dielectric materials

6. Conclusion

To summarize, high Al-composition $Al_xGa_{1-x}N$ devices are a new technology that has become a research area of significant interest. The key feature that has propelled these devices into prominence is their higher critical breakdown fields relative to GaN, which is buttressed by high electron saturated velocity comparable to GaN. Due to these desirable properties, these materials can be used to fabricate high performance RF devices which are highly scaled and can enable unprecedented levels of power density scaling. However, there are a few issues in these materials, such as high contact resistance and low mobility. Although significant strides remain to be made, very encouraging progress has been made in recent years in these devices. It is expected that with further technology improvement $Al_xGa_{1-x}N$ can position itself as a key competitor for usage in next generation RF electronics.

Acknowledgments

The authors acknowledge funding from Air Force Office of Scientific Research (AFOSR Grant FA9550-17-1-0227, Program Manager Kenneth Goretta) and the DARPA DREaM program (ONR N00014-18-1-2033, Program Manager Dr. Young-Kai Chen, monitored by Office of Naval Research, Program Manager Dr. Paul Maki).

Sandia National Laboratories is a multi-mission laboratory managed and operated by National Technology and Engineering Solutions of Sandia, LLC., a wholly owned subsidiary of Honeywell International, Inc., for the U.S. Department of Energy's National Nuclear Security Administration under contract DE-NA-0003525. The views expressed in the article do not necessarily represent the views of the U.S. Department of Energy or the United States Government.

References

1. Andrew M Armstrong, Brianna A Klein, Albert Colon, Andrew A Allerman, Erica A Douglas, Albert G Baca, Torben R Fortune, Vincent M Abate, Sanyam Bajaj, and Siddharth Rajan, Japanese Journal of Applied Physics **57** (7), 074103 (2018).
2. Albert G Baca, Andrew M Armstrong, Andrew A Allerman, Erica A Douglas, Carlos A Sanchez, Michael P King, Michael E Coltrin, Torben R Fortune, and Robert J Kaplar, Applied Physics Letters **109** (3), 033509 (2016).
3. Albert G Baca, Andrew M Armstrong, Andrew A Allerman, Erica A Douglas, Carlos A Sanchez, Michael P King, Michael E Coltrin, Christopher D Nordquist, Torben R Fortune, and Robert J Kaplar, presented at the Device Research Conference (DRC), 2016 74th Annual, 2016 (unpublished).
4. Sanyam Bajaj, Fatih Akyol, Sriram Krishnamoorthy, Yuewei Zhang, and Siddharth Rajan, Applied Physics Letters **109** (13), 133508 (2016).
5. Sanyam Bajaj, Andrew Allerman, Andrew Armstrong, Towhidur Razzak, Vishank Talesara, Wenyuan Sun, Shahadat H Sohel, Yuewei Zhang, Wu Lu, and Aaron R Arehart, IEEE Electron Device Letters **39** (2), 256 (2018).

6. Sanyam Bajaj, Ting-Hsiang Hung, Fatih Akyol, Digbijoy Nath, and Siddharth Rajan, Applied Physics Letters **105** (26), 263503 (2014).
7. EA Douglas, S Reza, C Sanchez, D Koleske, A Allerman, B Klein, AM Armstrong, RJ Kaplar, and AG Baca, physica status solidi (a) **214** (8), 1600842 (2017).
8. Sanyam Bajaj, The Ohio State University, 2018.
9. AFM Anwar, Shangli Wu, and Richard T Webster, IEEE Transactions on Electron devices **48** (3), 567 (2001).
10. Maziar Farahmand, Carlo Garetto, Enrico Bellotti, Kevin F Brennan, Michele Goano, Enrico Ghillino, Giovani Ghione, John D Albrecht, and P Paul Ruden, IEEE Transactions on electron devices **48** (3), 535 (2001).
11. T Razzak, S Hwang, A Coleman, S Bajaj, H Xue, Y Zhang, Z Jamal-Eddine, SH Sohel, W Lu, and A Khan, Electronics Letters **54** (23), 1351 (2018).
12. Towhidur Razzak, Seongmo Hwang, Antwon Coleman, Hao Xue, Shahadat H. Sohel, Sanyam Bajaj, Yuewei Zhang, Wu Lu, Asif Khan, and Siddharth Rajan, Applied Physics Letters **115** (4), 043502 (2019).
13. Towhidur Razzak, Hao Xue, Zhanbo Xia, Seongmo Hwang, Asif Khan, Wu Lu, and Siddharth Rajan, presented at the 2018 IEEE MTT-S International Microwave Workshop Series on Advanced Materials and Processes for RF and THz Applications (IMWS-AMP), 2018 (unpublished).
14. RJ Kaplar, Andrew A Allerman, AM Armstrong, Mary H Crawford, Jeramy Ray Dickerson, Arthur J Fischer, AG Baca, and EA Douglas, ECS Journal of Solid State Science and Technology **6** (2), Q3061 (2017).
15. Hao Xue, Choong Hee Lee, Kamal Hussian, Towhidur Razzak, Mamun Abdullah, Zhanbo Xia, Shahadat Hasan Sohel, Asif Khan, Siddharth Rajan, and Wu Lu, Applied Physics Express **12** (6), 066502 (2019).
16. Albert G Baca, Brianna A Klein, Joel R Wendt, Stefan M Lepkowski, Christopher D Nordquist, Andrew M Armstrong, Andrew A Allerman, Erica A Douglas, and Robert J Kaplar, IEEE Electron Device Letters **40** (1), 17 (2018).
17. Atsushi Nishikawa, Kazuhide Kumakura, and Toshiki Makimoto, Japanese journal of applied physics **46** (4S), 2316 (2007).
18. Xuhong Hu, Seongmo Hwang, Kamal Hussain, Richard Floyd, Shahab Mollah, Fatima Asif, Grigory Simin, and Asif Khan, IEEE Electron Device Letters **39** (10), 1568 (2018).
19. QZ Liu and SS Lau, Solid-State Electronics **42** (5), 677 (1998).
20. Dieter K Schroder, *Semiconductor material and device characterization.* (John Wiley & Sons, 2015).
21. Brianna Alexandra Klein, Albert G Baca, Andrew M Armstrong, Andrew A Allerman, Carlos Anthony Sanchez, Erica Ann Douglas, Mary H Crawford, Mary A Miller, Paul G Kotula, and Torben Ray Fortune, ECS Journal of Solid State Science and Technology **6** (11), S3067 (2017).
22. Hirokuni Tokuda, Maiko Hatano, Norimasa Yafune, Shin Hashimoto, Katsushi Akita, Yoshiyuki Yamamoto, and Masaaki Kuzuhara, Applied physics express **3** (12), 121003 (2010).
23. Alexei Vertiatchikh, Ed Kaminsky, Julie Teetsov, and Kevin Robinson, Solid-State Electronics **50** (7-8), 1425 (2006).
24. A Soltani, A BenMoussa, S Touati, V Hoël, J-C De Jaeger, J Laureyns, Y Cordier, C Marhic, MA Djouadi, and C Dua, Diamond and related materials **16** (2), 262 (2007).

25. Kazuki Mori, Kunihiro Takeda, Toshiki Kusafuka, Motoaki Iwaya, Tetsuya Takeuchi, Satoshi Kamiyama, Isamu Akasaki, and Hiroshi Amano, Japanese Journal of Applied Physics **55** (5S), 05FL03 (2016).
26. Ryan France, Tao Xu, Papo Chen, R Chandrasekaran, and TD Moustakas, Applied physics letters **90** (6), 062115 (2007).
27. G Vanko, T Lalinský, Ž Mozolová, J Liday, P Vogrinčič, A Vincze, F Uherek, Š Haščík, and I Kostič, Vacuum **82** (2), 193 (2007).
28. Norimasa Yafune, Shin Hashimoto, Katsushi Akita, Yoshiyuki Yamamoto, and Masaaki Kuzuhara, Japanese Journal of Applied Physics **50** (10R), 100202 (2011).
29. Hyung-Seok Lee, Dong Seup Lee, and Tomas Palacios, IEEE Electron Device Letters **32** (5), 623 (2011).
30. Liang Wang, Fitih M Mohammed, and Ilesanmi Adesida, Journal of applied physics **101** (1), 013702 (2007).
31. Deepak Selvanathan, Fitih M Mohammed, Asrat Tesfayesus, and Ilesanmi Adesida, Journal of Vacuum Science & Technology B: Microelectronics and Nanometer Structures Processing, Measurement, and Phenomena **22** (5), 2409 (2004).
32. Kevin W Kobayashi, Dan Denninghoff, and Dain Miller, presented at the 2015 IEEE Compound Semiconductor Integrated Circuit Symposium (CSICS), 2015 (unpublished).
33. D Qiao, LS Yu, SS Lau, Joan Marie Redwing, JY Lin, and HX Jiang, Journal of Applied Physics **87** (2), 801 (2000).
34. Felix Recht, L McCarthy, S Rajan, A Chakraborty, C Poblenz, A Corrion, JS Speck, and UK Mishra, (2006).
35. Colin Wood and Debdeep Jena, *Polarization effects in semiconductors: from ab initio theory to device applications.* (Springer Science & Business Media, 2007).
36. Sakib Muhtadi, S Hwang, Antwon Coleman, Fatima Asif, A Lunev, MVS Chandrashekhar, and Asif Khan, Applied Physics Letters **110** (17), 171104 (2017).
37. Albert G Baca, Andrew M Armstrong, Andrew A Allerman, Brianna A Klein, Erica A Douglas, Carlos A Sanchez, and Torben R Fortune, presented at the 2017 75th Annual Device Research Conference (DRC), 2017 (unpublished).
38. Takuma Nanjo, Akifumi Imai, Yosuke Suzuki, Yuji Abe, Toshiyuki Oishi, Muneyoshi Suita, Eiji Yagyu, and Yasunori Tokuda, IEEE Transactions on Electron Devices **60** (3), 1046 (2013).
39. T Nanjo, M Takeuchi, A Imai, M Suita, T Oishi, Y Abe, E Yagyu, T Kurata, Y Tokuda, and Y Aoyagi, Electronics letters **45** (25), 1346 (2009).
40. Ajay Raman, Sansaptak Dasgupta, Siddharth Rajan, James S Speck, and Umesh K Mishra, Japanese Journal of Applied Physics **47** (5R), 3359 (2008).
41. Norimasa Yafune, Shin Hashimoto, Katsushi Akita, Yoshiyuki Yamamoto, Hirokuni Tokuda, and Masaaki Kuzuhara, Electronics Letters **50** (3), 211 (2014).
42. Takuma Nanjo, Misaichi Takeuchi, Muneyoshi Suita, Yuji Abe, Toshiyuki Oishi, Yasunori Tokuda, and Yoshinobu Aoyagi, Applied physics express **1** (1), 011101 (2007).
43. Sakib Muhtadi, Seong Mo Hwang, Antwon Coleman, Fatima Asif, Grigory Simin, MVS Chandrashekhar, and Asif Khan, IEEE Electron Device Letters **38** (7), 914 (2017).
44. Takuma Nanjo, Misaichi Takeuchi, Muneyoshi Suita, Toshiyuki Oishi, Yuji Abe, Yasunori Tokuda, and Yoshinobu Aoyagi, Applied Physics Letters **92** (26), 263502 (2008).
45. Tian Fang, Ronghua Wang, Huili Xing, Siddharth Rajan, and Debdeep Jena, IEEE Electron Device Letters **33** (5), 709 (2012).

46. Zhanbo Xia, Caiyu Wang, Nidhin Kurian Kalarickal, Susanne Stemmer, and Siddharth Rajan, IEEE Transactions on Electron Devices **66** (2), 896 (2019).
47. Hao Xue, Towhidur Razzak, Seongmo Hwang, Antwon Coleman, Sanyam Bajaj, Yuewei Zhang, Zane Jamal-Eddin, Shahadat Hasan Sohel, Asif Khan, and Siddharth Rajan, presented at the 2018 76th Device Research Conference (DRC), 2018 (unpublished).

On the Progress Made in GaN Vertical Device Technology

Dong Ji[*] and Srabanti Chowdhury[†]

Department of Electrical Engineering, Stanford University,
Stanford, CA, USA
[*]dongji@stanford.edu
[†]srabanti@stanford.edu

Silicon technology enabled most of the electronics we witness today, including power electronics. However, wide bandgap semiconductors are capable of addressing high-power electronics more efficiently compared to Silicon, where higher power density is a key driver. Among the wide bandgap semiconductors, silicon carbide (SiC) and gallium nitride (GaN) are in the forefront in power electronics. GaN is promising in its vertical device topology. From CAVETs to MOSFETs, GaN has addressed voltage requirements over a wide range. Our current research in GaN offers a promising view of GaN that forms the theme of this article. CAVETs and OGFETs (a type of MOSFET) in GaN are picked to sketch the key achievements made in GaN vertical device over the last decade.

Keywords: GaN power devices; MOSFET; CAVET, OGFET; vertical devices.

1. Introduction

Power electronics has been the cornerstone of several technological advancements of our times. From its generation to delivery, power is handled, conditioned, and converted using electronics that are built with circuits that primarily rely on "switches".

With the development of semiconductors, mechanical and electro-mechanical switches gave way to electrical switches. Silicon technology enabled most of the electronics we witness today, including power electronics. However, wide bandgap semiconductors are capable of addressing high-power electronics more efficiently compared to Silicon, where higher power density is a key driver. Among the wide bandgap semiconductors, silicon carbide (SiC) and gallium nitride (GaN) are in the forefront of power electronics. Within GaN technology, lateral or HEMT-based technology complements vertical GaN technology; where the former is providing medium voltage devices (600/650 V) and the latter looks promising for higher voltage (>1.2 kV) needs. In this article we will cover vertical GaN technology that has manifested itself in various flavors, like Current aperture vertical electron transistor (CAVET) and vertical Metal-Oxide-Semiconductor Field Effect transistor (V-MOSFET). We will also address role of TCAD models that were extremely valuable to predict device performance during our research.

[†] Corresponding author.

2. Advantage of Vertical GaN Devices Over Others

The two most critical questions typically asked about GaN vertical devices are 1) How well can it perform compared to SiC vertical devices 2) How practical is vertical GaN device technology compared to lateral GaN, particularly in the presence of SiC.

While the answer to the first question is elaborated throughout this article with simulated and experimental results, the answer to the second question stems from the same reason why vertical devices in Si were adopted over and above lateral MOSFETs in power electronics. High electric field close to GaN's theoretical limits are reported in vertical p-n diodes [1–9], leading to avalanche in many cases [1, 6–9]. In HEMT based lateral devices, such avalanching is not possible due to highly non-uniform electric fields set up by the field plates between the source (or gate) and drain. Therefore, a performance at an electric field close to the critical value in GaN (~3 MV/cm) is not possible in lateral devices. Moreover, vertical devices offer higher current density compared to lateral HEMTs. The increment in the current density in vertical compared to lateral devices becomes significantly more with higher voltage-class designs, - a feature which also stems from higher "volts-per-micron" typically obtained in a vertical design. These two factors set a limit to the power density that a lateral device can offer at its best, which is less than a vertical device. Higher voltage-per-micron is a key advantage of a vertical configuration that helps it avoid device level overdesigns to achieve a stable performance.

The answer to the first question relies on several factors such as performance, learning curve, cost, reliability and field performance. Particularly cost, reliability and field performance are time bound data that are not easy to predict at this early stage of development.

When it comes to performance, the free carrier mobility in the bulk of the crystal is an extremely important marker. In the past decade the improvement in GaN substrate technology has been compelling. The growth of GaN on single crystalline GaN has been rewarded with electron mobility of over 1100 $cm^2/Vsec$ [10]. The mobility of the channel region in a GaN-based material system can be always designed using AlGaN-GaN (as used in a CAVET) to have 10x or higher mobility than any MOS channel we know to date in Si, SiC or even GaN. The effect of mobility (bulk and channel) was best captured and conveyed using TCAD modeling that was developed using Silvaco's mixed mode platforms [11–12].

3. Current Aperture Vertical Electron Transistors

A CAVET embodies the merging of a lateral and a vertical device where the channel is formed using an AlGaN/GaN layer structure and vertical conduction takes place in the bulk (drift region) of the device. The role of mobility is reflected by the device's on resistance (R_{on}) which is discussed in the following section.

The schematic of the CAVET is shown in Fig. 1. Typically, a thick n-GaN with 10^{16} cm^{-3} or less forms the drift region on a conductive GaN (10^{18} cm^{-3}) substrate. In our previously reported work, we designed 0.4μm-thick p-type current blocking layers (CBLs)

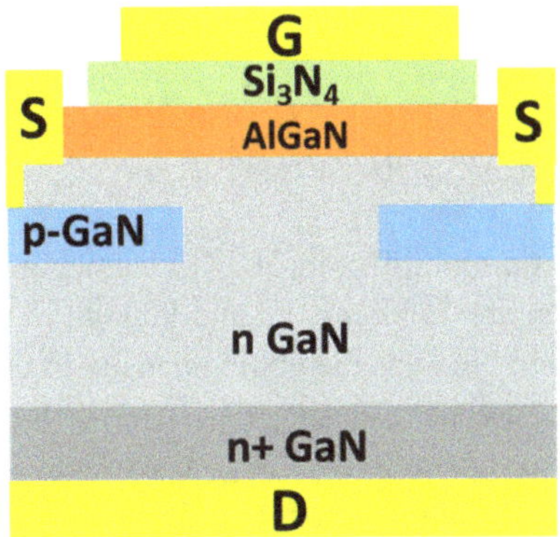

Fig. 1. Cross-section of an GaN CAVET.

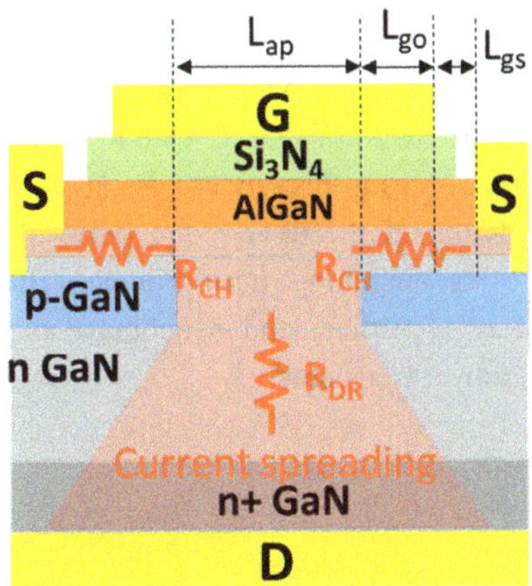

Fig. 2. On-state resistance distribution in the CAVET.

with a doping density of 3×10^{17} cm^{-3} to block current from flowing through any other path other than the designed aperture ($N_D = 3\times10^{17}$ cm^{-3}) [12]. AlGaN/GaN channel was defined on top of the CBL and the aperture to conduct electrons from the source through a high mobility 2D channel obtained at the hetero-interface. Sources contacting the 2D electron gas were designed on either side of the gate at a distance of 2 μm. The p-GaN CBLs were connected to the source to provide a discharge path for any electrons trapped in the p-type layer.

Considering the fact that the current spreading from the aperture into the drift region (shown in Fig. 2), the on-state resistance is also limited by the aperture region length. There are two major components that contribute to the device $R_{ON,TOT}$. The resistance offered by the AlGaN/GaN channel, R_{CH}, and drift region resistance, R_{DR}, can be written as

$$R_{CH} = \frac{L_{gs} + L_{go}}{q\mu_{2DEG}n_{2DEG}}\left(2\left(L_{gs} + L_{go}\right) + L_{ap}\right),$$

and

$$R_{DR} = \frac{T_{drift}\left(2\left(L_{gs}+L_{go}\right)+L_{ap}\right)}{q\mu_n N_{D,drift} 2\left(L_{gs}+L_{go}\right)} \ln\left(\frac{2\left(L_{gs}+L_{go}\right)+L_{ap}}{L_{ap}}\right),$$

here μ_{2DEG} and μ_n are electron mobilities in 2DEG channel and bulk GaN, the values of which used in the simulation are 1500 cm^2/Vs and 900 cm^2/Vs, n_{2DEG} is the 2D electron concentration the channel, and $N_{D,drift}$ is the doping density in the drift region.

Figure 3 shows the total on-state resistance ($R_{ON,TOT}$), drift region resistance (R_{DR}) and channel resistance (R_{CH}) versus aperture length L_{ap}. The lowest $R_{ON,TOT}$ is obtained between a L_{ap} of 4 μm and 10 μm, with a minimum of 1.5 mΩ•cm^2.

The trench gate CAVET structure is shown in Fig. 4. The electrons in the sidewall channel are depleted by the p-GaN base region, and the device has normally off behavior. The threshold voltage is dependent on the sidewall angle. From the simulation study, a > 1 V threshold voltage can be obtained for a straight sidewall (45 degrees).

Using the device-circuit-integrated model built in Silvaco's MixedMode platform [12, 13], the switching performance of the CAVET was analyzed. The simulation methodology is shown in Fig. 5. Using this device-circuit-integrated model, one can start with 2D drift-diffusion modeling of the device and build all the way up to its circuit implementation to evaluate its switching performance. The hybrid model gives an inexpensive and accurate way to project and benchmark the performance and can be extended to any GaN-based power transistors.

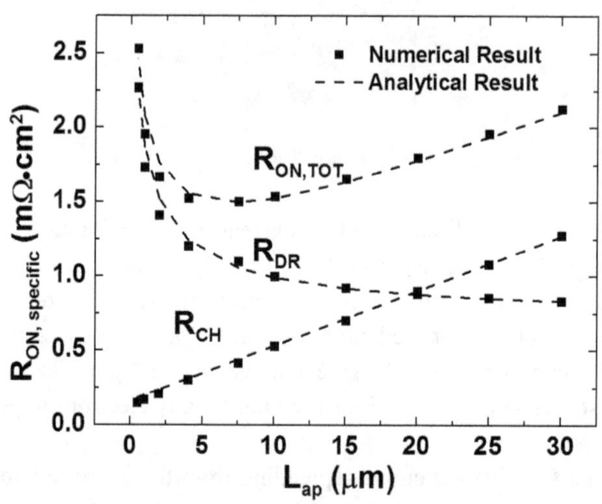

Fig. 3. $R_{ON,TOT}$, R_{DR} and R_{CH} as a function of L_{ap}. The solid squares indicate the results obtained numerically, while the dash curves show the analytical result (channel electron mobility: 1500 cm2/Vs; bulk GaN electron mobility: 900 cm2/Vs, L_{ap} and L_{go} were kept constant and only L_{ap} was varied).

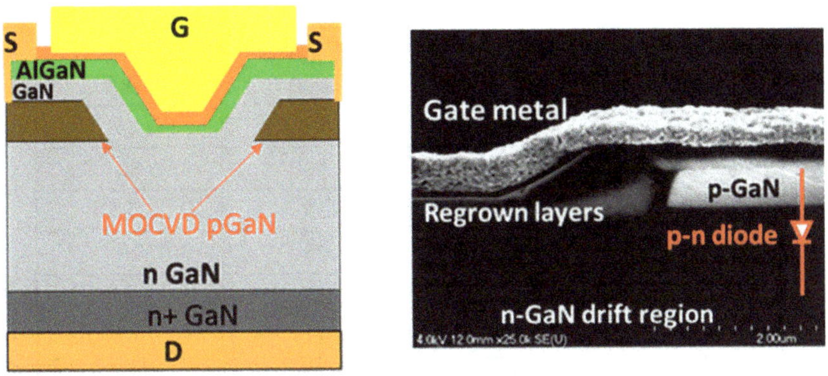

Fig. 4. Structure of a trench gate CAVET [13].

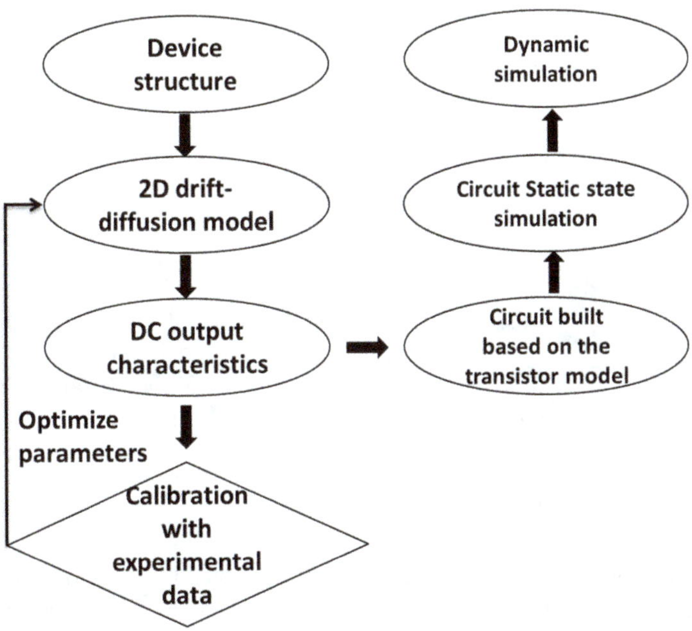

Fig. 5. Simulation methodology flowchart.

Since the CAVET is a normally on the device, a cascoded CAVET approach was adopted to achieve normally-off operation. The cascoded CAVET was subjected to switching operation and a detailed performance matrix was generated (shown in Table 1).

Figure 6 (a) and (b) show the turn-on and turn-off switching characteristics of a 1.2 kV cascoded CAVET, with a total on-state resistance R_{ON} of 80 mΩ. From the waveforms shown in Fig. 6, the turn-on delay time $T_{on\text{-}delay}$ is 2 ns, and rise time T_r is 16 ns, showing a turn-on Δv/Δt of 40 kV/μs. The total gate charge of the CAVET is 88 nC. The turn-off delay time $T_{off\text{-}delay}$ is 29.5 ns, and fall time T_f is 18 ns, giving a turn-off Δv/Δt = 35.6 kV/μs.

Table 1. Switching performance comparison of SiC MOSFET, and GaN CAVET.

Parameters	Cree SiC MOSFET	Simulated SiC MOSFET	Simulated GaN CAVET
V_{BR} (kV)	1.2	1.2	1.3
R_{on} (Ω)	80	80	80
Q_G (nC)	40	55	29.5
$t_{on\text{-}delay}$ (ns)	38	25	18
t_f (ns)	13	10	2
$t_{off\text{-}delay}$ (ns)	24	27	16
t_r (ns)	90.8	157	88
E_{on} (μJ)	305	312	57
E_{off} (μJ)	305	304	152
E_{ts} (μJ)	610	616	209

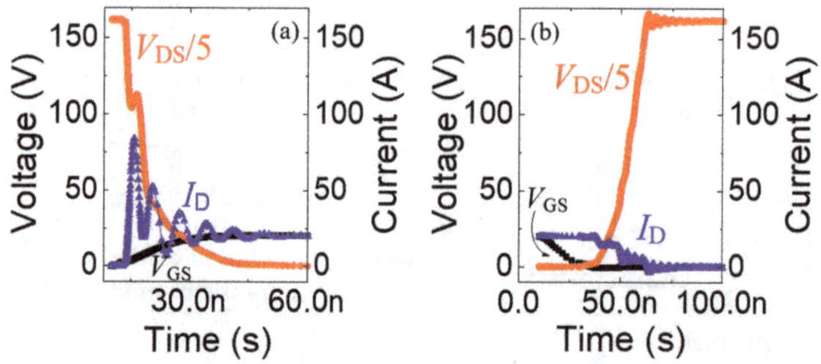

Fig. 6. (a) Simulated turn-on switching waveforms of the cascoded CAVET; (b) Simulated turn-off switching waveforms of the cascoded CAVET [12].

Table 1 summaries the performance matrix of simulated GaN cascode CAVET, Cree SiC MOSFET and simulated SiC MOSFET with the same voltage and current rating. The switching performance agreement between CREE and simulated SiC MOSFETs provides the competency of the model. When compared to standard SiC MOSFET, cascoded CAVET has faster switching time, and lower switching loss, hence enabling higher frequency operation.

The high mobility in the channel (1500 cm^2/Vs) and in the drift-region (900−1100 cm^2/Vs) led to such superior performance in CAVETs.

For higher voltage design the mobility of the drift region becomes a bigger differentiator. As shown in the following plot (Fig. 7), R_{on} of an 8 kV GaN device offers lower conduction losses than SiC if higher mobility (> 1100 cm^2/Vs) can be maintained in the drift region throughout. The percentage -improvement in performance due to mobility increases with the voltage range as evident by the simulated results in Fig. 7.

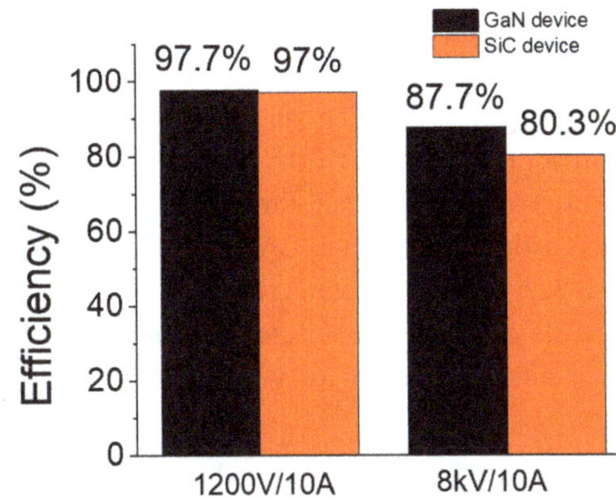

Fig. 7. Comparison of efficiencies in power converters using GaN and SiC MOSFETs as power switches.

The first generation CAVETs [14] depended on implanted Mg-based current blocking layers. The performance was limited by the efficacy of activation of the implanted Mg, ultimately limiting the overall blocking capability.

The trench-CAVET looked promising with the growth-based approach, where Mg is introduced by doping [11]. This technique depended on regrowth on trenches that comes with some other issues but can be improved with more research efforts. Some of our results discussed here gives the necessary evidence on the development and performance of Trench-CAVETs.

Figure 8 (a) shows the I_D-V_{DS} characteristics of the fabricated trench CAVET. The calculated $R_{on,sp}$ was 2.7 mΩ·cm^2. The double-swept I_D-V_{GS} transfer characteristics of the fabricated trench CAVET with a is shown in Fig. 8 (b). The threshold voltage was observed

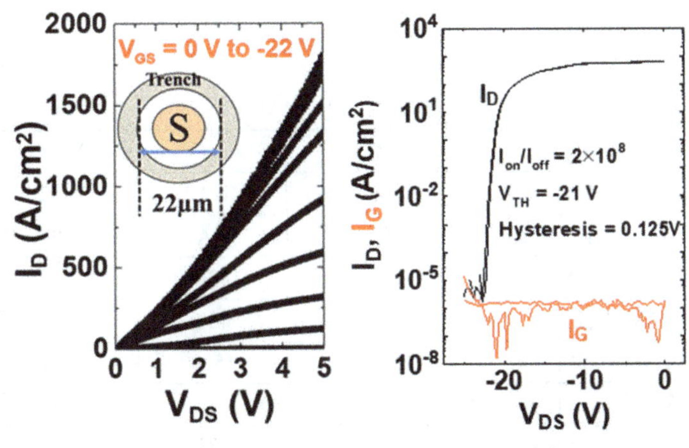

Fig. 8. (a) I_D-V_{DS} characteristics of the fabricated trench CAVET; (b) Double-swept I_D-V_{GS} transfer characteristics [15].

to be –21 V. Trench CAVET can be designed to provide normally off solution but the threshold voltage heavily relies on the accuracy of the angle of the sidewalls. The subthreshold leakage was as low as in the order of 10^{-6} A/cm^2 and a high I_{on}/I_{off} value of 2×10^8 was observed. In addition, due to the high quality of the in-situ grown Si_3N_4, the hysteresis of the device was ~ 0.1 V. The sub-threshold swing was as low as 180 mV/dec.

The off-state characteristics of the trench CAVETs with diffeent gate dielectric thicknesses are shown in Fig. 9. A breakdown voltage of 220 V was obtained in the fabricated trench CAVET with a gate dielectric thickness of 30 nm. While, a higher breakdown voltage of 880 V was obtained in the device with a gate dielectric thickness of 60 nm.

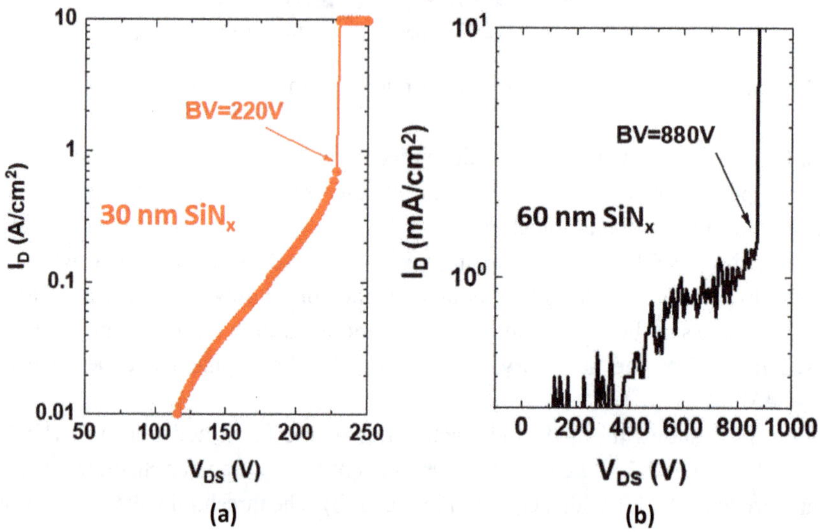

Fig. 9. (a) Breakdown characteristics of the fabricated trench CAVET with 30 nm gate dielectric; and (b) the breakdown characteristics of the trench CAVET with 60 nm gate dielectric [11, 15].

Table 2. Reported Vertical GaN Transistors with the AlGaN/GaN channel.

Device	Ron,sp (mΩ·cm^2)	V_{BR} (V)	Year
CAVET [17]	13Ω-mm	NA	2002
CAVET [14]	1.5	NA	2008
CAVET [18]	2.2	300	2012
CAVET [19]	NA	450	2018
VHFET [20]	2.2	1500	2014
VHFET [21]	7.6	672	2010
VHFET [16]	1.0	1700	2016
T-CAVET [13]	NA	225	2016
T-CAVET [15]	2.7	880	2018

Optimized designs of the trench etching and gate dielectric can improve the breakdown voltage further.

Table 2 summaries the reported AlGaN/GaN-based vertical GaN transistors. Excellent performance with a high breakdown voltage of 1700 V and a low $R_{on,sp}$ of 1 mΩ·cm^2 has been achieved simultaneously in the same AlGaN/GaN vertical device [16], which exceeded the material limit of SiC.

4. GaN MOSFETs

Although CAVETs bring the best of lateral and vertical designs in GaN by utilizing the best mobilities in channel and drift, MOSFETs are important to power electronics in multiple ways.

GaN MOSFETs are therefore other branch of devices that are being developed for superior performance. The MOSFETs are attractive particularly for offering a normally off solution which is a significant drawback of any GaN HEMT-based design, including CAVETs.

There are three types of vertical MOSFET in GaN reported so far 1) Nonregrowth-based trench MOSFET [22–26]; 2) Regrowth based trench MOSFET [27–32]; and 3) fin MOSFET [33–35]. In this paper, a regrowth-based trench MOSFET called the OGFET will be mainly discussed.

Figure 10 (a) shows the structure of a vertical GaN UMOSFET. A main *p-n* junction between source and drain is formed by *p*-base and *n*-drift regions. The device breakdown voltage is determined by the reverse characteristics of the main *p-n* junction. A *n*+ source region is formed partially on top of the *p*-base region, the junction between the *n*+ source region and the *p*-base region is shorted by source contact to improve the breakdown voltage by eliminating the *n-p-n* open base effect. The channel locates at the etched sidewall, formed by the inversion layer of the MOS structure.

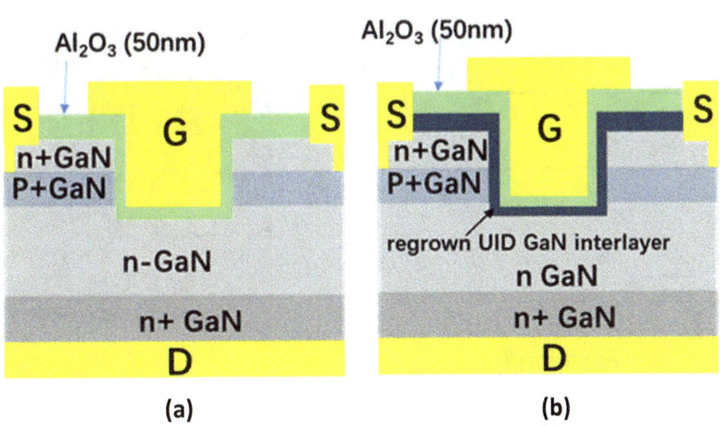

Fig. 10. Schematics of the (a) GaN UMOSFET, and (b) the OGFET.

Compared to the CAVET structure, there are two basic advantages of the UMOSFET: 1) the UMOSFET is a reliable normally off device with a high threshold voltage over 2 V; 2) the absence of the regrowth makes the process less challenging, reducing the cost. These advantages of the UMOSFET makes it an attractive design for vertical GaN transistors. However, for GaN UMOSFET, the biggest challenge lies in the channel electron mobility of the device. During on-state, the electrons flow through the inversion layer of the sidewall MOS structure, the channel electron mobility is limited by the surface roughness and impurity scattering. Another issue along with the poor channel property is the device reliability. The GaN UMOSFET cannot be widely recognized without a strong reliability.

GaN OGFET is a modified structure based on the conventional UMOSFET, as shown in Fig. 10 (b). Compared to the conventional UMOSFET, the OGFET has two features: (1) an unintentional doped (UID) GaN interlayer is used as the channel region, which enhance the channel electron mobility to reduce the Coulomb scattering by the dopants; (2) the oxide is in-situ grown by MOCVD, which reduces the interface states, and improves the gate oxide reliability. The novelty of the OGFET is enhancing the channel electron mobility without scarifying the normally off behavior.

The working principle of the OGFET is similar to the UMOSFET. Under a zero V_{GS}, the electrons in the GaN insert layer (shaded region in Fig. 11) is depleted by the p-GaN base region. the OGFET is in the off-state. The energy band diagram and the simulated electron contour are shown in Fig. 11. The p-n diode formed by the p-GaN base region and the n-GaN drift region is used to hold the high off-state blocking voltage. The electric field distribution along the p-GaN base region and the n-GaN drift region is shown in Fig. 12.

Under a positive V_{GS} (15 V), electrons are accumulated in the UID GaN insert layer, and the transistor is in its on-state. The energy band diagram and the electron concentration are shown in Fig. 13. Because of the enhanced channel electron mobility, the OGFET has smaller $R_{on,sp}$ compared to the conventional UMOSFET.

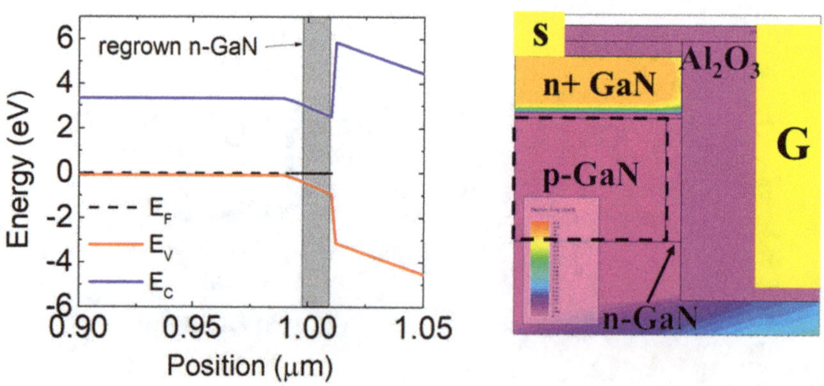

Fig. 11. Energy band diagram and the electron distribution of the OGFET in off-state.

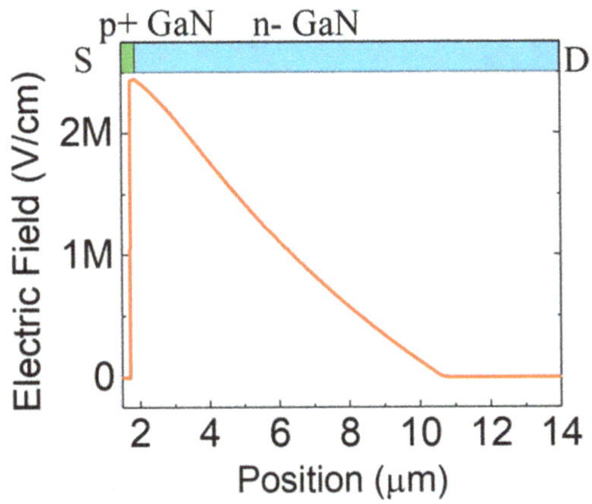

Fig. 12. Electric field distribution along the drift region.

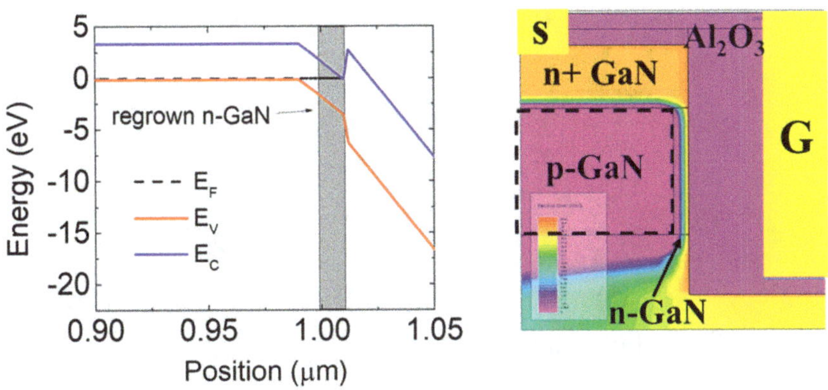

Fig. 13. Energy band diagram and the electron distribution of the OGFET in on-state.

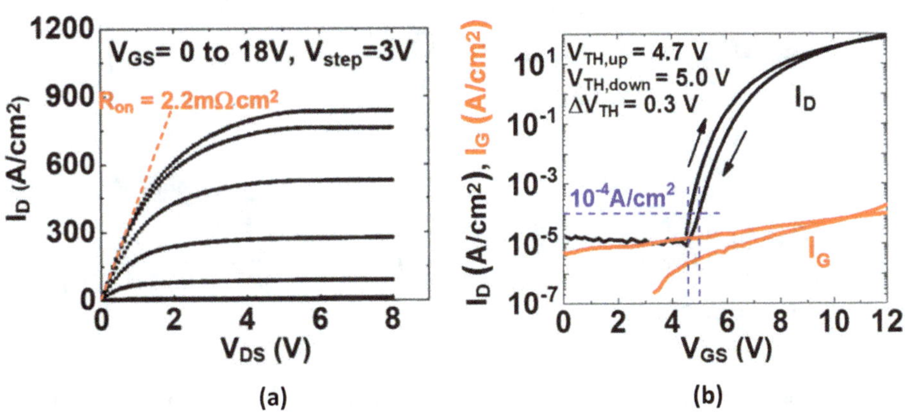

Fig. 14. I-V and transfer characteristics of the fabricated OGFET [30].

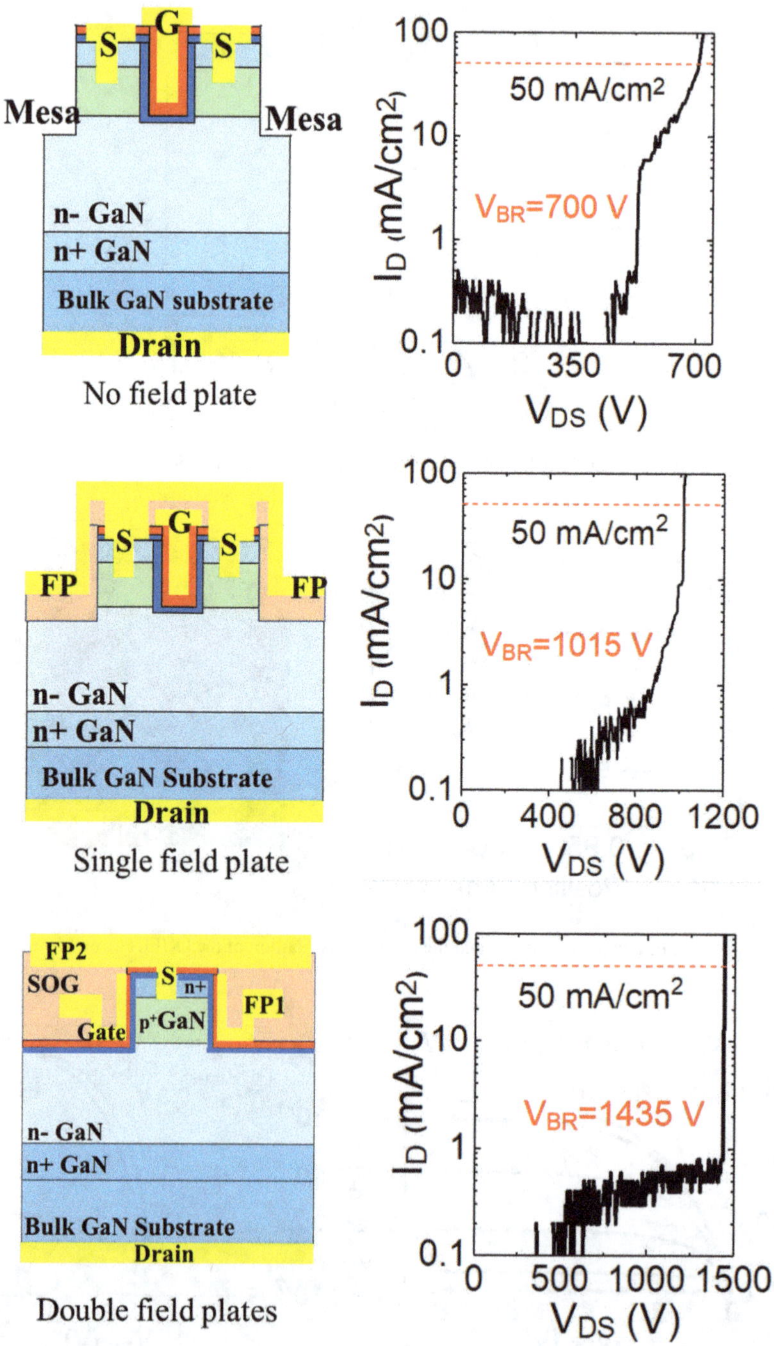

Fig. 15. Off-state characteristics of the fabricated OGFETs with no field plate (mesa etch only), single field plate, and double field plates. With the help of the double-field-plate structure, a 1.4 kV breakdown voltage was achieved [30, 31].

Some of our results discussed here gives the necessary evidence on the development and performance of OGFETs.

Figure 14 shows the I-V and transfer characteristics of a fabricated OGFET. The saturated current density was 850 A/cm^2 and $R_{ON,SP}$ was 2.2 mΩ·cm^2, which demonstrated an excellent on-state performance. The threshold voltage, V_{TH}, defined at a current level of 10^{-4} A/cm^2 (I_{on} / I_{off} = 10^6), obtained was 4.7 V (when V_{GS} sweeps up). A clockwise hysteresis of ΔV_{TH} of 0.3 V was observed. Low sub-threshold slope of 283 mV/decade was measured from $I_D = 10^{-5}$ A/cm^2 to 10^{-2} A/cm^2.

Figure 15 shows the off-state characteristics of the fabricated OGFETs with different edge termination designs: mesa etching, the single field plate, and double field plates. The breakdown voltage can be improved from 700 V to 1435 V by optimize the edge termination design.

Table 3 summaries the reported MOS channel based vertical GaN transistors. To date, excellent performances have been achieved in GaN vertical MOSFETs with high blocking voltage larger than 1200 V and low $R_{on,sp}$ smaller than 3 mΩ·cm^2. Recently, the demonstration of 100 A device indicated a significant step toward commercialization [36].

Table 3. Reported Vertical GaN Transistors with the MOS gate structure.

Device	Channel mobility (cm^2/VS)	$R_{on,sp}$ (mΩ·cm^2)	BV (V)	V_{TH} (V)	Year
UMOS [22]	120	33Ω-mm	NA	5.1	2007
UMOS [23]	131	9.3	NA	3.7	2008
UMOS [24]	NA	12.1	1600	7	2014
UMOS [25]	NA	1.8	1200	3.5	2015
UMOS [26]	NA	8.5	600	4.8	2016
OGFET [27]	NA	3.8	190	2	2016
OGFET [28]	NA	2.6	990	3	2017
OGFET [29]	NA	0.98	700	1	2017
OGFET [30]	185	2.2	1435	4.7	2017
V-MOS [32]	NA	300	600	18	2018
Fin-MOS [33]	10	200	NA	4.5	2015
Fin-MOS [34]	150	0.36	800	1	2017
Fin-MOS [35]	NA	0.2	1200	1	2017

5. Conclusion

In this article we discussed the progress made in vertical GaN device technology through two key examples – CAVETs and OGFETs (a variant of MOSFET). These two devices have so far demonstrated encouraging performance in two different voltage ranges. While

CAVETs are probably best suited for voltages up to 900 V or so, OGFETs definitely push the limits to 1.2 kV and up. One key achievement that has been discussed here is the ability to achieve a channel mobility in GaN OGFET up to 185 cm^2/Vs, using a regrowth-based technology. Higher channel mobility is also a key strength in a CAVET. The article describes the role of mobility throughout and how it impacts the switching loss through converter examples.

Acknowledgment

We would like to acknowledge the funding support by ARPA-E, ONR and NSF for accomplishing various stages of this work. We appreciatively extend our thanks to our co-authors of related work cited here, particularly, Prof. Mishra and Drs. S. Mandal, C. Gupta and S. Keller.

References

[1] I. C. Kizilyalli et al., IEEE Transactions on Electron Devices, 62, 414, 2015.
[2] J. Dickerson et al., IEEE Transactions on Electron Devices, 63, 419, 2016.
[3] H. Ohta et al., IEEE Electron Device Letters, 36, 1180, 2015.
[4] Z. Hu et al., Appl. Phys. Lett., 107, 243501, 2015.
[5] J. Wang et al., Appl. Phys. Lett., 113, 023502, 2018.
[6] K. Nomoto et al., IEEE Electron Device Letters, 37, 161, 2015.
[7] T. Maeda et al., IEEE International Electron Devices Meeting, 689, 2018.
[8] D. Ji et al., International Conference of Nitride Semiconductors, Seattle, July 10, 2019.
[9] H. Fukushima et al., Appl. Phys. Exp. 12, 026502, 2019.
[10] P. Kruszewski, et al., Int. Workshop Nitride Semiconductor, 2014.
[11] D. Ji et al., IEEE Workshop on Wide Bandgap Power Devices and Applications, 174-179, 2015.
[12] D. Ji et al., IEEE Transactions on Electron Devices, 63, 10, 4011-4017, 2016.
[13] D. Ji et al., IEEE Transactions on Electron Devices, 64, 805-808, 2016.
[14] S. Chowdhury et al., IEEE Electron Device Letters, vol. 29, no. 6, pp. 543-545, 2008.
[15] D. Ji et al., IEEE Electron Device Letters, vol. 39, no. 6, pp. 863-865, 2018.
[16] D. Shibata et al., IEEE International Electron Devices Meeting, pp. 248-251, 2016.
[17] I. Ben-Yaacov et al., J. Appl. Phys. 95, 4, 2073, 2004.
[18] S. Chowdhury et al., IEEE Electron Device Letters, vol. 33, no. 1, 41-43, 2012.
[19] S. Mandal et al., IEEE Electron Device Letters, vol. 38, no. 7, 933-936, 2017.
[20] H. Nie et al., IEEE Electron Device Letters, vol. 35, no. 9, pp. 939-941, 2014.
[21] M. Okada et al., Appl. Phys. Exp. 3, 054201, 2010.
[22] H. Otake, et al., J. J. Appl. Phys., vol. 46, no. 25, pp. L599-L601, 2007.
[23] H. Otake, et al., Appl. Phys. Ex., 1, 011105, 2008.
[24] T. Oka, et al, Appl. Phys. Ex., 7, 021002, 2014.
[25] T. Oka, et al, Appl. Phys. Ex., 8, 05401, 2015.
[26] R. Li et al. IEEE Electron Device Letters, 37, 11, 1466-1469, 2016.
[27] C. Gupta et al., IEEE Electron Device Letters, vol. 37, no. 12, 1601-1604, 2016.
[28] C. Gupta et al., IEEE Electron Device Letters, 38, 3, 353-355, 2017.
[29] D. Ji, Device Research Conference, p. 1-2, 2017.

[30] D. Ji et al., IEEE International Electron Devices Meeting, pp. 223-226, 2017.

[31] D. Ji, “Design and development of GaN-based vertical transistors for increased power density in power electronics applications,” Doctoral dissertation (University of California Davis, 2017).

[32] W. Li et al., IEEE Transactions on Electron Devices, vol. 65, no. 6, pp. 2558-2564, 2018.

[33] W. Li, Physics status solidi (a), vol. 213, no. 10, 2714-2720, 2016.

[34] M. Sun et al., IEEE Electron Device Letters, vol. 38, no. 4, 509-512, 2017.

[35] Y. Zhang, IEEE International Electron Devices Meeting, pp. 215-218, 2017.

[36] T. Oka, T. Ina, Y. Ueno, J. Nishii, in *Proc. IEEE 31st International Symposium on Power Semiconductor Devices and ICs (ISPSD)*, pp. 303-306, 2019.

Modeling and Simulation of Quasi-Ballistic III-Nitride Transistors for RF and Digital Applications

Kexin Li* and Shaloo Rakheja†

Holonyak Micro and Nanotechnology Laboratory,
Department of Electrical and Computer Engineering,
University of Illinois at Urbana-Champaign, Champaign, IL 61820, USA
* *kexinli4@illinois.edu*
† *Rakheja@illinois.edu*

This paper presents a self-consistent analytic model to describe the current-voltage (I-V) and charge-voltage (Q-V) behavior of quasi-ballistic III-nitride transistors. We focus on two types of transistor geometries: (i) high electron mobility transistors (HEMTs) suitable for radio frequency (RF) applications and (ii) nanowire field-effect transistors (FETs) for digital applications. Our core model is based on Landauer transport theory which is combined with the calculation of charge density and velocity of charges at the top-of-the-barrier in the transistor. The effect of extrinsic device features, such as the nonlinearity of access regions and Joule heating at high currents, are included in the static I-V model. In the case of the dynamic Q-V model, we calculate intrinsic terminal charges by approximating the solution of the 2D Poisson equation in the channel over a broad bias range. The effect of fringing capacitances, prominently inner-fringing capacitance that varies nonlinearly with the gate bias, is included in our Q-V model. We amend the model electrostatics and the description of source/drain rectifying contacts in our core model to represent the I-V characteristics of III-nitride nanowire FETs. The model shows excellent match against experimentally and numerically measured characteristics of GaN transistors with gate lengths ranging from 42 nm to 274 nm. With only 38 input parameters, most of which are extracted based on straightforward device characterization, our model can be used for device-circuit co-design and optimization using a standard hierarchical circuit simulator.

Keywords: Gallium nitride; compact modeling; surface potential; ballistic transport; self-consistent charge-current.

1. Introduction

The high electron velocity in III-nitride semiconductors ($\sim 2 \times 10^7$ cm/s) combined with the large material breakdown voltage (3 MV/cm) are the key enablers for radio frequency (RF) electronics including high-speed power amplifiers, ultra-linear mixers [1], and high-output power digital-to-analog converters [2]. Unlike the current transistor technologies in silicon and GaAs which can provide high speed only with unacceptably low voltage swings, GaN can achieve a Johnson Figure of Merit (JFoM) $\sim$10 THz·V [3], thus, lifting the technology constraints for developing crucial commercial and military electronics.

Due to their spontaneous and piezoelectric polarization effects, III-nitride high electron mobility transistors (HEMTs) feature a high density of the two-dimensional electron gas (2DEG). Experiments have reported values of the 2DEG density in excess of 10^{13} cm^{-2} at room temperature in III-nitride HEMTs [24]. At the same time, the intrinsic mobility of the 2DEG in III-nitride HEMTs is $\sim$(1000-2000) cm^2/Vs [25]. This combination of high 2DEG density, high carrier mobility, and a large breakdown field enable III-nitride HEMTs the ability to outperform Si and GaAs HEMTs for high frequency high power applications [26, 27]. Over the last decade, there has been a significant improvement in the performance metrics of III-nitride-based RF HEMTs. As shown in Fig. 1, the operation frequency of power amplifiers entered the W-band, delivering an output power of 6.7 W/mm at a power added efficiency (PAE) of 15% [11, 28, 29]. With appropriate device scaling, the III-nitride-based power amplifiers can achieve f_T ($f_{\max}$) of 400 (582) GHz [4, 30]. Here, f_T and $f_{\max}$ denote the cut-off frequency and maximum oscillation frequency, respectively.

Given that the optical phonon energy of $\approx$ 92 meV in III-nitride materials is 4$\times$ larger than that in III-V and group-IV semiconductors, room temperature (RT) ballistic transport of electrons is feasible in III-nitride materials [31]. Indeed, such ballistic transport has been experimentally observed at 300 K in nanoscale Gallium Nitride (GaN) crosses with widths ranging from 315 nm down to 33 nm [32]. The phonon-dominated scattering rate in other semiconducting materials, including InSb/AlInSb, InGaAs/InP, and Si/SiGe, is significant since their optical phonon energy is $\approx$ (20 − 35) meV, which is comparable to the thermal energy at RT. The RT thermal energy k_BT = 26 meV, where k_B is Boltzmann constant and T is the

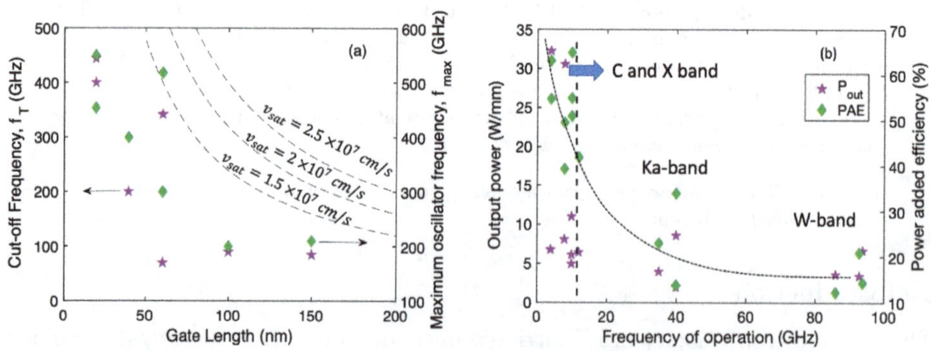

Fig. 1. Improvement of (a) cut-off frequency and maximum oscillation frequency [4–10]. Here, v_{sat} indicates the saturation velocity of electrons in the III-nitride channel material. (b) Output power (W/mm) and power added efficiency, of power amplifier design based on GaN HEMTs technology [11–23].

lattice temperature. In these materials, electrons are typically scattered due to the crystal lattice potential via the emission of optical phonons.

More recently, III-V materials have been investigated for digital computing applications where their steep sub-threshold behavior can lead to lower leakage, and therefore, low-power operation [34]. Measured maximum transconductance, g_m, versus subthreshold swing, SS, for various semiconducting materials is shown in Fig. 2. Compared to Si, Ge, and InAs devices, GaN nanowire (NW) field-effect transistors (FETs) exhibit a higher value of the ratio g_m/SS (denoted as Q), clearly showing the superior potential of III-nitride devices for digital logic applications. For the same geometry and dimensions, GaN NW FETs offer 17% higher Q compared to their silicon counterparts. Moreover, GaN NW FETs with Al_2O_3 gate oxide have a near-ideal SS of 68 mV/dec [35] and a high on-off current ratio of 10^6 [33], which is desirable for low-power and fast digital switching.

For the design, simulation, and optimization of RF circuits using III-nitride technology, it is imperative to develop a compact model of III-nitride HEMTs with capability to self-consistently handle the statics and dynamics of carrier transport in the device. In this work, we present the physics of operation of III-nitride HEMTs which lays the foundation for our compact model that is valid for ultra-scaled quasi-ballistic (QB) III-nitride HEMTs. The model comprehends the physics pertinent to III-nitride HEMTs, such as polarization-induced channel charge, nonlinear access region resistance, self-heating, and cutting off of negative momenta states from the drain contact under off-equilibrium conditions. The core of the model is based on the Landauer transport theory for QB channels [36], which is combined with the Ward-Dutton charge partitioning scheme to calculate the nodal charges and inter-nodal capacitances. The model respects charge continuity and can faithfully reproduce the current-voltage (I-V) and capacitance-voltage (C-V) data of devices with gate lengths ranging from 42 nm to 105 nm. We amend the model to capture

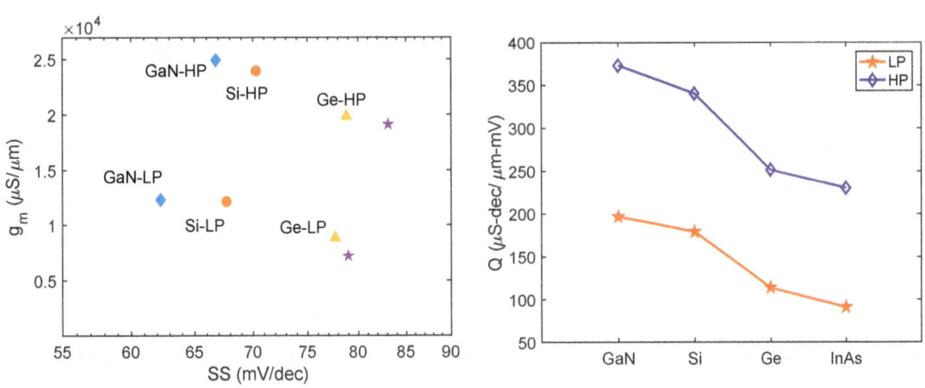

Fig. 2. (a) g_m versus SS of various nanowire FETs for low standy power (LP) and high standby power (HP) operation. (b) $Q = g_m/SS$ of various semiconducting technologies using values reported in the left figure [33].

the behavior of III-nitride NW FETs, which are currently being investigated for digital switching applications.

Several analytic I-V models of III-nitride HEMTs have been reported in the literature [37–42]. Most previous models are empirical in nature and focus only on the drift-diffusive (DD) transport, which is relevant for long-channel devices For example, the model presented in Ref. 37 uses smoothing functions to stitch together the piece-wise continuous I-V model of AlGaN-GaN HEMTs. As such, the accuracy and mathematical robustness of the model are expected to be poor. A compact model based on a polynomial expansion to determine the Fermi level as a function of the 2DEG density is reported in Refs. 38, 39. The use of polynomial expansion not only increases the empiricism in the model but also requires complex integration of the 2DEG density with the channel potential to obtain the I-V relationship. A surface-potential-based compact model relying on the DD transport theory was recently proposed in Ref. 41. A threshold-voltage based I-V model that is mathematically more robust is reported in Ref. 42. The drawback of the threshold-voltage-based model is the empirical formulation of the 2DEG density that ignores the effect of voltage-dependent density of states and quantum-mechanical capacitance of the inversion layer on the overall gate capacitance. The explicit use of the gradual-channel approximation (GCA) throughout the channel in Ref. 40 limits the model use to GaN HEMTs in which the channel length is significantly larger than the electron mean free path (MFP).

There are relatively fewer models available for the dynamic charge-voltage (Q-V) model in III-nitride HEMTs. The Angelov model is empirical in nature and contains a large number of fitting parameters [43]. Moreover, in the Angelov model, internodal capacitances are not obtained from terminal charges, which could potentially lead to a violation of the charge conservation law in the device. The Advance SPICE model for GaN (ASM-GaN) HEMTs calculates the channel charge from position-dependent surface potential using the GCA [44]. The validity of GCA becomes questionable in short-channel transistors with QB transport, for which the the GCA is typically only valid near the virtual source (VS) point. The model in Ref. 42 cannot comprehend quantum effects present in scaled HEMTs and is, therefore, inapplicable to short channel devices considered in this work. Additionally, there is no compact I-V model presently available to simulate the behavior of digital III-nitride NW FETs.

This paper is organized as follow. In the Section 2, the design criteria of III-nitride devices for both RF and digital applications is explained. In Section 3, we present an analytic device model to capture the static and dynamic behavior of III-nitride HEMTs over a broad bias and temperature range. Parameter extraction and verification against experimental data are discussed in Section 4. We summarize the key findings of our work in Section 5.

2. Design of III-Nitride Transistors

2.1. *RF device design*

Device metrics that are relevant for high-frequency and high-power applications include breakdown voltage, maximum current carrying capability, and cut-off frequency. In the case of RF power amplifiers, metrics such as efficiency, RF linearity, output power, PAE, and signal noise are crucial and must be optimized at the materials and device design level.

Even in the absence of external doping, GaN-based heterostructures provide high polarization-induced carrier concentration and high carrier velocities [45, 46]. Taking $Al_xGa_{1-x}N/GaN$ heterojunction as an example, the spontaneous (P_{sp}) and piezoelectric (P_{pz}) polarizations result in a 2DEG with a high carrier concentration on the GaN side of the heterojunction. The 2DEG formed due to polarization effects at the heterointerface is shown in Fig. 3. The spontaneous polarization results from an intrinsic asymmetry of the bonding in the equilibrium wurtzite crystal structure of the III-nitride material. The value of P_{sp} is negative and increases from GaN to InN to AlN, as the crystal structure becomes more non-ideal [47]. On the other hand, piezoelectric polarization results from the electromechanical interaction between the mechanical and the electrical state in crystalline materials without any inversion symmetry. Piezoelectricity is a reversible process in that the materials exhbiting direct piezoelectric effect (generation of charges due to mechanical strain) also have inverse piezoelectric effect (generation of strain due to electric field). The sign of P_{pz} in a material depends on whether the material is under the effects of compressive or tensile strain during its growth on a substrate. For AlGaN grown on GaN, experi-

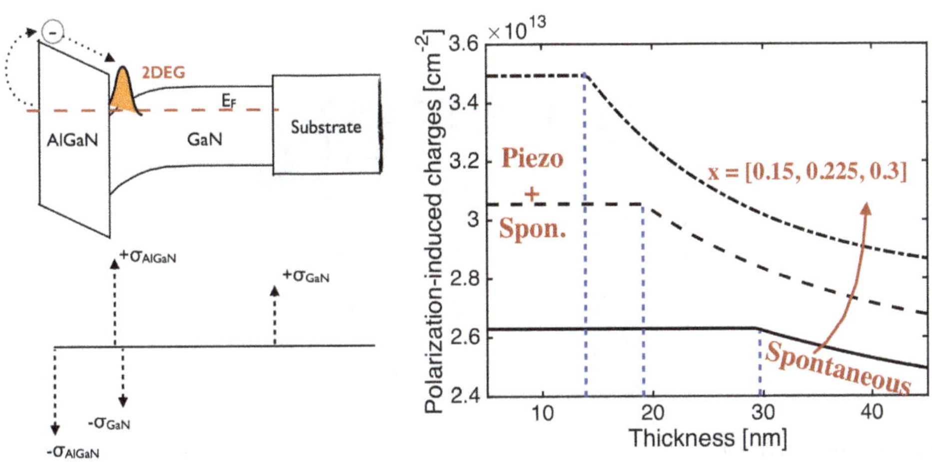

Fig. 3. (left) Polarization-induced charge in a typical AlGaN/GaN heterostructure. Here, E_f denotes the Fermi level from the gate to the substrate. (right) Effect of alloy composition and AlGaN barrier thickness on polarization-induced charges in GaN. The vertical dashed lines correspond to the critical thickness of AlGaN for strain relaxation.

ments have shown that the AlGaN layer is always under tensile strain, which makes the sign of P_{pz} negative–similar to P_{sp} [48]. Due to their similar signs, the piezoelectric and spontaneous polarization fields can be simply added in the AlGaN/GaN heterostructure. While spontaneous polarization is not affected by the thickness of the barrier (AlGaN) layer, piezoelectric polarization is strongly dependent on the AlGaN layer thickness and strain relaxation. The total polarization charge induced at the AlGaN/GaN interface is plotted in Fig. 3 for various alloy compositions and thickness of the AlGaN barrier. For a fixed thickness of the AlGaN layer, the 2DEG increases from $\approx 6 \times 10^{12}$ cm^{-2} to 2×10^{13} cm^{-2} as the Al composition x increases from 0.15 to 0.31 [46]. Beyond a certain critical layer thickness, the strain energy stored in the AlGaN layer leads to a rapid formation of dislocations that relaxes the strain in the layer. The phenomena of strain relaxation leads to plastic deformations and surface irregularities. As the Al composition in AlGaN increases, the critical layer thickness for strain relaxation reduces, and eventually the structure may be rendered useless for device applications [49].

While large polarization effects are desirable to improve the 2DEG density, they also lead to lattice mismatch issues, making the fabrication of high-quality and stress-free GaN-based epilayers challenging. Alternative materials and novel device structures, such as lattice-matched InAlN/GaN, have been proposed to optimize the HEMT performance. The InAlN/GaN heterostructure provides charge densities in excess of 2.8×10^{13} cm^{-2} [24, 50].

AlGaN/AlN/GaN HEMTs with insertion of the very thin AlN interfacial layer ($\sim$1 nm) maintains high mobility at high sheet charge densities by increasing the effective conduction band offset and decreasing alloy scattering [51]. InGaN back barrier can be added to enhance the confinement of the 2DEG, which has shown to result in $\approx$ 50% increase in output resistance in some reports [52]. AlGaN/GaN

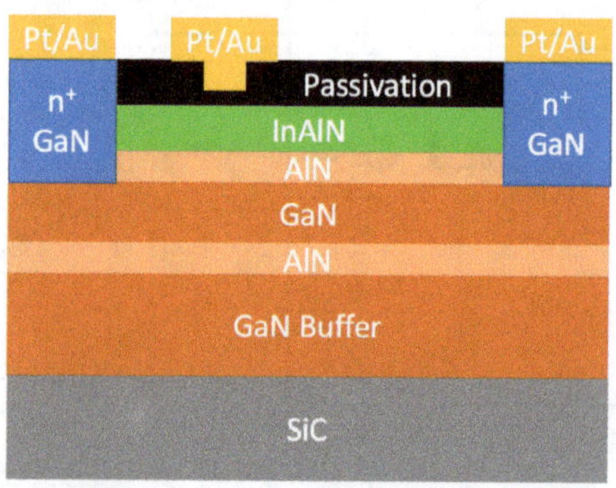

Fig. 4. Cross-section of a III-nitride HEMT for superior RF performance. Device innovations include (i) using thin AlGaN barrier with high Al content to improve the 2DEG density in the channel, (ii) thinning GaN channel for superior electrostatics, (iii) passivating the surface (SiN) to remove surface traps, (iv) addition of n+ capping layer at the source and drain contacts to reduce access region resistance, and (v) introduction of high bandgap AlN barrier above and below the channel.

double-channel HEMTs based on a multilayer structure exhibit high current drive, low buffer leakage, and fast frequency response [53]. Enhancement-mode GaN transistors have also been implemented using dual-gate AlGaN/GaN designs [54]. Other innovative designs that can help improve RF linearity include (i) implementing n-doped GaN capping layer above the AlGaN barrier [55]; (ii) moderate GaN thickness ($\sim$20 nm) to confine the 2DEG for improved electrostatics; (iii) using field plates for electric field modulation in the channel [56]; (iv) modifying the design of the access regions to lower their resistance [57]; and (v) inserting an AlGaN layer with low-Al composition in the channel region [58]. Additionally, passivation layers can be added to the surface of GaN HEMTs to reduce surface traps and overcome current collapse issues [59, 60]. Figure 4 shows the cross-section of a III-nitride HEMT with various design innovations to yield superior RF performance.

2.2. *Digital device design*

Due to their higher on-off current ratio and lower standby power dissipation, NW FETs shown in Fig. 5 are suited for future wireless internet-of-things applications [61] that place stringent demands on both performance and energy efficiency. While Si and C-based NW FETs with excellent on-off performance have been demonstrated [62, 63], gallium nitride (GaN) is an alternative channel material for NW FETs that offers some unique features. The wide band gap of GaN can suppress the device off-current, while the high electron mobility and comparatively large density of states mass enables high current density in on state to achieve high switching speed. Recently, GaN NW FETs based on selectively-grown GaN NWs exhibiting both appreciable g_m desirable for high-frequency operations as well as near-ideal SS for low power have been fabricated [35].

Some outstanding challenges of GaN digital technology include difficulties in fabricating (i) fast p-type GaN FETs, (ii) enhancement-mode FETs, and (iii) temperature stable, low-resistance ohmic source and drain contacts. Most GaN-based FETs currently designed operate as depletion-mode devices. That is, they are normally-on and a large negative gate voltage is required to turn them off. However, to lower device leakage current and improve its reliability, enhancement-mode devices will be needed to meet the performance targets of digital applications.

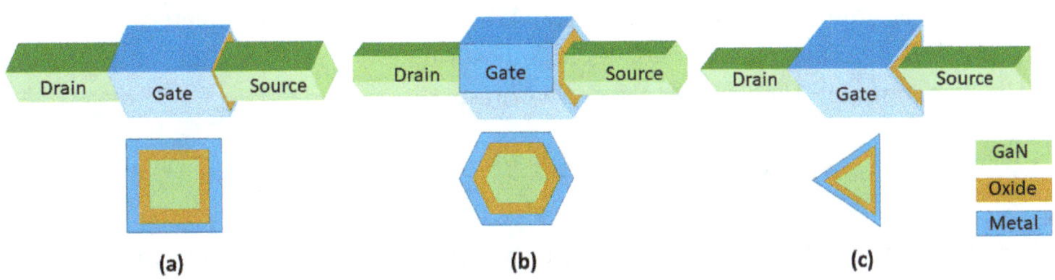

Fig. 5. Nanowire GaN FETs with various types of cross-sections suitable for digital applications.

3. Analytic Device Modeling

In this section, we present an analytic model to describe the unified current-voltage (static) and capacitance-voltage (dynamic) behaviour of scaled III-nitride HEMTs in which transport within the channel is considered QB in nature. Our model combines the Landauer transport theory with carrier injection at the top-of-the-barrier (ToB) or the VS point in the channel to obtain the I-V relationship. The ToB is defined as the point in the channel at which the conduction band profile achieves its peak. When carrier concentration is low in the channel, the ToB is located in the middle of the channel, but it gradually moves toward the source terminal as the drain-source bias increases. This reflects the asymmetry in the source and drain contacts under off-equilibrium conditions. We calculate the charge density, Q_{x0}, and the velocity of charges, v_{x0}, at the ToB to define the channel current according to $I = WQ_{\mathrm{x0}}v_{\mathrm{x0}}$. Here, W is the device width. The model comprehends the effect of nonlinear access region resistance and self-heating during nominal operation of the device. The potential profile in the channel, needed for the Q-V model formulation, is obtained by solving the 2D Poisson equation accounting for the effects of gate and drain bias. The Ward-Dutton charge partitioning scheme under quasi-static conditions is used to obtain an analytic description of the terminal charges in terms of applied terminal voltages. We also include the effect of inner- and outer-fringing capacitances to obtain the broad bias C-V profile of the device. Amendments to the core model to handle digital switching applications are also discussed.

3.1. *Charge modeling*

3.1.1. *One-dimensional (1D) electrostatics*

We will begin with a simplified description of electrostatics under equilibrium conditions in the device. That is, we assume the applied drain source bias, $V_{\mathrm{ds}} = 0$ V. Here, the electrostatics at the ToB can be described within a one-dimensional (1D) treatment, implying that the effect of drain can be neglected. To obtain the charge at the ToB with applied gate bias, we consider the energy-band diagram along the gate-substrate direction as shown in Fig. 6. While this energy band diagram is illustrated for an AlGaN/GaN heterostructure, the calculations shown hereafter are applicable to other barrier and channel materials by using appropriate material-dependent parameters. Assuming that the AlGaN barrier is fully depleted, such that the gate depletion and the channel depletion overlap, and by balancing the potential in the device under 1D electrostatics, we obtain

$$q(\Psi_{b0} - V_g) = \frac{q(\sigma_{\mathrm{pol}} - qn_s)}{C_{\mathrm{ins}}} + \Delta E_c - E_{\mathrm{fs}}, \tag{1}$$

where Ψ_{b0} is the Schottky barrier height (SBH) at the metal/AlGaN interface, V_g is the applied gate voltage, ΔE_c is the conduction-band discontinuity at the hetero-interface (AlGaN/GaN interface), E_{fs} is the Fermi level at the hetero-interface

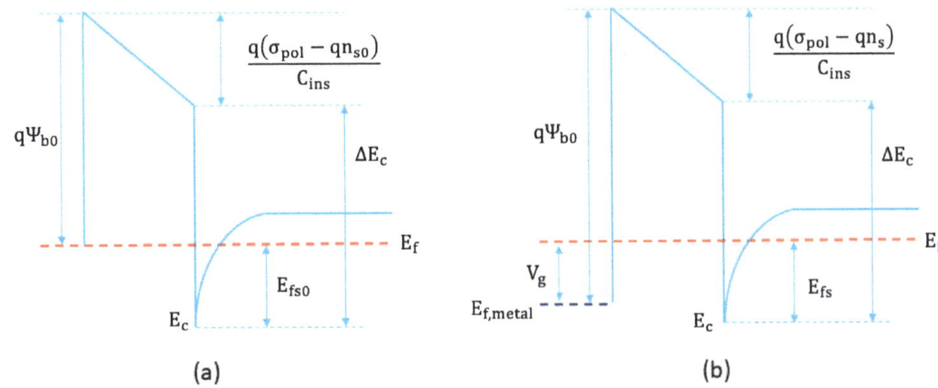

Fig. 6. Band-bending in the vertical direction of AlGaN/GaN HEMTs by considering only 1D electrostatics. Similar band-diagram can be drawn for different heterostructures. Meaning of symbols is explained in the text.

measured relative to the bottom of the conduction band edge in GaN, σ_{pol} is the polarization-induced charge density, n_s is the sheet carrier concentration of 2DEG in GaN, and $C_{\text{ins}} = \epsilon_{\text{ins}}/d_{\text{ins}}$ is the barrier layer capacitance density. Here, ϵ_{ins} and d_{ins} are the barrier permittivity and thickness, respectively. A re-arrangement of the terms in Eq. (1) yields

$$\frac{E_{\text{fs}}}{q} + \frac{qn_s}{C_{\text{ins}}} = V_g - \left(\Psi_{\text{b0}} - \frac{\sigma_{\text{pol}}}{C_{\text{ins}}} - \frac{\Delta E_c}{q}\right). \tag{2}$$

When $V_g = 0$ V, the carrier concentration $n_s = n_{\text{s0}}$ and the Fermi level $E_{\text{fs}} = E_{\text{fs0}}$, which satisfy the following condition:

$$\frac{E_{\text{fs0}}}{q} + \frac{qn_{\text{s0}}}{C_{\text{ins}}} = -\left(\Psi_{\text{b0}} - \frac{\sigma_{\text{pol}}}{C_{\text{ins}}} - \frac{\Delta E_c}{q}\right). \tag{3}$$

Using Eq. (3), we can re-write Eq. (2) as

$$\psi_s = V_g - \left(-\frac{qn_{\text{s0}}}{C_{\text{ins}}}\right) - \frac{qn_s}{C_{\text{ins}}} = V_g - V_{\text{off}} - \frac{qn_s}{C_{\text{ins}}}, \tag{4}$$

where ψ_s is the potential change in the channel due to applied gate bias, and $V_{\text{off}} = -qn_{\text{s0}}/C_{\text{ins}}$.

3.1.2. *Two-dimensional (2D) electrostatics*

Under finite V_{ds}, the effect of the drain terminal on the ToB charge in the channel is incorporated by updating the 1D potential balance equation to handle 2D electrostatics as shown in Fig. 7. At the ToB, the change in surface potential, $\psi_{\text{s,2D}}$, due to applied bias can be obtained by examining the potential divider network,

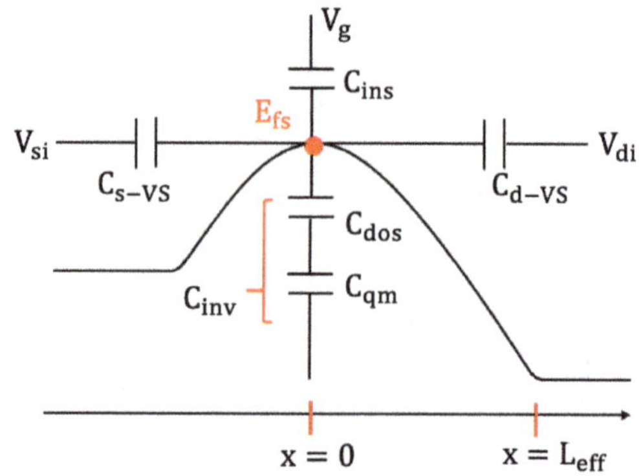

Fig. 7. 2D capacitance network, which relates the surface potential at the ToB to terminal voltages. V_{si}, and V_{di} are the intrinsic source and drain voltage, C_{ins} is capacitance of AlGaN barrier, $C_{\mathrm{d-VS}}$ models the effects of drain on VS electrostatics, $C_{\mathrm{s-VS}} = 0$. Inversion layer capacitance, C_{inv}, is obtained simply as $\partial Q_{\mathrm{x0}}/\partial \psi_s$. C_{inv} consists of quantum mechanical part, C_{qm}, and DOS capacitance, C_{dos}.

which yields

$$\psi_{\mathrm{s,2D}} = \frac{C_{\mathrm{ins}}}{C_{\Sigma}} \left(V_{\mathrm{gsi}} - V_{\mathrm{off}} + \frac{C_{\mathrm{d-VS}}}{C_{\mathrm{ins}}} V_{\mathrm{dsi}} - \frac{q n_s}{C_{\mathrm{ins}}} \right), \tag{5}$$

where V_{gsi} and V_{dsi} are the intrinsic gate-source and drain-source voltages, respectively. The intrinsic voltages are obtained by removing the potential drop across the access regions from externally applied bias voltages. The capacitance $C_{\mathrm{d-VS}}$ ($C_{\mathrm{s-VS}}$) captures the effect of drain (source) voltage on the surface potential at the ToB, and $C_{\Sigma} = C_{\mathrm{ins}} + C_{\mathrm{d-VS}} + C_{\mathrm{s-VS}}$.

The channel, or inversion layer capacitance, C_{inv}, is comprised of the series combination of density of states (DOS) capacitance, C_{dos}, and the quantum-mechanical capacitance, C_{qm}. C_{dos} reflects a finite band bending required to increase n_s due to the finite density of states in the band [64], while $C_{\mathrm{qm}} = \frac{\partial q^2 n_s}{\Delta E_0}$, where ΔE_0 refers to the energy of the first sub-band in the channel due to quantum confinement of carriers. C_{qm} is determined by the self-consistent solution of the Schrödinger-Poisson equations and is related to the shape of the confinement potential and the confinement electron mass [65, 66]. C_{inv} represents the variation in n_s with $\psi_{\mathrm{s,2D}}$ and is given as

$$C_{\mathrm{inv}}^{-1} = C_{\mathrm{dos}}^{-1} + C_{\mathrm{qm}}^{-1}. \tag{6}$$

Given that our model is source-referenced, we consider $C_{\mathrm{s-VS}} = 0$. The value of $C_{\mathrm{d-VS}}$ can be known from the measured values of the non-ideality factor, n, and the drain-induced barrier lowering (DIBL), δ. That is, $n = C_{\Sigma}/C_{\mathrm{ins}}$ and $\delta = C_{\mathrm{d-VS}}/C_{\mathrm{ins}}$. Therefore, Eq. (5) can be re-written as

$$\psi_{\mathrm{s,2D}} = \frac{1}{n} \left(V_{\mathrm{gsi}} - V_{\mathrm{off}} + \delta V_{\mathrm{dsi}} - \frac{q n_s}{C_{\mathrm{ins}}} \right). \tag{7}$$

The non-ideality factor, n, is related to SS according to $SS = 2.3n\phi_t$ ($\phi_t = k_B T/q$ is the thermal voltage). To account for finite punch-through effects at high $V_{\rm dsi}$, we consider $n = n_0 + n_d V_{\rm dsi}$, where n_d is the first-order punch-through factor [36].

3.1.3. *Two-dimensional electron gas (2DEG)*

The 2DEG density in the channel, n_s, is found from using the 2D density of states ($D_{\rm 2D}(E)$) in the triangular potential well formed at the heterointerface and applying the Fermi-Dirac statistics ($f_{\rm FD}(E)$). At zero applied drain-source bias,

$$n_s = \int_{E_c+\Delta E_0}^{\infty} D_{\rm 2D}(E) f_{\rm FD}(E) dE, \tag{8}$$

where ΔE_0 is the subband energy of the first quantized energy level referenced to the bottom of the conduction band at the heterointerface. By including the effect of the first- and the second-order conduction-band non-parabolicity in $D_{\rm 2D}(E)$, the integration in Eq. (8) yields

$$n_s = \frac{m_{\rm eff} k_B T}{\pi \hbar^2} \mathcal{F}_0(\eta_{\rm fs}), \tag{9a}$$

$$m_{\rm eff} = m_e \left(1 + 2\alpha k_B T \frac{\mathcal{F}_1(\eta_{\rm fs})}{\mathcal{F}_0(\eta_{\rm fs})} + 3\beta (k_B T)^2 \frac{\mathcal{F}_2(\eta_{\rm fs})}{\mathcal{F}_0(\eta_{\rm fs})}\right). \tag{9b}$$

In the above set of equations, the parameters α and β denote the first- and the second-order non-parabolicities, respectively, of the conduction band, $m_{\rm eff}$ is the effective DOS mass of carriers in channel. The function $\mathcal{F}_j$ is the Blakemore Fermi-Dirac integral of order j [67]. The factor $\eta_{\rm fs}$ is defined as $\frac{E_{\rm fs}-\Delta E_0}{k_B T}$, and ΔE_0 is obtained by solving the Schrödinger equation in the triangular potential well formed at the heterointerface. The wave-functions of electrons confined in a triangular potential well are Airy functions with eigenenergies given as [68]

$$\Delta E_i = \left(\frac{\hbar^2}{2m_{\rm eff}}\right)^{1/3} \left(\frac{3\pi}{2}\frac{q^2 n_s}{\epsilon_{\rm GaN}}\right)^{2/3} (i + 3/4)^{2/3}. \tag{10}$$

Substituting $i = 0$ in the above equation, we get

$$\Delta E_0 = \left(\frac{\hbar^2}{2m_{\rm eff}}\right)^{1/3} \left(\frac{9\pi}{8}\frac{q^2 n_s}{\epsilon_{\rm GaN}}\right)^{2/3} = \gamma_0 n_s^{2/3}, \tag{11}$$

where γ_0 can be adjusted to match experimental and numerical device data. Here, we consider only the first quantized energy level in the triangular well, since the second quantized energy level is at least $4k_B T$ higher in energy and its population by free carriers can be neglected [69]. To account for the quantum-mechanical tunneling of electron wave-function into the barrier, we amend Eq. (11) as

$$\Delta E_0 = \gamma_0 n_s^{2/3} \left(\frac{n_s}{n_{s0}}\right)^e, \tag{12}$$

where the empirical parameter e depends on the barrier thickness and composition.

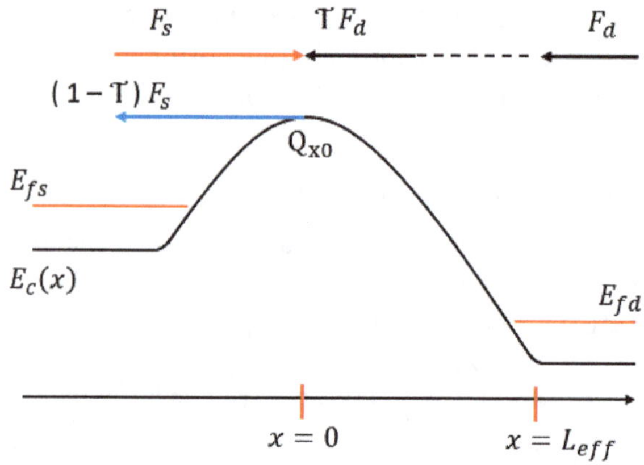

Fig. 8. Electron states at the ToB under non-equilibrium condition will be filled differently from both source and drain contacts. The total charge at the ToB is $Q_{\rm x0} = qn_s$. The source side flux depends on its Fermi level, $E_{\rm fs}$. Likewise, the drain side flux depends on its Fermi level, $E_{\rm fd}$. For a given intrinsic drain-source bias $V_{\rm dsi}$, $E_{\rm fs} - E_{\rm fd} = qV_{\rm dsi}$. $\mathcal{T}$ is the transmission coefficient which accounts for elastic scatterings in the channel. $\mathcal{T}$ approaches zero for drift diffusive transport, while for ballistic transport T approaches unity.

Under non-equilibrium transport conditions ($V_{\rm ds} \neq 0$), the channel charge, n_s must be obtained by accounting for the filling of states at the ToB by both the source and the drain contacts. We model the source and drain fluxes in terms of the Landauer channel transmission coefficient, $\mathcal{T}$. For DD transport in the channel, $\mathcal{T} \to 0$, while for ballistic transport, $\mathcal{T} \to 1$. As illustrated in Fig. 8, the total charge at the ToB under finite $V_{\rm ds}$ is given as

$$n_s = \frac{(2-\mathcal{T})F_s + \mathcal{T}F_d}{v_T}, \tag{13a}$$

$$F_s = v_T \frac{m_{\rm eff} k_B T}{2\pi\hbar^2} \mathcal{F}_0(\eta_{\rm fs}), \tag{13b}$$

$$F_d = v_T \frac{m_{\rm eff} k_B T}{2\pi\hbar^2} \mathcal{F}_0(\eta_{\rm fd}), \tag{13c}$$

where F_s and F_d are the fluxes from the source and the drain contacts, respectively. The thermal velocity at the ToB is $v_T = v_{\rm T0}\mathcal{F}_{1/2}(\eta_{\rm fs})/\mathcal{F}_0(\eta_{\rm fs})$, which includes the effect of degeneracy in the channel for high gate-source voltages [70, 71]. Here, $v_{\rm T0}$ is the thermal velocity of carriers in the non-degenerate regime [72]. The reduced energy $\eta_{\rm fd} = \eta_{\rm fs} - V_{\rm dsi}/\phi_t$. From Eq. (13a) we can infer that when $V_{\rm dsi} \approx 0$ or when the transport in the channel is DD (i.e. $\mathcal{T} \to 0$), n_s will simply be given as Eq. (9a). Hence, the 1D charge model is a limiting case of the 2D charge model under equilibrium channel conditions or when the channel length is significantly greater than the carrier MFP.

The transmission coefficient in the channel is mathematically given as [36]

$$\mathcal{T} = \frac{\lambda}{\lambda + L_{\mathrm{crit}}}. \tag{14}$$

Here, λ is the MFP of carriers associated with backscattering in the channel, and L_{crit} denotes the critical length for backscattering of carriers in the low-field region near the source end. We refer to L_{crit} as the critical length hereafter. In general, under low V_{dsi} conditions, $L_{\mathrm{crit}} \to L_{\mathrm{eff}}$. Here, L_{eff} is the effective channel length, which is assumed equal to the gate length. Under non-equilibrium conditions, the critical length is only a fraction of the effective channel length and is, therefore, given as $L_{\mathrm{crit}} = \xi L_{\mathrm{eff}}$, where $\xi < 1$ [73]. We use a semi-empirical formulation for L_{crit} given as

$$L_{\mathrm{crit}} = L_{\mathrm{eff}} \times (1 - f_2) + \xi L_{\mathrm{eff}} \times f_2, \tag{15a}$$

$$f_2 = \frac{V_{\mathrm{dsi}}/(\theta\phi_t)}{(1 + (V_{\mathrm{dsi}}/(\theta\phi_t))^{\beta_{\mathrm{crit}}})^{1/\beta_{\mathrm{crit}}}}, \tag{15b}$$

where the parameters ξ and θ are independent of the channel length for a given HEMT technology. The parameter β_{crit} controls the rate at which L_{crit} changes from L_{eff} to ξL_{eff} as V_{dsi} increase. The parameter β_{crit} depends on the channel length and must, therefore, be fitted for each device in the technology. The parameter ξ depends on the applied bias and junction temperature and is modeled as

$$\xi = \frac{1}{1 + \eta_{v0}\,(T/T_{\mathrm{nom}})^{\eta_e}\,(V_{\mathrm{gsi}} - V_{\mathrm{off}})^{0.5}\,V_{\mathrm{dsi}}^{0.25}}. \tag{16}$$

The MFP, λ, is obtained directly from the measurement of scattering-limited mobility, μ_{eff}, of a long-channel device according to [74]

$$\lambda = 2\phi_t \frac{\mu_{\mathrm{eff}}}{v_{\mathrm{T0}}} \frac{\mathcal{F}_0(\eta_{\mathrm{fs}})}{\mathcal{F}_{-1/2}(\eta_{\mathrm{fs}})}. \tag{17}$$

Due to self-heating in the device under high drain-source current, the scattering-limited mobility degrades [75, 76]. We account for the reduction in μ_{eff} due to increase in the device temperature according to

$$\mu_{\mathrm{eff}}(T) = \mu_{\mathrm{eff,0}} \left(\frac{300\mathrm{K}}{T}\right)^{\zeta}, \tag{18}$$

where $\mu_{\mathrm{eff,0}}$ is experimentally measured at 300 K, and ζ is a fitting parameter in the model [77]. Experimental data in III-nitride HEMTs typically show a degradation in mobility due to an increase in gate bias [78]. This is because under high gate bias, carriers from the channel may spill into the barrier layer, leading to an increase in carrier scatterings from the interface roughness [78]. To model the observed behaviour of $\mu_{\mathrm{eff,0}}$ on V_{gs}, we consider that the mobility can be expressed as

$$\frac{1}{\mu_{\mathrm{eff,0}}} = \frac{1}{\mu_{\mathrm{eff,00}}} \left(\left(\frac{Q_{\mathrm{x0}}}{Q_{01}}\right)^{v1} + 1\right) \tag{19}$$

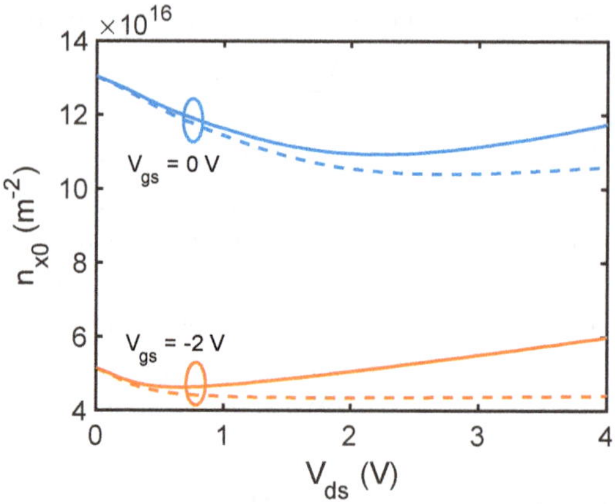

Fig. 9. ToB charge versus V_{ds} at fixed V_{gs} for various DIBL values. Here, channel transmission is unity to illustrate non-monotonicity of ToB charge with drain bias in ballistic transistors.

where $\mu_{\mathrm{eff},00}$ is the maximum carrier mobility for a given 2DEG density, while Q_{01} and $v_1 \approx 2.0$ are fitting parameters.

Typically, the drain bias affects the carrier concentration at the ToB through the DIBL parameter as in Eq. (7). However, in the case of QB transport in which case $L_{\mathrm{eff}} \approx \lambda$, drain bias also affects the carrier concentration through the transmission coefficient as in Eq. (13a). For $V_{\mathrm{dsi}} \to 0$ V and $\mathcal{T} \to 1$, the net charge at the ToB is contributed equally by the source and the drain contacts. However, as V_{dsi} in a QB channel increases, the negative momenta states directed from the drain contact toward the source are cut-off and the net charge at the ToB actually reduces. The effects of DIBL and cutting-off of drain momenta states on the charge at the ToB are examined in Fig. 9. In the absence of DIBL, the ToB charge for a fully ballistic device decreases with an increase in V_{ds} indicating that the drain flux no longer fills the ToB states. This effect is counteracted by a finite DIBL, which causes the ToB charge to increase with V_{ds} beyond a certain value of V_{ds}. The minimum point in the profile shown in Fig. 9 corresponds to the V_{ds} value at which DIBL balances the loss of drain flux at the TOB.

3.1.4. *One-dimensional electron gas (1DEG)*

For nanowire GaN FETs suitable for digital applications, we calculate the carrier concentration in the channel using 1D density of states according to

$$n_s = \int_{E_c+\Delta E_0}^{\infty} D_{1\mathrm{D}} f_{\mathrm{FD}}(E)dE = \frac{1}{\hbar}\sqrt{\frac{2m_{\mathrm{eff}}k_BT}{\pi}}\mathcal{F}_{-1/2}(\eta_{\mathrm{fs,fd}}). \tag{20}$$

Similar to the case of HEMTs with a 2DEG in the channel, the total charge at the ToB in NW FETs depends on the flux from the source and drain contacts. The flux is defined as

$$F_{s,d} = \frac{v_T}{2\hbar}\sqrt{\frac{2m_{\text{eff}}k_BT}{\pi}}\mathcal{F}_{-1/2}(\eta_{\text{fs,fd}}). \tag{21}$$

The barrier capacitance per length is calculated as [79]

$$C_{\text{ins}} = \frac{2\pi\epsilon_{\text{ins}}}{\cosh^{-1}\left(\frac{r+d_{\text{ins}}}{r}\right)}, \tag{22}$$

where r is the radius of the NW channel.

3.2. *Velocity modeling*

The carrier velocity, v_{x0}, at the ToB under QB transport in the channel is modeled as

$$v_{\text{x0}} = \begin{cases} \mu_{\text{app}}\frac{V_{\text{dsi}}}{L_{\text{eff}}} & \text{if } V_{\text{dsi}} < V_{\text{dsat}}, \\ v_{\text{inj}} & \text{if } V_{\text{dsi}} > V_{\text{dsat}}, \end{cases} \tag{23}$$

where V_{dsat} is the saturation drain-source voltage, v_{inj} is the source-side injection velocity under QB transport conditions, and μ_{app} is the apparent mobility of carriers in the channel [80]. Apparent mobility is given by the Matthiessen's sum of long-channel diffusive mobility (μ_{eff}) and ballistic mobility (μ_{ball}) [80]. That is,

$$\mu_{\text{app}}^{-1} = \mu_{\text{eff}}^{-1} + \mu_{\text{ball}}^{-1}, \tag{24a}$$

$$\mu_{\text{ball}} = \frac{v_{\text{T0}}L_{\text{eff}}}{2\phi_t}\frac{\mathcal{F}_{-1/2}(\eta_{\text{fs}})}{\mathcal{F}_0(\eta_{\text{fs}})}. \tag{24b}$$

$$\mu_{\text{eff}} = \frac{v_{\text{T0}}\lambda}{2\phi_t}\frac{\mathcal{F}_{-1/2}(\eta_{\text{fs}})}{\mathcal{F}_0(\eta_{\text{fs}})}. \tag{24c}$$

For low drain-source bias, the velocity of carriers at the ToB increases linearly with V_{dsi}. Beyond a certain saturation drain-source voltage (V_{dsat}) the velocity is limited by the injection velocity (v_{inj}) that is proportional to the unidirectional thermal velocity. The profile of v_{x0} for the entire range of V_{dsi} is modeled as [81]

$$v_{x0} = v_{\text{inj}}F_{\text{sat}}, \tag{25a}$$

$$F_{\text{sat}} = \frac{V_{\text{dsi}}/V_{\text{dsat}}}{(1 + (V_{\text{dsi}}/V_{\text{dsat}})^{\beta})^{1/\beta}}, \tag{25b}$$

where β is an empirical parameter to control the sharpness of transition of v_{x0} from the linear to the saturation regimes. The source-side injection velocity is given as

$$v_{\text{inj}} = v_{\text{T0}}\frac{\mathcal{F}_{1/2}(\eta_{\text{fs}})}{\mathcal{F}_0(\eta_{\text{fs}})}\left[\frac{\mathcal{T}_{\text{sat}}}{2-\mathcal{T}_{\text{sat}}}\right] \tag{26}$$

where $v_{\text{T0}}\frac{\mathcal{F}_{1/2}(\eta_{\text{fs}})}{\mathcal{F}_0(\eta_{\text{fs}})}$ gives the degenerate thermal velocity v_T, $\mathcal{T}_{\text{sat}}$ is the channel transmission coefficient evaluated under saturation, i.e., for $L_{\text{crit}} = \xi(V_{\text{dsi}}, V_{\text{gsi}})L_{\text{eff}}$.

Equation (26) correctly predicts the increase in both injection velocity and thermal velocity of carriers with an increase in carrier concentration [36].

The saturation drain-source voltage for QB transport is given as [82]

$$V_{\text{dsat}} = \frac{v_{\text{inj}} L_{\text{eff}}}{\mu_{\text{app}}}. \tag{27}$$

Substituting the value of saturation velocity v_{inj} and apparent mobility μ_{app}, in Eq. (27), V_{dsat} is simplified as

$$V_{\text{dsat}} = 2\phi_t \frac{\mathcal{F}_{1/2}(\eta_{\text{fs}})}{\mathcal{F}_{-1/2}(\eta_{\text{fs}})} \frac{\lambda + L_{\text{eff}}}{\lambda + 2\xi(V_{\text{dsi}}, V_{\text{gsi}}) L_{\text{eff}}}. \tag{28}$$

3.3. *Secondary effects*

3.3.1. *Parasitic source/drain Schottky contact*

Electrical characteristics of Schottky contacts formed by various metal contacts with n-type GaN have been discussed in early publications [83–86]. Table 1 summarizes the electrical properties of metal-GaN Schottky contacts measured experimentally using current-density-temperature and capacitance-voltage methods.

For optimal performance, FETs require thermally stable and low resistance ohmic contacts. Metals, such as Au and Ti, are preferred for GaN transistors since the Schottky contact formed by these materials tends to have a low reverse breakdown voltage. Once the reverse bias across the barrier is greater than the breakdown voltage, the current increases rapidly, which leads to low contact resistance. The nonlinear current-voltage characteristics of rectifying Schottky contacts is due to a very low value of the reverse breakdown voltage [87]. In the case of GaN-based NW structures suitable for digital applications, a Schottky barrier is indeed formed between the n-type semiconductor and the metal source and drain contacts (Fig. 10). This combined with the small cross-sectional area of the structure generates a large parasitic contact resistance that will limit the drain current. The Schottky contact at the source side is always reverse biased, while the Schottky contact on the drain side is forward biased. As a result of the rectifying nature of the Schottky contacts, the output I-V characteristics of GaN NW FETs measured experimentally display nonlinear behavior under low drain bias. To capture this nonlinearity, we model

Table 1. Summary of metal-GaN Schottky characteristics from current-density-temperature and capacitance-voltage experimental measurements [84].

	Work function ϕ_m	Metal electronegativity	Non-ideality factor n	Breakdown voltage (V)	A ($\text{Acm}^{-2}\text{K}^{-2}$)	ϕ_{bn} (eV)
Pt	5.65	2.2	1.21	18.1	6.61	1.03
Pd	5.12	2.2	1.14	17.5	0.44	0.91
Au	5.1	2.4	1.03	...	0.006	0.84
Ti	4.33	1.5	1.28	...	...	0.59
Ni	5.15	1.8	1.15	9.6	0.11	0.66

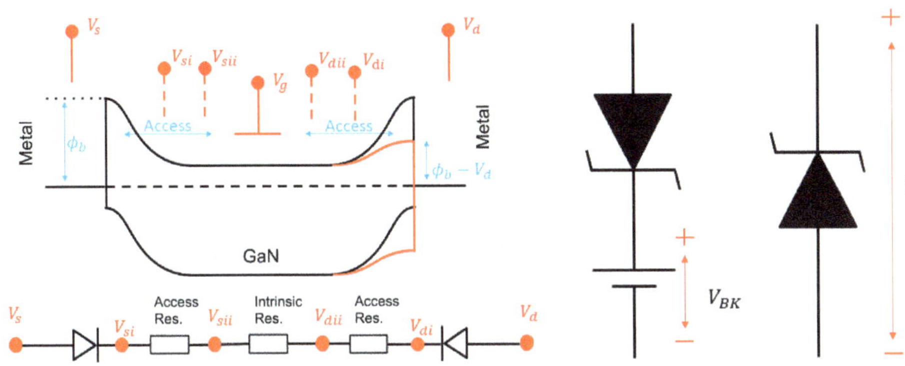

Fig. 10. Equivalent circuit model of the nanowire device with Schottky barriers on both source and drain side under low longitudinal bias. Constant contact resistance R_c is not shown in the figure. The constant voltage drop (CVD) Zener model for Schottky contact with a very low breakdown field is shown on the right.

the source side Schottky contact resistance that is reverse biased using the constant voltage drop (CVD) Zener model [88]. Accordingly,

$$R_{\text{SB,s}} = \frac{V_{\text{si,s}} - V_{\text{BK}}}{I_0\left(\exp\left(\frac{V_{\text{si,s}} - V_{\text{BK}}}{K_s\phi_t} - 1\right)\right)}, \tag{29}$$

where $V_{\text{si,s}} = V_{\text{si}} - V_{\text{s}}$ as shown in Fig. 10, V_{BK} is the reverse breakdown voltage, K_s is fitting parameter that controls the decay rate of $R_{\text{SB,s}}$ due to breakdown, and I_0 is the reverse saturation current of thermal emission and is found by fitting the model to measured data. The resistance of the drain side Schottky contact that is forward biased is calculated as [89]

$$R_{\text{SB,d}} = \frac{V_{\text{d,di}}}{I_0}\frac{1}{\exp\left(\frac{V_{\text{d,di}}}{K_d\phi_t}\right) - 1}, \tag{30}$$

which decreases exponentially with $V_{\text{d,di}}(= V_{\text{d}} - V_{\text{di}})$. The parameter K_d is the non-ideality factor of the drain schottky contact. Since this resistance is exponentially decreasing with the voltage drop across it, compare with $R_{\text{SB,s}}$, $R_{\text{SB,d}}$ is neglectable.

3.3.2. *Access region resistance*

Most III-nitride HEMTs are non-self-aligned transistors due to process constraints. As the drive current of the HEMT increases, the resistance associated with access regions also increases, which ultimately limits the maximum achievable transconductance and the linearity of the device [90, 91]. The nonlinearity of access regions has been exposed experimentally by various groups using transmission length measurements (TLMs) [42, 57]. Typically, the nonlinearity of access regions is attributed to a combination of Joule heating, channel pinch-off, and velocity saturation. Since scaled access regions are preferred for RF applications, we restrict ourselves to velocity-saturation-limited resistance model of access regions. This is in agreement with previously published reports in which RF nonlinearity in scaled GaN HEMTs is attributed to velocity saturation in access regions [90–94]. For high-voltage applications demanding long access regions, current saturation occurs due to pinch off at the drain end. The effect of pinch-off can be captured by representing access regions as implicitly gated transistors with length-dependent saturation [95].

To model the nonlinearity of access resistances, we apply the DD formalism considering that the typical length of access regions will be few hundreds of nanometers even in ultra-scaled GaN HEMTs to support a high breakdown field. For a given voltage drop V_{acc} across the access regions, the large-signal access-region resistance is modeled as

$$R_{\mathrm{acc}} = \frac{V_{\mathrm{acc}}}{I_{\mathrm{acc}}}, \tag{31a}$$

$$\frac{I_{\mathrm{acc}}}{W} = Q_{\mathrm{acc}} v_{\mathrm{acc}}, \tag{31b}$$

$$v_{\mathrm{acc}} = v_{\mathrm{sat,acc}} F_{\mathrm{sat,acc}}, \tag{31c}$$

where Q_{acc} is the fixed, temperature-independent 2DEG charge in the access region, v_{acc} is the velocity of carriers in the access region, $v_{\mathrm{sat,acc}}$ is the saturation velocity of carriers, and $F_{\mathrm{sat,acc}}$ is an empirical function that models the transition of current from linear to saturation regimes. This function is similar to that in Eq. (25b) with with an updated value of saturation voltage denoted as $V_{\mathrm{dsat,acc}}$. This function is given as

$$F_{\mathrm{sat,acc}} = \frac{V_{\mathrm{acc}}/V_{\mathrm{dsat,acc}}}{\left(1 + (V_{\mathrm{acc}}/V_{\mathrm{dsat,acc}})^{\beta_{\mathrm{acc}}}\right)^{1/\beta_{\mathrm{acc}}}}, \tag{32a}$$

$$V_{\mathrm{dsat,acc}} = \frac{v_{\mathrm{sat,acc}} L_{\mathrm{acc}}}{\mu_{\mathrm{eff,acc}}}, \tag{32b}$$

where L_{acc} is the length of the access region, β_{acc} is an empirical parameter to control the rate of current saturation in access regions, and $\mu_{\mathrm{eff,acc}}$ is the effective mobility of carriers in the access region. The value of $\mu_{\mathrm{eff,acc}}$ and its temperature and carrier concentration dependence are assumed to be the same as μ_{eff} that is defined in the main channel.

Under the DD transport formalism, the saturation velocity of carriers in access regions is modeled as

$$v_{\mathrm{sat,acc}} = \mu_{\mathrm{eff,acc}} E_{\mathrm{crit}}, \tag{33}$$

where E_{crit} is the critical electric field beyond which carrier velocity is saturated. The critical electric field increases with an increase in junction temperature and is modeled according to [96]

$$E_{\mathrm{crit}} = E_{\mathrm{crit,0}} \left[\frac{T}{T_{\mathrm{nom}}}\right]^{\epsilon_{\mathrm{crit}}}, \tag{34}$$

where $E_{\text{crit},0}$ is measured at $T = T_{\text{nom}}$ (nominal operating temperature of 300 K), and ϵ_{crit} is the temperature sensitivity of the critical electric field. By substituting Eqs. (32)-(34) in Eq. (31), the access region resistance simplifies to

$$R_{\text{acc},(\text{s,d})} = \underbrace{\frac{1}{Q_{\text{acc}}\mu_{\text{eff,acc}}}}_{R_{\text{sh}}} \frac{L_{\text{acc}}}{W} \left[1 + \left(\frac{E_{s,d}}{E_{\text{crit}}}\right)^{\beta_{\text{acc}}}\right]^{1/\beta_{\text{acc}}}, \tag{35}$$

where $E_{s,d} = V_{\text{acc(s,d)}}/L_{acc(s,d)}$ is the electric field in the access region, and R_{sh} is the sheet resistance of the access region. The temperature dependence of R_{sh} arises mainly due to the temperature dependence of $\mu_{\text{eff,acc}}$ as given in Eq. (18). Access-region resistance model introduces three additional fitting parameters: (i) Q_{acc}, (ii) $E_{\text{crit},0}$, and (iii) ϵ_{crit}. The voltage drop across the access regions is calculated based on the current flow in the device. The voltage drop is then used within Eq. (35) to self-consistently obtain the nonlinear resistance of access regions and the corresponding voltage drop is updated simultaneously.

3.3.3. *Self heating*

Measurement of temperature using micro-raman spectroscopy [97] and gate-end-to-end method [98], which is sensitive to the temperature rise adjacent to the gate contact on the drain side, show temperature rise of $\sim 100°$C in GaN HEMTs. This temperature rise of the junction, known as self heating, is defined as $\Delta T = T - T_{\text{nom}} = R_{\text{th}}P_{\text{diss}}$, where R_{th} is the thermal resistance of the device, and P_{diss} is the power dissipation. Our model assumes a unique temperature for the entire device and effects of temperature gradient are ignored in favor of model simplicity. For one-dimensional heat flow and constant thermal conductivity (κ_0), $R_{\text{th}} = t/(A\kappa_0)$, where t is the thickness along the heat flow direction, and A is the cross-sectional area.

The thermal resistance $R_{\text{th,material}}$ is calculated for a rectangular heat source at the gate-drain region of the heterointerface between the AlGaN barrier and the GaN channel. The width of the heat source is equal to the width of the device, and the length of the heat source is independent of gate length by assuming that the heating is concentrated around the gate-drain edge. The heat is taken to flow down uniformly in a volume made by extending the edges of the heat source downwards and outwards by an angle α, known as the heat spreading angle. The thermal resistance is calculated based on a three dimensional integration for each material layer (in this case, GaN and SiC) separately, before adding each layers contribution to obtain the total thermal resistance. Figure 11 shows the 3D heat flow in the GaN layer from the top heat source. Based on the 3D heat flow diagram in Fig. 11, the differential thermal resistance of the GaN layer is given as

$$dR_{\text{th,GaN}} = \frac{dz}{\kappa_{\text{GaN}} 2(t_{\text{ins}} + t_{\text{GaN}} - z)\tan\alpha[W + 2(t_{\text{GaN}} - z)\tan\alpha]}, \tag{36}$$

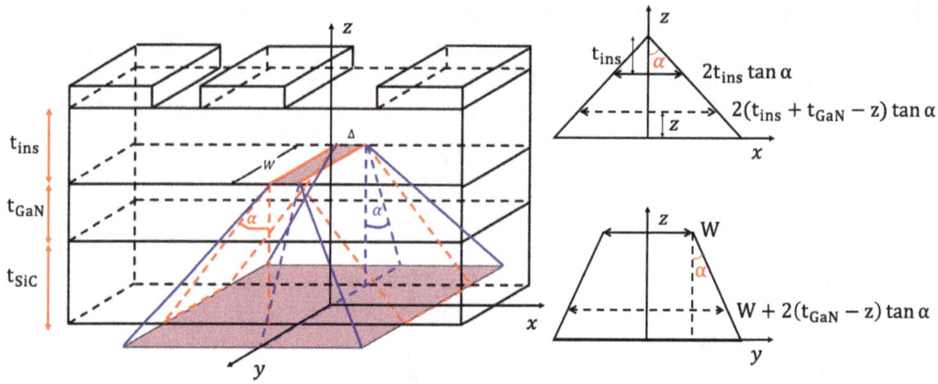

Fig. 11. The 3D heat flow in the GaN layer. The right top and right bottom figures are the front view and left view of the highlighted heat spreading area. Cross-sectional area perpendicular to the z-axis direction is calculated as $2(t_{\text{ins}} + t_{\text{GaN}} - z)\tan\alpha \times (W + 2(t_{\text{GaN}} - z)\tan\alpha)$.

which is integrated along the thickness of GaN, t_{GaN}, to yield the total resistance according to

$$R_{\text{th,GaN}} = \int_0^{t_{\text{GaN}}} dR_{\text{th,GaN}} = \frac{1/(4\kappa_{\text{GaN}}\tan^2(\alpha))}{t_{\text{ins}} - W/(2\tan(\alpha))} \ln\left(\frac{t_{\text{ins}}(t_{\text{GaN}} + W/(2\tan(\alpha)))}{W/(2\tan(\alpha))(t_{\text{GaN}} + t_{\text{ins}})}\right), \tag{37}$$

where W is the width of the device structure. A similar procedure is applied to obtain the thermal resistance of the substrate. The thermal resistance for the SiC layer is given as

$$R_{\text{th,SiC}} = \frac{1/(4\kappa_{\text{SiC}}\tan^2(\alpha))}{t_{\text{ins}} - W/(2\tan(\alpha))} \times \ln\left(\frac{(t_{\text{ins}} + t_{\text{GaN}})}{(t_{\text{GaN}} + W/(2\tan(\alpha)))} \frac{(t_{\text{GaN}} + t_{\text{SiC}} + W/(2\tan(\alpha)))}{(t_{\text{GaN}} + t_{\text{ins}} + t_{\text{SiC}})}\right). \tag{38}$$

In the more general case, thermal conductivity follows the relationship: $\kappa(T) = \kappa_0(T/T_{\text{nom}})^{-k_e}$, where κ_0 is the thermal conductivity defined at T_{nom}, and $k_e > 0$ is the temperature coefficient of thermal conductivity [99, 100]. In this case, the junction temperature is obtained by applying Kirchoff's transformation to solve the nonlinear heat equation. The junction temperature, T, for the nonlinear heat flow is given as [101, 102]

$$\widetilde{T} - T_{\text{nom}} = \frac{1}{\kappa_0} \int_{T_{\text{nom}}}^{T} \kappa(T) dT, \tag{39a}$$

$$\widetilde{T} = T_{\text{nom}} + R_{\text{th},0} P_{\text{diss}}, \tag{39b}$$

where R_{th0} is the thermal resistance calculated at T_{nom}. Substituting $\kappa(T)$ in

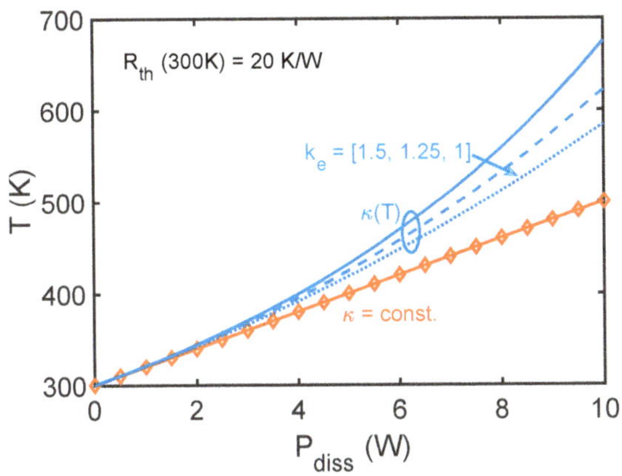

Fig. 12. Difference in junction temperature due to temperature dependent thermal conductivity. The junction temperature increases more rapidly with dissipated power as the exponent k_e increases.

Eq. (39a) and solving for the junction temperature T yields

$$T = \left[\frac{\widetilde{T} - k_e(\widetilde{T} - T_{\mathrm{nom}})^{k_e}}{T_{\mathrm{nom}}^{k_e}}\right]^{1/(1-k_e)}. \tag{40}$$

Figure 12 shows that a temperature-independent thermal conductivity model will underestimate the junction temperature rise significantly compared to a realistic scenario assuming temperature-dependent thermal conductivity. An accurate estimate of junction temperature is critical for analyzing the device performance over its lifetime.

3.4. *Dynamic effects*

In addition to the steady state DC current, the device also supports displacement currents resulting from the charges stored in device capacitances. The capacitive characteristics are a sum of intrinsic (channel region) and extrinsic (access regions) capacitances and must be modeled to simulate the large-signal and small-signal response of the device for frequencies greater than about 1 KHz. In our model, nodal charges are formulated as functions of applied voltages. The inter-nodal capacitances are obtained by differentiating the nodal charges with respect to the appropriate terminal voltage.

The channel charges in the device vary non-linearly with the applied gate- and drain-bias from both linear-to-saturation regimes and from off-to-on state of the device, resulting in non-linear large-signal charges and, hence, nonlinear small-signal capacitances. To partition the position-dependent channel charge into the source and drain terminals, we use the linear Ward-Dutton (WD) charge partitioning scheme [103]. The WD partitioning scheme is used extensively to develop transient and high frequency advanced compact models in MOSFET analysis [103, 104]. To apply the WD scheme, we assume quasi-static conditions in the device according to which the potential and charge density at any location in the channel follow the variation in applied bias without any delay. This assumption is valid when the transit time of carriers through the channel is greater than the transit time of the bias variation [105]. According to the WD scheme, the terminal charges Q_s (source), Q_d (drain), and Q_g (gate) are given as

$$Q_s = W \int_0^{L_{\text{eff}}} \left(1 - \frac{x}{L_{\text{eff}}}\right) Q_{\text{ch}}(x)dx, \tag{41a}$$

$$Q_d = W \int_0^{L_{\text{eff}}} \left(\frac{x}{L_{\text{eff}}}\right) Q_{\text{ch}}(x)dx, \tag{41b}$$

$$Q_g = -(Q_s + Q_d). \tag{41c}$$

The appearance of the factor $(1-x/L_{\text{eff}})$ and (x/L_{eff}) can be interpreted physically as the linear grading of the charge along the channel length. At the ToB, channel charge $Q_{\text{ch}}(x_0) = Q_{\text{x0}}$. Equation (41c) is based on charge conservation in the transistor. Using the voltage-dependent terminal charges, the inter-nodal capacitances are given as

$$C_{ij} = \begin{cases} -\partial Q_i/\partial V_j \text{ if } i \neq j, \\ \partial Q_i/\partial V_j \text{ if } i = j. \end{cases} \tag{42}$$

In the case of ultra-scaled channel lengths, the transport is QB in nature, which precludes the use of GCA, along the channel (the voltages vary gradually along the channel from the drain to the source) when the transistor is operating in the saturation mode. This necessitates modeling the channel charge separately in the linear and saturation regimes for QB transistors.

To obtain inter-nodal capacitances according to Eqs. (41) and (42), the potential profile in the gated region of the HEMT must be obtained by solving the 2D Poisson equation given as

$$\frac{d^2u(x,y)}{dx^2} + \frac{d^2u(x,y)}{dy^2} = \frac{\rho}{\epsilon_{\text{ch}}}. \tag{43}$$

Here, $u(x, y)$ is the channel potential referenced to the ToB potential, $x-$ is along the length of the channel, $y-$ is in the direction from the gate to the substrate, ρ is the channel charge density, and ϵ_{ch} is the permittivity of the channel. To obtain a closed-form solution of the potential profile, the 2D Poisson equation is converted into its 1D form by assuming negligible free carrier density below subthreshold and neglecting the vertical electric field from the gate to the substrate above threshold [106, 107]. At the hetero-interface, we model the change in surface potential, $\psi_0(x)$, with respect to the potential at the ToB, $\psi_0(x_0)$, according to

$$\underbrace{\psi_0(x) - \psi_0(x = x_0)}_{u(x)} = \underbrace{\eta_s \frac{\sinh\left(\frac{L_{\text{eff}}-x}{\lambda_{\text{sl}}}\right)}{\sinh\left(\frac{L_{\text{eff}}}{\lambda_{\text{sl}}}\right)}}_{u_l(x)} + \underbrace{\eta_d \frac{\sinh\left(\frac{x}{\lambda_{\text{sl}}}\right)}{\sinh\left(\frac{L_{\text{eff}}}{\lambda_{\text{sl}}}\right)}}_{u_r(x)}. \tag{44}$$

Here, λ_{sl} is the characteristic length of the exponential decay of the lateral electric field lines in the channel. That is, it represents the spread of the potential in x- direction and is referred to as the "natural length" or the "electrostatic length" of the transistor [108]. The value of λ_{sl} depends strongly on the material and geometrical properties of the transistor. Transistors are designed to reduce λ_{sl} to reduce short-channel effects including lower threshold voltage, DIBL, and punch-through. With higher permittivity of the barrier and lower thicknesses of the channel and barrier, λ_{sl} can be reduced. While here we consider λ_{sl} as a technology-dependent empirical parameter that is independent of the channel length for a specific technology, its interpretation is physics inspired. The value of λ_{sl} can also be changed separately for the source and the drain terminals (i.e. $\lambda_{\text{sl,s}}$ and $\lambda_{\text{sl,d}}$) depending on the device geometry and doping to account for off-equilibrium conditions although its impact on C-V in QB regime is not expected to be quite significant. This is because in QB conditions, the channel charge decreases rapidly as we move away from the ToB; therefore, the charge near the ToB dominates the total channel carrier density and the terminal charges. By choosing $\eta_s = -\psi_s$ and $\eta_d = V_{\text{dsi}} - \psi_s$ in Eq. (44), we get the expected potential drop across the channel, i.e. $\psi_0(x = L_{\text{eff}}) - \psi_0(x = 0) = V_{\text{dsi}}$. The sensitivity of intrinsic gate capacitance on the numerical value of λ_{sl} is examined in Sec. 4.

In the linear region, $\eta_s \approx \eta_d$ and the channel potential is nearly symmetric about the channel center ($L_{\text{eff}}/2$). In this case,

$$u(x) \approx u(0) + (u(L_{\text{eff}}) - u(0))\left(1 - \exp\left(\frac{x}{-\lambda_{\text{sl}}}\right)\right). \tag{45}$$

3.4.1. *Q-V model in the saturation region*

In the saturation region, we apply current continuity and energy conservation equations to derive the position-dependent channel charge, $Q_{\text{ch}}(x)$. The total energy of carriers in the channel is the sum of their kinetic and potential energy. At the ToB ($x = x_0$), the kinetic energy of carriers is $\frac{1}{2}mv_{\text{x0}}^2$. At any general location x along the channel length, the change of carrier potential energy is $u(x)$. Under QB transport, the potential energy transferred to the kinetic energy of carriers is $\eta_{\text{ball}}u(x)$, where $\eta_{\text{ball}} \in (0, 1)$ is the ballisticity parameter and is chosen equal to $\mathcal{T}_{\text{sat}}$. Using energy conservation, the position-dependent velocity of carriers in the channel is given as

$$v(x) = \sqrt{v_{\text{x0}}^2 + \frac{2q}{m_{\text{eff}}}\eta_{\text{ball}}u(x)}. \tag{46}$$

Based on current continuity in the channel, the position-dependent channel charge, $Q_{\text{ch}}(x)$, is given as

$$Q_{\text{ch}}(x) = qn(x) = q\frac{n_{\text{x0}}v_{\text{x0}}}{v(x)} = \frac{qn_{\text{x0}}v_{\text{x0}}}{\sqrt{v_{\text{x0}}^2 + \frac{2q}{m_{\text{eff}}}\eta_{\text{ball}}u(x)}}, \tag{47}$$

where $n(x)$ is the channel carrier concentration. Using Eq. (44) in Eq. (47) and carrying out the integral in Eq. (41), the gate and drain terminal charges (Q_{GQB} and Q_{DQB}) under QB transport conditions can be obtained (results not shown here for brevity).

The analytic $v(x)$ model in Eq. (46) is strictly valid in saturation transport conditions as it predicts that the velocity of charges will increase along the channel from the source-end toward the drain-end even for V_{ds} in the linear regime, unless η_{ball} is chosen to be nearly zero.

3.4.2. *Q-V model in the linear region*

In the linear transport region, when V_{ds} is low, the charge concentration under the gated region is nearly flat except for an increase near the source/drain edges. This non-uniform charge distribution in the channel can be modeled as

$$\begin{aligned} Q_{\text{ch}}(x) &= Q_{\text{ch}}(x=0) + \\ &\left(Q_{\text{ch}}\left(x = \frac{L_{\text{eff}}}{2}\right) - Q_{\text{ch}}(x=0)\right) \times \left(1 - \exp\left(\frac{\left|\frac{L_{\text{eff}}}{2} - x\right| - \frac{L_{\text{eff}}}{2}}{\lambda_{\text{sl}}}\right)\right). \end{aligned} \tag{48}$$

The above charge profile can be also explained by considering that in the linear region, the profile of $Q_{\text{ch}}(x)$ is expected to be similar to that of $u(x)$ due to the validity of the GCA throughout the channel (see Eq. (45)).

Substituting Eq. (48) in Eq. (41) and carrying out the integration, the terminal charges in the linear regime at source (Q_{SLIN}), drain (Q_{DLIN}) and gate (Q_{GLIN}) are given as

$$Q_{\text{SLIN}} = -qP_{\text{ne,s}}WL_{\text{eff}}n_{\text{x0}}, \tag{49a}$$

$$Q_{\text{DLIN}} = -qP_{\text{ne,d}}WL_{\text{eff}}n_{\text{x0}}, \tag{49b}$$

$$Q_{\text{GLIN}} = -Q_{\text{SLIN}} - Q_{\text{DLIN}}, \tag{49c}$$

where the coefficients $P_{\text{ne,s}}$ and $P_{\text{ne,d}}$ account for the effects of electric field penetration from the source and drain access regions, respectively, into the channel. In the case of symmetric source/drain access regions, we expect $P_{\text{ne,s}} = P_{\text{ne,d}} = 1/2P_{\text{ne}}$. The total penetration coefficient $P_{\text{ne}} = P_{\text{ne,s}} + P_{\text{ne,d}}$ is given as

$$\begin{aligned} P_{\text{ne}} &= \frac{\int_0^{L_{\text{eff}}} Q_{\text{ch}}(x)dx}{L_{\text{eff}}Q_{\text{ch}}(x = L_{\text{eff}}/2)} \\ &= 1 - \frac{2\lambda_{\text{sl}}}{L_{\text{eff}}}\left(1 - \exp\left(\frac{-L_{\text{eff}}}{2\lambda_{\text{sl}}}\right)\right). \end{aligned} \tag{50}$$

For $L_{\text{eff}} \gg \lambda_{\text{sl}}$, P_{ne} is close to unity, indicating that the penetration of electric field lines from access regions into the channel can be ignored. For the purpose of model calibration against measured data, P_{ne} is tuned within ±10% of its value given by Eq. (50).

3.4.3. *Fringing capacitances*

In addition to the terminal charges due to intrinsic channel charges, there are fringing field charges that must be included in the Q-V model (see Fig. 13). The outer fringing field lines extend from the gate metal to the 2DEG in the access regions. These field lines exist in both on- and off- regimes of the HEMT. The outer fringing charges are given as

$$Q_{\text{ofs}} = -C_{\text{ofs}} V_{\text{gsi}}, \tag{51a}$$

$$Q_{\text{ofd}} = -C_{\text{ofd}} V_{\text{gdi}}, \tag{51b}$$

where C_{ofs} and C_{ofd} are bias-independent outer-fringing capacitances of the source and the drain terminals, respectively.

The inner fringing field lines extend from the gate metal to the access regions through the channel region. In the on-state there is a significant 2DEG in the channel, which screens the inner fringing field lines. Therefore, inner-fringing capacitance, C_{if}, will decrease rapidly as V_{gs} becomes greater than a certain off-voltage. However, in off-state, C_{if} is also expected to decrease as V_{gs} becomes highly negative. This is due to the widening of the depletion region under the gate, which reduces the net inner-fringing charge, Q_{if}.

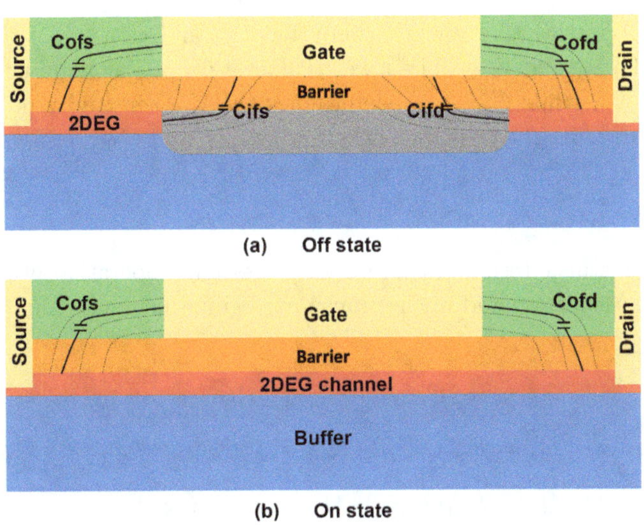

Fig. 13. Electric field lines from the gate to the access region in off-state and on-state. In on-state, inner fringing field lines are screened out due to the finite 2DEG in the channel.

Based on this, we expect C_{if} to vary non-monotonically with V_{gs} and display a peak at a specific V_{gs} value in the vicinity of off-voltage. These physical features of the inner-fringing capacitance can be captured through the following model of $Q_{\mathrm{if(s/d)}}$:

$$Q_{\mathrm{if(s/d)}} = \underbrace{-\left(C_{\mathrm{on(s/d)}} - C_{\mathrm{corr(s/d)}} V_{\mathrm{dsi}}\right) F_{1(s/d)}}_{\text{On-state}} - \underbrace{\left(C_{\mathrm{on(s/d)}} - C_{\mathrm{if0(s/d)}} - \left(C_{\mathrm{corr(s/d)}} + C_{\mathrm{corr0(s/d)}}\right) V_{\mathrm{dsi}}\right) F_{2(s/d)}}_{\text{Off-state}}, \tag{52}$$

where $F_{1(s/d)}$ and $F_{2(s/d)}$ are logistic functions given as

$$F_{1(s/d)} = V_{\mathrm{g(si/di)}} - n_1 \phi_t \ln\left[1 + \exp\left(\frac{V_{\mathrm{g(si/di)}} - (V_{\mathrm{off1}} - \delta_{\mathrm{1d}} V_{\mathrm{dsi}})}{n_1 \phi_t}\right)\right]. \tag{53}$$

$$F_{2(s/d)} = -V_{\mathrm{g(si/di)}} + n_2 \phi_t \ln\left[1 + \exp\left(\frac{V_{\mathrm{g(si/di)}} - (V_{\mathrm{off2}} - \delta_{\mathrm{2d}} V_{\mathrm{dsi}})}{n_2 \phi_t}\right)\right]. \tag{54}$$

The parameter n_1 in $F_{1(s/d)}$ models the rate at which the inner-fringing charges are screened out by the channel charge in on-state, while the parameter n_2 in $F_{2(s/d)}$ models the rate at which the depletion region widens in off-state. The voltage $(V_{\mathrm{off2}} - \delta_{\mathrm{2d}} V_{\mathrm{dsi}})$ gives the maximum $V_{\mathrm{g(si/di)}}$ voltage below which the depletion region widening dominates the inner-fringing charges. The inner fringing capacitance will peak at $V_{\mathrm{g(si/di)}} = (V_{\mathrm{off1}} - \delta_{\mathrm{1d}} V_{\mathrm{dsi}})$ and will decrease with further increase in $V_{\mathrm{g(si/di)}}$ due to screening. The value of V_{off1} is similar in magnitude to V_{off}, where V_{off} is the off-state voltage corresponding to intrinsic channel charges. The peak value of $C_{\mathrm{if(s/d)}}$ is set by $C_{\mathrm{on(s/d)}}$, while $C_{\mathrm{corr(s/d)}}$ gives the V_{dsi} dependency of $C_{\mathrm{on(s/d)}}$. The parameters δ_{1d} and δ_{2d} are non-zero only for C_{ifd} and can model the drain-bias dependence of V_{off1} and V_{off2}. The lowest value of $C_{\mathrm{if(s/d)}}$ at large negative V_{gsi} is given by $C_{\mathrm{if0(s/d)}}$, while $C_{\mathrm{corr0(s/d)}}$ determines the V_{dsi} sensitivity of $C_{\mathrm{if0(s/d)}}$. For symmetric source/drain access regions, $C_{\mathrm{ons}} = C_{\mathrm{ond}}$, and $C_{\mathrm{if0s}} = C_{\mathrm{if0d}}$.

3.4.4. *Unified Q-V model*

To obtain the intrinsic terminal charges for large-signal operation, we combine the charges in the saturation and linear transport regimes using an empirical function F_{sat}:

$$Q_{\mathrm{gi}} = F_{\mathrm{sat}} Q_{\mathrm{GQB}} + (1 - F_{\mathrm{sat}}) Q_{\mathrm{GLIN}}, \tag{55a}$$

$$Q_{\mathrm{si}} = F_{\mathrm{sat}} Q_{\mathrm{SQB}} + (1 - F_{\mathrm{sat}}) Q_{\mathrm{SLIN}}, \tag{55b}$$

$$Q_{\mathrm{di}} = F_{\mathrm{sat}} Q_{\mathrm{DQB}} + (1 - F_{\mathrm{sat}}) Q_{\mathrm{DLIN}}. \tag{55c}$$

The function F_{sat} is the same as that used for modeling the static I-V transport (see Eq. 25b).

The total terminal charge is sum of the intrinsic and fringing charges. That is,

$$Q_s = Q_{\text{si}} + Q_{\text{ifs}} + Q_{\text{ofs}}, \tag{56a}$$

$$Q_d = Q_{\text{di}} + Q_{\text{ifd}} + Q_{\text{ofd}}, \tag{56b}$$

$$Q_g = -\left(Q_d + Q_s\right). \tag{56c}$$

Once the terminal charges are known, the inter-nodal capacitances and the full capacitance matrix of the HEMT can be constructed using Eq. (42).

We develop a unified large-signal compact model to describe static (I-V) and dynamic (Q-V) behavior of ultra-scaled III-nitride HEMTs in which transport is QB in nature.

4. Model Validation

The unified model to describe I-V and Q-V behavior of QB III-nitride HEMTs has a total of 38 parameters, most of which have a physical interpretation and can be extracted from device characterization. A systematic procedure for parameter extraction using this model has been outlined in our previous work [109]. We validate the model against measured data of 42-nm and 105-nm gate-length III-nitride HEMTs, as well as a 50-nm gate-length GaN HEMT that was simulated in Sentaurus. The static model for digital GaN NW FETs contains 16 parameters and is validated by comparing model-generated results against experimentally measured I-V characteristics of a 274-nm gate length FET. Extracted model parameters are listed in Tables 2 and 3.

4.1. *105 nm and 42 nm gate-length InAlN/GaN HEMTs (experimental data)*

We use data corresponding to 25-µm wide InAlN/GaN RF devices with gate lengths of 105 nm and 42 nm, fabricated at MIT [42] to validate our HEMT model. The heterostructure of the devices comprises $In_{0.17}Al_{0.83}N/AlN/GaN$ (7.5 nm/1 nm/26 nm), with AlN as the interlayer. A 2DEG density of appriximately 1.2×10^{13} cm^{-2} and diffusive electron mobility of 1650 cm^2/Vs are reported for the GaN channel layer at 300 K. A back barrier of 3.3 nm is inserted between the channel layer and thick GaN buffer. Contact resistance of ohmic contacts formed with Si/Ge/Ti/Al/Ni/Au metal is found to be 0.1 Ω.mm. The 105-nm gate length device has symmetric access regions measuring 0.46 µm each. For the 42-nm gate-length device, the source access region length is 0.56 µm, while the drain access region length is 0.42 µm. The flat-band voltage of both 42-nm and 105-nm gate-length devices is ≈ 3.0 V. Hall measurements indicate sheet resistance of the access regions to be around 200 Ω/□.

Table 2. Input parameters for the unified static and dynamic model. Devices with effective channel length of 42 nm and 105 nm are fabricated at MIT. The device with effective channel length of 50 nm is simulated in Sentaurus. We assume that the effective channel length is the same as the gate length.

Model Parameter	L_{eff} **42 nm**	L_{eff} **105 nm**	L_{eff} **50 nm**	**Parameter meaning**
Geometry parameters				
W (µm)	25	25	10	Device width
L_s (nm)	560	460	400	Length of the source access region
L_d (nm)	420	460	400	Length of the drain access region
Static model parameters exclusive to access regions				
R_c (ohm-mm)	0.1	0.1	0.1	Contact resistance
Q_{acc} (C/m^2)	0.05	0.025	0.0067	Uniform charge density in access regions
$E_{\text{crit},0}$ (V/m)	4.0e5	4.8e5	1.0e6	Critical field for current saturation in access regions at 300 K
ϵ_{crit} (unit-less)	-1.8	-1.6	-1.5	Temperature coefficient for $E_{\text{crit},0}$
β_{acc} (unit-less)	1.7	1.8	1.8	Shape parameter for F_{sat} function to merge linear and saturation regimes in access regions
Static model parameters for the entire device				
$\mu_{\text{eff},0}$ (cm^2/Vs)	1650	1650	1650	Diffusive carrier mobility in the channel at 300 K
ζ	1.5	2	2	Temperature exponent coefficient of $\mu_{\text{eff},0}$
R_{th} (K/W)	686	686	800	Thermal resistance of the device
C_{ins} (F/m^2)	0.0069	0.0069	0.01	Areal barrier capacitance
n_{s0} ($\times 10^{17}$m^{-2})	1.32	1.3	0.48	Off-state carrier concentration in the channel
γ_0 (Vm$^{-2/3}$)	10^{-12}	10^{-12}	10^{-12}	Quantization of the first sub-band energy in conduction band
e (unit-less)	-0.62	-0.65	-0.45	Parameter for wave-function tunneling in barrier
θ (unit-less)	5	5	5	Parameter to model saturation of L_{crit} with V_{ds}
n_0 (unit-less)	3.3	1.7	2.3	Non-ideality parameter
n_d (V^{-1})	0.48	0.0535	0.7	Punch-through factor
δ (mV/V)	260	140	120	Drain-induced barrier lowering

β (unit-less)	1.7	1.8	1.8	Shape parameter for F_{sat} to merge linear and saturation regimes in the channel
β_{crit} (unit-less)	1.7	1.8	1.8	Shape parameter for F_{sat} used in L_{crit} formulation
η_{v0} ($\text{V}^{-3/4}$)	6.7	4.0	5.0	Parameter to model the bias dependence of L_{crit}
η_e (unit-less)	-0.6	-0.8	-0.5	Parameter to model temperature dependence of L_{crit}
Q_{01} (C/m^2)	0.004	0.007	0.006	Parameter to capture the degradation of μ_{eff} with carrier concentration
v_1 (unit-less)	2	1.6	2.9	Exponent coefficient to capture the degradation of μ_{eff} with carrier concentration
Dynamic model parameters				
$C_{\text{on(s/d)}}$ (fF/mm)	70/10	60/60	45/45	Maximum C_{if} before screening via mobile channel charges
$C_{\text{corr(s/d)}}$ (fF/mmV)	-5/45	15/70	-20/30	V_{dsi} dependency of C_{on}
$C_{\text{if0(s/d)}}$ (fF/mm)	5/5	15/15	3/3	Inner fringing capacitance extracted from C_{gg}-L_g plot
$C_{\text{corr0(s/d)}}$ (fF/mmV)	7/7	-25/55	-8/17	V_{dsi} dependency of C_{if0}
n_1 (unit-less)	4	3	10	Rate of C_{if} screening by channel charge
n_2 (unit-less)	10	12	10	Rate of C_{if} reduction due to the extended depletion region
V_{off1} (V)	-3.8	-3.4	-0.98	Minimum voltage beyond which the screening effect dominates
V_{off2} (V)	-4.8	-4.2	-1.8	Maximum voltage below which extension of depletion region dominates
δ_{1d} (unit-less)	0	0.3	0.3	V_{dsi} dependency of drain side V_{off1}
δ_{2d} (unit-less)	0	0	0.15	V_{dsi} dependency of drain side V_{off2}
$P_{\text{ne(s/d)}}$ (unit-less)	0.37/0.37	0.43/0.43	0.245/0.245	Electric field penetration from access regions into the channel
λ_{sl} (nm)	4.0	4.0	12	Electrostatic scaling length
$C_{\text{of(s/d)}}$ (fF/mm)	132/132	132/132	95/95	Outer fringing capacitance

Table 3. Input parameters for the static I-V model for the GaN NW FET. Here, we ignore thermal effects, as well as the resistance of the forward biased drain side Schottky contact. The resistance of the access region is considered to be voltage-independent.

Model Parameter	**Nanowire 274 nm**	**Parameter meaning**
	Geometry parameters	
r (nm)	73	Device width
L_s (nm)	150	Length of the source access region
L_d (nm)	150	Length of the drain access region
	Static model parameters exclusive to access regions	
R_{acc} (Ω)	5400	Resistance of access region
V_{BK} (V)	0.4	Breakdown voltage of Schottky barrier
I_0 (A)	1×10^{-6}	Reverse saturation current of Schottky barrier
K_s (unit-less)	1	fitting parameter that controls the decay rate of $R_{\text{SB,s}}$
	Static model parameters for the entire device	
$\mu_{\text{eff,0}}$ (cm^2/Vs)	130	Diffusive carrier mobility in the channel at 300 K
C_{ins} (nF/m^{-1})	1.83	Areal barrier capacitance
n_{s0} ($\times10^{10}$m^{-1})	1.89	Off-state carrier concentration in the channel
γ_0 (Vm$^{-2/3}$)	10^{-12}	Quantization of the first sub-band energy in conduction band
e (unit-less)	0	Parameter for wave-function tunneling in barrier
θ (unit-less)	5	Parameter to model saturation of L_{crit} with V_{ds}
n_0 (unit-less)	1.13	Non-ideality parameter
n_d (V^{-1})	0	Punch-through factor
δ (mV/V)	27	Drain-induced barrier lowering
β (unit-less)	2	Shape parameter for F_{sat} to merge linear and saturation regimes in the channel
β_{crit} (unit-less)	2	Shape parameter for F_{sat} used in L_{crit} formulation
η_{v0} (V$^{-3/4}$)	0.65	Parameter to model bias dependence of L_{crit}
Q_{01} (C/m)	2.2×10^{-10}	Parameter to capture the degradation of μ_{eff} with carrier concentration
v_1 (unit-less)	0.3	Exponent coefficient to capture the degradation of μ_{eff} with carrier concentration

Model fits to the static I-V experimental data of the devices are shown in Fig. 15 (a, c). The model predicts the I-V characteristics accurately over a broad range of drain and gate bias voltages. The model does not include gate leakage which is why the fit of I-V curves at large negative gate voltages ($V_{gs} < -4$ V) is poor. The dynamic C-V model is also validated against experimental data as shown in Fig. 15 (b, d). The capacitive elements are obtained from s-parameter measurements, which are de-embedded (pad capacitances, terminal resistance and inductance removed) and converted into Y-parameters at each bias point [110]. We see that the model provides a good match for gate capacitance, C_{gg}, versus V_{gs} for various V_{ds} values, ranging from the linear to saturation regimes of operation. The poor fitting quality of I-V curves at large negative V_{gs} does not affect the fitting quality of C-V curves in this bias range. This is because in off-state the contribution of intrinsic capacitance to the total capacitance is negligible, while parasitic capacitances are dominant.

4.2. *50-nm gate-length AlGaN/GaN HEMT (numerically simulated)*

The model is validated against hydrodynamic numerical simulations using the TCAD tool Sentaurus [111, 112]. Here, we consider a heterostructure of 12 nm $Al_{0.3}Ga_{0.7}N$ barrier layer with 100 nm GaN channel. A back barrier of 1 nm $Al_{0.04}Ga_{0.96}N$ is inserted between the channel and GaN buffer layer. The device length is fixed to 50 nm. The thickness of GaN buffer is set at 2.5 µm, and the substrate is chosen as SiC. The device is passivated with 50 nm Si_3N_4. Ohmic contacts are simulated by introducing a highly doped region under the source and drain terminals with phosphorus doping (10^{20} cm^{-3}).

The Schottky barrier height of the gate contact is fixed at 1.45 eV. To correctly model the carrier temperature in the device, the hydrodynamic transport model is used for both electrons and holes in Sentaurus [112]. The temperature dependence of bandgap and the density of states is incorporated [113–115]. The low-field ballistic mobility model is applied to incorporate the reduction of low-field mobility due to ballistic effects [116]. The temperature dependence of mobility is modeled with constant mobility model [117]. The net carrier mobility due to ballistic effects and temperature degradation is obtained using the Matthiessen's rule. Interface traps are assumed to be fully occupied, while dynamic traps due to deep defect levels are modeled using Shockley-Read-Hall recombination [118]. The C-V data is obtained from Sentaurus AC simulations at a frequency of 100 MHz.

The value of λ_{sl} for this device is obtained by fitting the numerically extracted potential profile at the hetero-interface along the channel length with the analytic model in Eq. (44). The penetration of electric field lines from access regions is expected to be more significant in this device since L_{eff} is comparable to λ_{sl}.

We also study the gate-length scaling of C_{gd} of the device. Results are shown in Fig. 14 for $(V_{gs} - V_{off}) = 1$ V and $V_{ds} = 1$ V. The total C_{gd}, which is the sum of intrinsic and fringing gate-drain capacitances, increases with gate length in the range of (15 − 90) nm. Because of the reduction in intrinsic gate charge with respect to V_{ds}, the intrinsic C_{gd} is positive in QB transport. When added to the parasitic capacitances, the total C_{gd} shows an increase with the gate length,

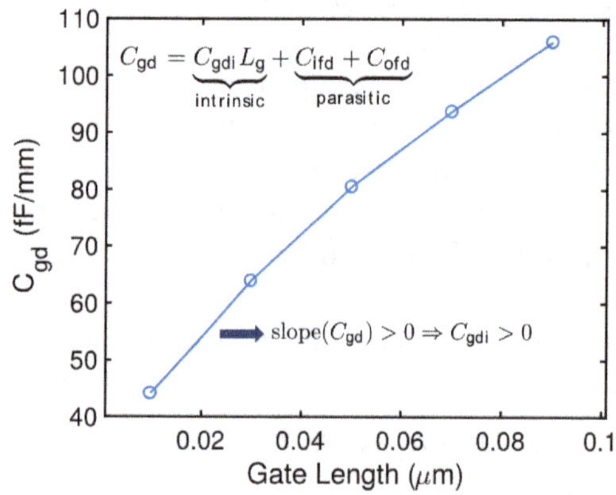

Fig. 14. Scaling of gate-drain capacitance with gate length for the numerically simulated heterostructure. Results reported at $V_{\mathrm{ds}} = 1$ V (saturation) and $V_{\mathrm{gs}} - V_{\mathrm{off}} = 1$ V. Reprinted from [Li, Kexin, and Shaloo Rakheja. "A unified static-dynamic analytic model for ultra-scaled III-nitride high electron mobility transistors." Journal of Applied Physics 125.13 (2019): 134503.], with the permission of AIP Publishing.

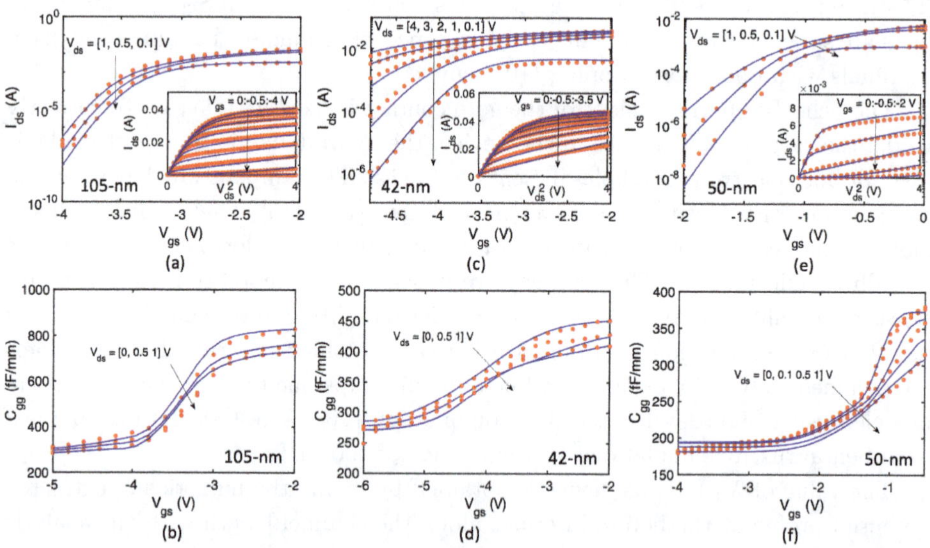

Fig. 15. Model fit to the static I-V and C-V experimental data of 105-nm, 42-nm, and 50-nm gate-length HEMTs. Solid lines are model fits, while symbols corresponds to numerical data. Reprinted from [Li, Kexin, and Shaloo Rakheja. "A unified static-dynamic analytic model for ultra-scaled III-nitride high electron mobility transistors." Journal of Applied Physics 125.13 (2019): 134503.], with the permission of AIP Publishing.

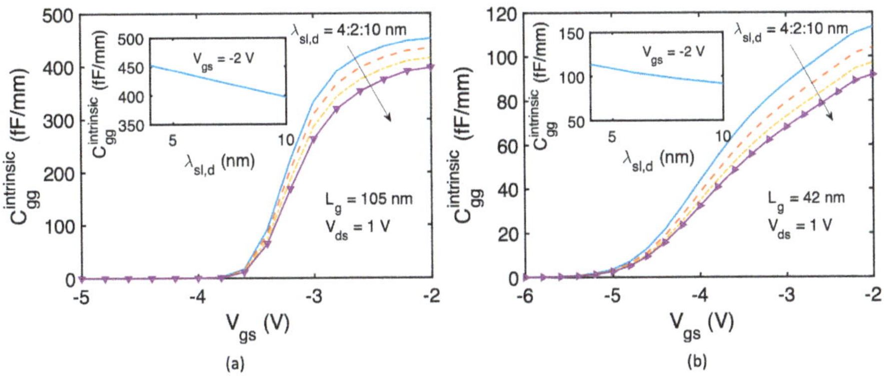

Fig. 16. Intrinsic gate capacitance for different values of drain-side electrostatic screening length, $\lambda_{sl,d}$ for (a) 105-nm gate-length device and (b) 42-nm gate-length device. The insert shows that the net charge in intrinsic gate capacitance in the on-state (V_{gs} = -2V, V_{ds} = 1V) as $\lambda_{sl,d}$ increases from 4 nm to 10 nm. Reprinted from [Li, Kexin, and Shaloo Rakheja. "A unified static-dynamic analytic model for ultra-scaled III-nitride high electron mobility transistors." Journal of Applied Physics 125.13 (2019): 134503.], with the permission of AIP Publishing.

which is confirmed by the hydrodynamic simulations of the heterostructure. On the other hand, devices with DD transport display the reverse scaling of C_{gd} with gate length. This difference in the scaling trend of C_{gd} versus gate length for QB and DD transport has also been confirmed for III-V HEMTs in prior work [119].

Under off-equilibrium conditions, it is possible for λ_{sl} to be different on the source and the drain sides. In Fig. 16, the sensitivity of the intrinsic gate capacitance to distinct values of $\lambda_{sl,s}$ and $\lambda_{sl,d}$ is shown. We see that the impact of changing λ_{sl} separately for the source and drain terminals affects the intrinsic gate capacitance by as much as 10%. However, due to finite parasitic (outer-fringing) capacitance, the effect of $\lambda_{sl,s} \neq \lambda_{sl,d}$ on the total gate capacitance will be lower than 10%. Here, we consider $\lambda_{sl,s} = \lambda_{sl,d}$ without sacrificing the quality of C-V model fits.

4.3. *GaN NW n-FET (experimentally measured)*

The GaN NW device used for model validation has a gate length of 274 nm, and the diameter of the NW is 146 nm. The conformal Al_2O_3 oxide is 16 nm thick and is formed at the hexagonal sidewall surface of the channel [35]. The channel region is uniformly doped with Si at a concentration of $1 \pm 0.1 \times 10^{18}$ cm^{-3}. The access region at the drain and source sides are symmetric, and measure approximately 150 nm. The access region contributes a series resistance of $\sim$5.4 kΩ per side. A sharp turn-on behavior that exhibits an SS of 68 mV/dec and DIBL of 27 mV/V are observed from the transfer characteristics of the device. The non-linear increase of current under low drain bias as shown in the output characteristics indicates the presence of a reverse-biased Schottky barrier at the source contact. Based on temperature-dependent measurements of the drain current at low-V_{ds}, a Schottky barrier height of 0.29 eV is extracted for the contacts. As explained earlier, the drain side Schottky contact is forward biased, while the source Schottky contact is reverse biased. Therefore, the drain contact resistance is assumed to be negligible relative to the source side contribution. Model fits to the measured data are shown in Fig. 17. The model can capture the I-V characteristics accurately over a board range of drain and gate bias voltage. The nonlinear increase of drain current under low drain bias in the output characteristics is also accurately represented. The nonlinearity of source side Schottky barrier diminishes at high V_{ds} when the voltage exceeds the breakdown voltage. The NW FET is normally on and a V_{gs} < -4 V is needed to turn it off (see Fig. 17b). The model-fitted low-field diffusive carrier mobility agrees with the experimentally reported value [78]. Joule heating effects are not included for the NW FET.

The effect of Schottky contacts on output characteristics of the NW FET is examined in Fig. 18. Considering ohmic contacts overestimates the on-current of

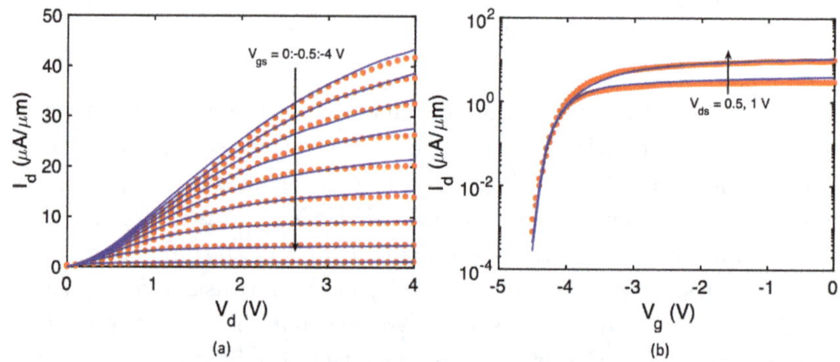

Fig. 17. Model fit to the experimental (a) output characteristics and (b) transfer characteristics of GaN NW n-FET. Solid lines are model fits, while symbols correspond to the experimental data [35].

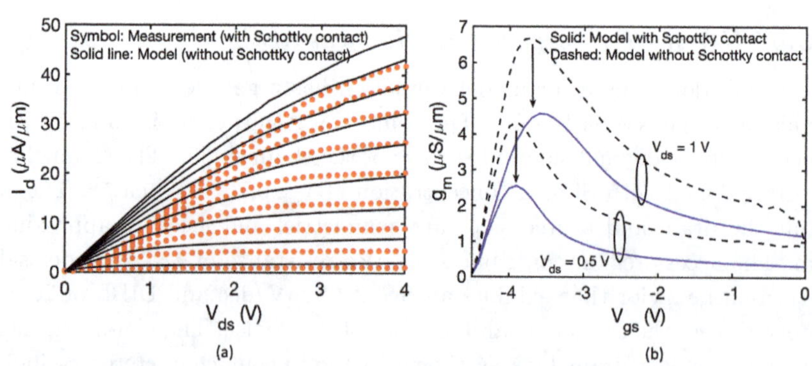

Fig. 18. Effect of Schottky contacts on the (a) output characteristics and (b) transconductance of the GaN NW FET.

the device by 14.5% (V_{ds} = 4 V, V_{gs} = 0 V). Likewise the peak value of g_m achieved at V_{gs} = -3.6 V, V_{ds} = 1 V is overestimated by 45.1% if the impact of Schottky contacts is ignored.

5. Summary

We develop a unified large-signal compact model to describe static (I-V) and dynamic (Q-V) behavior of ultra-scaled III-nitride HEMTs and nanowire FETs in which transport is quasi-ballistic in nature. The static and dynamic model are obtained self-consistently and can be used to simulate the dc and transient response of the device to time-varying inputs over a broad range of temperature and bias voltages. With a total of 38 parameters, most of which have a physical origin and can be obtained from device characterization, the model provides a physical and intuitive insight into device operation. The accuracy of the model is demonstrated by comparing model results against experimental and numerical simulation data of InAlN-on-GaN and AlGaN-on-GaN HEMTs with gate lengths of 42 nm, 50 nm, and 105 nm. With proper justification, the model is also capable of capture I-V characteristics of GaN based nanowire n-FET. The nonlinear increase of drain current under low drain bias as observed from output characterstics is due to Schottky cotacts formed on source and drain side. The model is flexible enough to capture the I-V behavior of NW FETs through modification of the charge calculation using 1D density of states and by incorporating the rectifying nature of Schottky source/drain contacts. The NW FET model is validated against measured I-V data of 274-nm gate-length GaN FET with Al_2O_3 gate oxide. We use model-generated results to elucidate the impact on device on-current and maximum g_m stemming from the large non-linear contact resistance in the NW geometry. Our model currently does not capture effects of electric field and temperature gradients in the device and time-dependent degradation mechanisms, such as Joule heating, trapping and de-trapping phenomena, that could limit the benefits of III-nitride technology in real commercial and military applications.

Acknowledgments

The authors acknowledge the support of The Boeing Company for sponsoring this research.

References

1. K. Andersson, V. Desmaris, J. Eriksson, N. Rorsman and H. Zirath, C-band linear resistive wide bandgap FET mixers, in *IEEE MTT-S International Microwave Symposium Digest, 2003*, (2003), pp. 1303–1306.
2. V. Bassoo, K. Tom, A. K. Mustafa, E. Cijvat, H. Sjoland and M. Faulkner, *EURASIP Journal on Wireless Communications and Networking* **2009**, 1 (2009).
3. U. K. Mishra, L. Shen, T. E. Kazior and Y.-F. Wu, *Proceedings of the IEEE* **96**, 287 (2008).

4. M. Micovic, D. Brown, D. Regan, J. Wong, Y. Tang, F. Herrault, D. Santos, S. Burnham, J. Tai, E. Prophet and I. Khalaf, High frequency GaN HEMTs for RF MMIC applications, in *2016 IEEE International Electron Devices Meeting (IEDM)*, (2016), pp. 3.3.1–3.3.4.
5. M. Micovic, A. Kurdoghlian, P. Hashimoto, M. Hu, M. Antcliffe, P. Willadsen, W. Wong, R. Bowen, I. Milosavljevic, A. Schmitz and M. Wetzel, GaN HFET for W-band power applications, in *2006 International Electron Devices Meeting (IEDM)*, (2006), pp. 1–3.
6. K. Makiyama, T. Ohki, M. Kanamura, K. Imanishi, N. Hara and T. Kikkawa, *physica status solidi (a)* **204**, 2054 (2007).
7. J. W. Chung, W. E. Hoke, E. M. Chumbes and T. Palacios, *IEEE Electron Device Letters* **31**, 195 (2010).
8. K. Shinohara, A. Corrion, D. Regan, I. Milosavljevic, D. Brown, S. Burnham, P. Willadsen, C. Butler, A. Schmitz, D. Wheeler and A. Fung, 220GHz f_T and 400GHz f_{max} in 40-nm GaN DH-HEMTs with re-grown ohmic, in *2010 International Electron Devices Meeting (IEDM)*, (2010), pp. 30–1.
9. K. Shinohara, D. Regan, A. Corrion, D. Brown, Y. Tang, J. Wong, G. Candia, A. Schmitz, H. Fung, S. Kim and M. Micovic, Self-aligned-gate GaN-HEMTs with heavily-doped n+-GaN ohmic contacts to 2DEG, in *2012 International Electron Devices Meeting (IEDM)*, (2012), pp. 27.2.1–27.2.4.
10. Y. Tang, K. Shinohara, D. Regan, A. Corrion, D. Brown, J. Wong, A. Schmitz, H. Fung, S. Kim and M. Micovic, *IEEE Electron Device Letters* **36**, 549 (2015).
11. S. Wienecke, B. Romanczyk, M. Guidry, H. Li, E. Ahmadi, K. Hestroffer, X. Zheng, S. Keller and U. K. Mishra, *IEEE Electron Device Letters* **38**, 359 (2017).
12. H.-Q. Tao, W. Hong, B. Zhang and X.-M. Yu, *IEEE Microwave and Wireless* **27**, 73 (2016).
13. Y.-F. Wu, A. Saxler, M. Moore, R. Smith, S. Sheppard, P. Chavarkar, T. Wisleder, U. Mishra and P. Parikh, *IEEE Electron Device Letters* **25**, 117 (2004).
14. D. Dumka and P. Saunier, *Electronics Letters* **46**, 946 (2010).
15. D. Marti, S. Tirelli, A. R. Alt, J. Roberts and C. Bolognesi, *IEEE Electron Device Letters* **33**, 1372 (2012).
16. D. Dumka, T. Chou, J. Jimenez, D. Fanning, D. Francis, F. Faili, F. Ejeckam, M. Bernardoni, J. Pomeroy and M. Kuball, Electrical and thermal performance of AlGaN/GaN HEMTs on diamond substrate for RF applications, in *2013 IEEE Compound Semiconductor Integrated Circuit Symposium (CSICS)*, (2013), pp. 1–4.
17. Y. Niida, Y. Kamada, T. Ohki, S. Ozaki, K. Makiyama, Y. Minoura, N. Okamoto, M. Sato, K. Joshin and K. Watanabe, 3.6 W/mm high power density W-band InAlGaN/GaN HEMT MMIC power amplifier, in *2016 IEEE Topical Conference on Power Amplifiers for Wireless and Radio Applications (PAWR)*, (2016), pp. 24–26.
18. P.-C. Chao, K. Chu, C. Creamer, J. Diaz, T. Yurovchak, M. Shur, R. Kallaher, C. McGray, G. D. Via and J. D. Blevins, *IEEE Transactions on Electron Devices* **62**, 3658 (2015).
19. M.-S. Lee, D. Kim, S. Eom, H.-Y. Cha and K.-S. Seo, *IEEE Electron Device Letters* **35**, 995 (2014).
20. X. Wang, S. Huang, Y. Zheng, K. Wei, X. Chen, G. Liu, T. Yuan, W. Luo, L. Pang, H. Jiang and J. Li, *IEEE Electron Device Letters* **36**, 666 (2015).
21. W. Shaobing, G. Jianfeng, W. Weibo and Z. Junyun, *IEEE Transactions on Electron Devices* **63**, 3882 (2016).
22. A. M. Darwish, K. Boutros, B. Luo, B. D. Huebschman, E. Viveiros and H. A. Hung, *IEEE Transactions on Microwave Theory and Techniques* **54**, 4456 (2006).

23. T. Palacios, A. Chakraborty, S. Rajan, C. Poblenz, S. Keller, S. DenBaars, J. Speck and U. Mishra, *IEEE Electron Device Letters* **26**, 781 (2005).
24. M. Gonschorek, J.-F. Carlin, E. Feltin, M. Py and N. Grandjean, *Applied physics letters* **89**, 062106 (2006).
25. U. K. Mishra, P. Parikh and Y.-F. Wu, *Proceedings of the IEEE* **90**, 1022 (2002).
26. B. P. Downey, D. J. Meyer, D. S. Katzer, J. A. Roussos, M. Pan and X. Gao, *IEEE Electron Device Letters* **35**, 527 (2014).
27. X. Zheng, M. Guidry, H. Li, E. Ahmadi, K. Hestroffer, B. Romanczyk, S. Wienecke, S. Keller and U. K. Mishra, *IEEE Electron Device Letters* **37**, 77 (2015).
28. B. Romanczyk, S. Wienecke, M. Guidry, H. Li, E. Ahmadi, X. Zheng, S. Keller and U. K. Mishra, *IEEE Transactions on Electron Devices* **65**, 45 (2017).
29. B. Romanczyk, M. Guidry, S. Wienecke, H. Li, E. Ahmadi, X. Zheng, S. Keller and U. Mishra, *Electronics Letters* **52**, 1813 (2016).
30. K. Shinohara, D. C. Regan, Y. Tang, A. L. Corrion, D. F. Brown, J. C. Wong, J. F. Robinson, H. H. Fung, A. Schmitz, T. C. Oh *et al.*, *IEEE Transactions on Electron Devices* **60**, 2982 (2013).
31. S. Fischer, C. Wetzel, E. Haller and B. Meyer, *Applied Physics Letters* **67**, 1298 (1995).
32. E. Matioli and T. Palacios, *Nano letters* **15**, 1070 (2015).
33. N. Chowdhury, G. Iannaccone, G. Fiori, D. A. Antoniadis and T. Palacios, *IEEE Electron Device Letters* **38**, 859 (2017).
34. G. Doornbos and M. Passlack, *IEEE Electron Device Letters* **31**, 1110 (2010).
35. W. Li, M. D. Brubaker, B. T. Spann, K. A. Bertness and P. Fay, *IEEE Electron Device Letters* **39**, 184 (2018).
36. S. Rakheja, M. S. Lundstrom and D. A. Antoniadis, *IEEE Transactions on Electron Devices* **62**, 2786 (2015).
37. A. Kranti, S. Haldar and R. Gupta, *Solid-State Electronics* **46**, 621 (2002).
38. X. Cheng, M. Li and Y. Wang, *IEEE Transactions on Electron Devices* **56**, 2881 (2009).
39. X. Cheng and Y. Wang, *IEEE Transactions on Electron Devices* **58**, 448 (2011).
40. N. Karumuri, G. Dutta, N. DasGupta and A. DasGupta, *IEEE Transactions on Electron Devices* **63**, 4226 (2016).
41. S. Khandelwal, C. Yadav, S. Agnihotri, Y. S. Chauhan, A. Curutchet, T. Zimmer, J.-C. De Jaeger, N. Defrance and T. A. Fjeldly, *IEEE Transactions on Electron Devices* **60**, 3216 (2013).
42. U. Radhakrishna, Modeling gallium-nitride based high electron mobility transistors: Linking device physics to high voltage and high frequency circuit design, PhD thesis, Massachusetts Institute of Technology (2016).
43. I. Angelov, H. Zirath and N. Rosman, *IEEE Transactions on Microwave Theory and Techniques* **40**, 2258 (1992).
44. S. Khandelwal and T. A. Fjeldly, A physics based compact model for drain current in AlGaN/GaN HEMT devices, in *2012 24th International Symposium on Power Semiconductor Devices and ICs (ISPSD)*, (2012), pp. 241–244.
45. A. Bykhovski, B. Gelmont and M. Shur, *Journal of Applied Physics* **81**, 6332 (1997).
46. O. Ambacher, B. Foutz, J. Smart, J. Shealy, N. Weimann, K. Chu, M. Murphy, A. Sierakowski, W. Schaff, L. Eastman and R. Dimitrov, *Journal of Applied Physics* **87**, 334 (2000).
47. O. Ambacher, J. Smart, J. Shealy, N. Weimann, K. Chu, M. Murphy, W. Schaff, L. Eastman, R. Dimitrov, L. Wittmer and M. Stutzmann, *Journal of Applied Physics* **85**, 3222 (1999).

48. B. Kang, S. Kim, J. Kim, F. Ren, K. Baik, S. Pearton, B. Gila, C. Abernathy, C.-C. Pan, G.-T. Chen and J. Chyi, *Applied Physics Letters* **83**, 4845 (2003).
49. A. Bykhovski, R. Gaska and M. Shur, *Applied Physics Letters* **73**, 3577 (1998).
50. Gonschorek, M and Carlin, J-F and Feltin, E and Py, MA and Grandjean, N and Darakchieva, Vanya and Monemar, Bo and Lorenz, M and Ramm, G, *Journal of Applied Physics* **103**, 093714 (2008).
51. L. Shen, S. Heikman, B. Moran, R. Coffie, N.-Q. Zhang, D. Buttari, I. Smorchkova, S. Keller, S. DenBaars and U. Mishra, *IEEE Electron Device Letters* **22**, 457 (2001).
52. T. Palacios, A. Chakraborty, S. Heikman, S. Keller, S. DenBaars and U. Mishra, *IEEE Electron device letters* **27**, 13 (2006).
53. R. Chu, Y. Zhou, J. Liu, D. Wang, K. J. Chen and K. M. Lau, *IEEE Transactions on Electron Devices* **52**, 438 (2005).
54. B. Lu, O. I. Saadat and T. Palacios, *IEEE Electron Device Letters* **31**, 990 (2010).
55. I. Khalil, E. Bahat-Treidel, F. Schnieder and J. Wurfl, *IEEE Transactions on Electron Devices* **56**, 361 (2009).
56. Y.-F. Wu, A. Saxler, T. Wisleder, M. Moore, R. Smith, S. Sheppard, P. Chavarkar and P. Parikh, Linearity performance of GaN HEMTs with field plates, in *Conference Digest [Includes' Late News Papers' volume] Device Research Conference, 2004. 62nd DRC.*, (2004), pp. 35–36.
57. T. Palacios, A. Chini, D. Buttari, S. Heikman, A. Chakraborty, S. Keller, S. P. DenBaars and U. K. Mishra, *IEEE Transactions on Electron Devices* **53**, 562 (2006).
58. J. Liu, Y. Zhou, R. Chu, Y. Cai, K. J. Chen and K. M. Lau, *IEEE Electron Device Letters* **26**, 145 (2005).
59. A. V. Vertiatchikh and L. F. Eastman, *IEEE Electron Device Letters* **24**, 535 (2003).
60. W. Lu, V. Kumar, R. Schwindt, E. Piner and I. Adesida, *Solid-State Electronics* **46**, 1441 (2002).
61. R. Aitken, V. Chandra, J. Myers, B. Sandhu, L. Shifren and G. Yeric, Device and technology implications of the internet of things, in *2014 Symposium on VLSI Technology (VLSI-Technology): Digest of Technical Papers*, (2014), pp. 1–4.
62. N. Singh, A. Agarwal, L. Bera, T. Liow, R. Yang, S. Rustagi, C. Tung, R. Kumar, G. Lo, N. Balasubramanian and D. Kwong, *IEEE Electron Device Letters* **27**, 383 (2006).
63. A. D. Franklin, M. Luisier, S.-J. Han, G. Tulevski, C. M. Breslin, L. Gignac, M. S. Lundstrom and W. Haensch, *Nano letters* **12**, 758 (2012).
64. S.-i. Takagi and A. Toriumi, *IEEE Transactions on Electron Devices* **42**, 2125 (1995).
65. H. S. Pal, K. D. Cantley, S. S. Ahmed and M. S. Lundstrom, *IEEE Transactions on Electron Devices* **55**, 904 (2008).
66. A. Majumdar, *Journal of Applied Physics* **117**, 205704 (2015).
67. J. Blakemore, *Solid-State Electronics* **25**, 1067 (1982).
68. F. Stern, *Physical Review B* **5**, 4891 (1972).
69. Q. Hao, H. Zhao, Y. Xiao and M. B. Kronenfeld, *International Journal of Heat and Mass Transfer* **116**, 496 (2018).
70. Y. Liu, M. Luisier, A. Majumdar, D. A. Antoniadis and M. S. Lundstrom, *IEEE Transactions on Electron Devices* **59**, 994 (2012).
71. A. Rahman and M. S. Lundstrom, *IEEE Transactions on Electron Devices* **49**, 481 (2002).
72. F. Assad, Z. Ren, D. Vasileska, S. Datta and M. Lundstrom, *IEEE Transactions on Electron Devices* **47**, 232 (2000).
73. A. Majumdar and D. A. Antoniadis, *IEEE Transactions on Electron Devices* **61**, 351 (2014).

74. M. Lundstrom and C. Jeong, *Near-Equilibrium Transport: Fundamentals and Applications* (World Scientific Publishing Company, 2012).
75. J. Albrecht, R. Wang, P. Ruden, M. Farahmand and K. Brennan, *Journal of Applied Physics* **83**, 4777 (1998).
76. Y. Zhang and J. Singh, *Journal of Applied Physics* **85**, 587 (1999).
77. I. Fishchuk, A. Kadashchuk, J. Genoe, M. Ullah, H. Sitter, T. B. Singh, N. Sariciftci and H. Bässler, *Physical Review B* **81**, 045202 (2010).
78. X. Dang, P. Asbeck, E. Yu, G. Sullivan, M. Chen, B. McDermott, K. Boutros and J. M. Redwing, *Applied Physics Letters* **74**, 3890 (1999).
79. L. Mansfield, K. A. Bertness, P. T. Blanchard, T. E. Harvey, A. W. Sanders and N. A. Sanford, *Journal of Electronic Materials* **38**, 495 (2009).
80. D. Antoniadis, *IEEE Transactions on Electron Devices* **63**, 2650 (2016).
81. C. Canali, G. Majni, R. Minder and G. Ottaviani, *IEEE Transactions on Electron Devices* **22**, 1045 (1975).
82. A. Khakifirooz, O. M. Nayfeh and D. Antoniadis, *IEEE Transactions on Electron Devices* **56**, 1674 (2009).
83. J. Foresi and T. Moustakas, *Applied Physics Letters* **62**, 2859 (1993).
84. J. Guo, F. Pan, M. Feng, R. Guo, P. Chou and C. Chang, *Journal of Applied Physics* **80**, 1623 (1996).
85. A. Schmitz, A. Ping, M. A. Khan, Q. Chen, J. Yang and I. Adesida, *Journal of Electronic Materials* **27**, 255 (1998).
86. H. Imadate, T. Mishima and K. Shiojima, *Japanese Journal of Applied Physics* **57**, 04FG13 (2018).
87. S. Miller and P. H. Holloway, *Journal of Electronic Materials* **25**, 1709 (1996).
88. R. C. Jaeger and T. N. Blalock, *Microelectronic circuit design* (McGraw-Hill New York, 1997).
89. E. H. Rhoderick, *IEE Proceedings I-Solid-State and Electron Devices* **129**, 1 (1982).
90. S. Ghosh, S. A. Ahsan, Y. S. Chauhan and S. Khandelwal, Modeling of source/drain access resistances and their temperature dependence in GaN HEMTs, in *2016 IEEE International Conference on Electron Devices and Solid-State Circuits (EDSSC)*, (2016), pp. 247–250.
91. W. Xing, Z. Liu, H. Qiu, G. I. Ng and T. Palacios, *IEEE Electron Device Letters* **38**, 619 (2017).
92. D. R. Greenberg and J. A. Del Alamo, *IEEE Transactions on Electron Devices* **41**, 1334 (1994).
93. M. Azize, A. L. Hsu, O. I. Saadat, M. Smith, X. Gao, S. Guo, S. Gradecak and T. Palacios, *IEEE Electron Device Letters* **32**, 1680 (2011).
94. R. J. Trew, Y. Liu, L. Bilbro, W. Kuang, R. Vetury and J. B. Shealy, *IEEE Transactions on Microwave Theory and Techniques* **54**, 2061 (2006).
95. U. Radhakrishna, T. Imada, T. Palacios and D. Antoniadis, *physica status solidi (c)* **11**, 848 (2014).
96. U. Radhakrishna, L. Wei, D.-S. Lee, T. Palacios and D. Antoniadis, Physics-based GaN HEMT transport and charge model: Experimental verification and performance projection, in *Electron Devices Meeting (IEDM), 2012 IEEE International*, (2012), pp. 13–6.
97. S. Choi, E. R. Heller, D. Dorsey, R. Vetury and S. Graham, *IEEE Transactions on Electron Devices* **60**, 1898 (2013).
98. B. M. Paine, T. Rust and E. A. Moore, *IEEE Transactions on Electron Devices* **63**, 590 (2016).
99. J. C. Freeman, Channel temperature model for microwave AlGaN/GaN power

HEMTs on SiC and sapphire, in *IEEE MTT-S International Microwave Symposium Digest*, (2004), pp. 2031–2034.
100. H. Hjelmgren, M. Thorsell, K. Andersson and N. Rorsman, *IEEE Transactions on Electron Devices* **59**, 3344 (2012).
101. W. Joyce, *Solid-State Electronics* **18**, 321 (1975).
102. J.-W. Lee and K. J. Webb, *IEEE Transactions on Microwave Theory and Techniques* **52**, 2 (2004).
103. S.-Y. Oh, D. E. Ward and R. W. Dutton, *IEEE Journal of Solid-State Circuits* **15**, 636 (1980).
104. A. S. Roy, C. C. Enz and J.-M. Sallese, *IEEE Transactions on Electron Devices* **54**, 1384 (2007).
105. A. B. Bhattacharyya, *Compact MOSFET models for VLSI design* (John Wiley & Sons, 2009).
106. Y. Tosaka, K. Suzuki and T. Sugii, *IEEE Electron Device Letters* **15**, 466 (1994).
107. R.-H. Yan, A. Ourmazd and K. F. Lee, *IEEE Transactions on Electron Devices* **39**, 1704 (1992).
108. D. J. Frank, Y. Taur and H.-S. Wong, *IEEE Electron Device Letters* **19**, 385 (1998).
109. K. Li and S. Rakheja, *Journal of Applied Physics* **125**, 134503 (2019).
110. J. W. Chung, Advanced technologies for improving high frequency performance of AlGaN/GaN high electron mobility transistors, PhD thesis, Massachusetts Institute of Technology (2008).
111. A. Majumdar, K. Fushinobu and K. Hijikata, *Journal of Applied Physics* **77**, 6686 (1995).
112. X.-D. Wang, W.-D. Hu, X.-S. Chen and W. Lu, *IEEE Transactions on Electron Devices* **59**, 1393 (2012).
113. J. Wu, W. Walukiewicz, W. Shan, K. Yu, J. Ager III, S. Li, E. Haller, H. Lu and W. J. Schaff, *Journal of Applied Physics* **94**, 4457 (2003).
114. I. Vurgaftman, J. Meyer and L. Ram-Mohan, *Journal of Applied Physics* **89**, 5815 (2001).
115. M. E. Levinshtein, S. L. Rumyantsev and M. S. Shur, *Properties of Advanced Semiconductor Materials: GaN, AIN, InN, BN, SiC, SiGe* (John Wiley & Sons, 2001).
116. M. S. Shur, *IEEE Electron Device Letters* **23**, 511 (2002).
117. C. Lombardi, S. Manzini, A. Saporito and M. Vanzi, *IEEE Transactions on Computer-Aided Design of Integrated Circuits and Systems* **7**, 1164 (1988).
118. J. M. Tirado, J. L. Sanchez-Rojas and J. I. Izpura, *IEEE Transactions on Electron Devices* **54**, 410 (2007).
119. L. Wei, O. Mysore and D. Antoniadis, *IEEE Transactions on Electron Devices* **59**, 1263 (2012).

Recent Progress in III-Nitride Tunnel Junction-Based Optoelectronics

Zane Jamal-Eddine[1,*], Yuewei Zhang[2], and Siddharth Rajan[1,3]

[1]*Department of Electrical and Computer Engineering, The Ohio State University, Columbus, Ohio 43210, USA*
[2]*Materials Department, University of California, Santa Barbara, California 93106, USA*
[3]*Department of Materials Science and Engineering, The Ohio State University, Columbus, Ohio 43210, USA*
[*]*jamal-eddine.1@osu.edu*

Tunnel junctions have garnered much interest from the III-Nitride optoelectronic research community within recent years. Tunnel junctions have seen applications in several material systems with relatively narrow bandgaps as compared to the III-Nitrides. Although they were initially dismissed as ineffective for commercial device applications due to high voltage penalty and on resistance owed to the wide bandgap nature of the III-Nitride material systems, recent development in the field has warranted further study of such tunnel junction enabled devices. They are of particular interest for applications in III-Nitride optoelectronic devices in which they can be used to enable novel device designs which could potentially address some of the most challenging physical obstacles presented with this unique material system. In this work we review the recent progress made on the study of III-Nitride tunnel junction-based optoelectronic devices and the challenges which are still faced in the field of study today.

Keywords: Light emitting diodes; tunnel junction; hole injection; polarization; cascaded LEDs.

1. Introduction

The III-nitride material system boasts a wide range of applications, owed to the unique material properties of the system. The GaInAlN alloys are direct gap semiconductors which have bandgap energies that span from the infrared wavelengths, all the way across the visible spectrum, and into the ultraviolet (UV) spectrum, including UVA, UVB, and UVC. As such they have a wide range of application in optoelectronic devices such as light emitting diodes (LEDs), laser diodes (LDs), vertical-cavity surface-emitting lasers (VCSELs), solar cells, and photodetectors [1]. Furthermore the III-nitride system is unique compared to other III-V material systems in the nature of its crystal structure. The wurtzite crystal structure gives rise to strong spontaneous polarization charge, at the same time strong piezoelectric polarization charge exists at heterojunction interfaces due to strain effects. The existence of such polarization dipoles provides extra design space for both electronic and optoelectronic devices which is not available in the zincblende structured III-V systems [2].

Just as the nitride material system offers a range of unique physical properties, it also hosts a variety of innate physical limitations, fundamental to the material system properties.

One of such limitations is the inefficient p-type doping of (Al)GaN. The high acceptor ionization energy of Mg, 140 meV for GaN [3], and 510 meV for AlN [4], is especially detrimental to III-nitride applications in UVLEDs, as the large acceptor ionization energy of the Mg doped p-AlGaN layers could lead to poor hole injection into the active region of the UVLED.

Such technical challenges are not limited to AlGaN based UVLEDs either. It is well known that visible wavelength LEDs suffer from a phenomenon known as efficiency droop. Efficiency droop occurs when LEDs are pushed to higher input powers. By increasing the driving current electrons saturate the quantum wells in the active region and non-radiative auger recombination processes take over as the dominant recombination process within the filled quantum well active region, thus leading to a reduction of the internal quantum efficiency of the LED as the driving current is increased [5]. As such the ability of visible wavelength LEDs to achieve high power output is severely limited by the efficiency droop phenomenon.

Recently III-nitride tunnel junction-based UVLEDs and cascaded visible LEDs have been proposed to address both of these issues discussed above. In fact, the issues discussed above are just two examples of problems faced in the design of efficient III-Nitride optoelectronic devices, and there are numerous other issues to be addressed, many more of which can be approached with novel tunnel junction-based device designs. Over the past decade there have been a variety of applications for tunnel junction-based III-nitride optoelectronic and electronic devices, including tunnel junction based multi-color visible LEDs [6, 7], cascaded active region LEDs [8], tunnel injected UVLEDs [9], tunnel injected laser diodes [10], tunnel junction-based current apertures for VCSELs [11], and even tunnel junction field-effect transistors [12]. Tunnel junction-based III-nitride devices have found a wide range of applications aimed at improving the efficiencies of conventional III-nitride devices, as well as to overcome some of the fundamental limitations of the III-nitride materials system and the device constraints associated with these limitations. These tunnel junction based devices will be discussed in further detail below.

2. History of III-Nitride Tunnel Junctions

The first tunnel junction device was realized by Leo Esaki in 1958 when studying a heavily doped Ge p-n junction [13]. Negative differential resistance (NDR) was observed when the device was forward biased. Such behavior is characteristic of inter-band tunneling. The current density was larger in reverse bias than it was in forward bias, which is another characteristic of tunnel junctions. It is this reverse bias operation of tunnel junctions which finds such wide application in III-Nitride emitter optoelectronics. Since Esaki first discovered the Ge tunnel junction there has been extensive study demonstrating the viability of novel tunnel junction-based optoelectronic devices in lower bandgap III-V material systems such as the III-As system. Forward biased p-GaAs/n-GaAs and p-$Al_xGa_{1-x}As$/n-GaAs tunnel junctions were implemented as tunneling interconnects for cascading solar cells into a multi-junction configuration by Miller *et al.* [14]. DeSalvo demonstrated GaAs and AlAs tunnel junctions with delta doping at the TJ interface that were shown to have ultra-low voltage drop across the tunnel junction of 0.02V and 0.1V respectively, when operated at 30A/cm^2 current density [15]. Later it was shown by Kim *et al.* that three VCSEL active region structures operating at 1.55μm wavelength could be cascaded via tunnel junctions. The tunnel junction-based cascaded VCSEL structure enabled differential quantum efficiency above 100% while maintaining a threshold current of 1kA/cm^2 and a threshold voltage V_{th} of 3.2V [16]. Such utilization of tunnel junction carrier regeneration between optical emitter active regions has still captured the research interest of the III-As optoelectronic community in recent years [17]. As mentioned in the introduction, similar cascaded active region structures have been proposed for tunnel junction-based III-nitride emitters as well. These studies will be discussed in more detail later on.

Although the tunnel junction has a wide variety of potential applications in III-nitride optoelectronics the initial attempts to realize such devices were not as fruitful as their III-As counterparts. Early experimental work on tunnel junction-based III-nitride opto-electronic devices began in 2001 when Jeon *et al.* demonstrated enhanced lateral current spreading in a blue LED [18]. The device consisted of a conventional blue LED structure comprised of an n-GaN bottom contact layer, an InGaN multi quantum well active region emitting at 455nm, and a p-GaN cladding layer, however instead of the standard p-type GaN contact formed by semitransparent ITO, there was a heavily doped p++GaN ([Mg]=3x10^{19} cm^{-3})/Si-δ doping ([Si]=3x10^{13} cm^{-2})/n++GaN ([Si]=3x10^{19} cm^{-3}) tunnel junction with an n-GaN current spreading layer on top. This novel device design circum-vented the necessity of covering the top surface of the LED with an absorbing ITO contact to the p-GaN top contact layer. In the conventional blue LED structure, an ITO top contact is required for efficient current spreading in the resistive p-GaN layer. By utilizing the tunnel junction structure the n-GaN top contact layer which contacts the tunnel junction acts as an efficient and transpaent current spreading layer and contact layer to the LED structure below. In the novel device design the tunnel junction is reverse biased as the LED is forward biased, enabling efficient hole injection into the p-GaN cladding layer of the LED as well. The removal of the absorbing ITO current spreading contact from the LED and the reduction in current crowding effect by efficient lateral current spreading in the n-GaN contact layer lead to an increase in the output power of the device. As compared to the standard ITO p-GaN based LED, the light output power was doubled in the TJLED. However, the forward voltage of the TJLED was 1.0V higher than that of the reference ITO LED at 16.3 A/cm^2. An increase in series resistance was also observed in the TJLED (45 Ω) as compared to the reference ITO LED (35 Ω). Tunneling probability depends directly on both the energy barrier height and the width of the energy barrier [19]. The increase in forward voltage and series resistance of the TJLED can be attributed to the inefficient tunneling process in the GaN based tunnel junction which has a large tunneling barrier due to the wide bandgap of GaN and the appreciable depletion width within the tunnel junction owed to the dopant solubility limits.

The III-nitride optoelectronics research community developed many novel tunnel junction-based devices designs over the next few years. Many of the new tunnel junction enabled devices employed a p++InGaN/n++GaN tunnel junction design in hopes of reducing the depletion width within the tunnel junction by achieving a higher hole concentration on the p-side of the tunnel junction. Some of these devices included a dual-wavelength LED [7], a vertical cavity violet LED [20], and a near-white cascaded LED which contained two LED active regions emitting at 455nm and 515nm grown in series and connected via tunnel junctions in order to achieve white light output [21]. In each case the TJLED had an additional forward voltage of at least 1V over the reference LEDs. Such a voltage penalty is too high to be of use commercially the research interest in the area of tunnel junction-based device structures waned over the next few years.

It was not until 2007 that the research interest in III-nitride tunnel junction-based optoelectronic devices was reignited when Grundman *et al.* demonstrated an increase in III-nitride tunnel junction efficiency by inserting a thin AlN layer within the GaN p-n tunnel junction structure [6]. Utilizing this novel heterostructure tunnel junction design authors were able to demonstrate a dual-wavelength LED and reduce the voltage dropped across each tunnel junction to roughly 0.6-0.75V per tunnel junction. The series resistance of the device was also lowered considerably from that of the "first generation" of tunnel junction-based III-nitride optoelectronics. This reduction in tunnel junction resistance was accomplished by utilizing the spontaneous and piezoelectric polarization charges at the heterojunction interface within the tunnel junction. These polarization charges give rise to a strong electric field within the tunnel junction thereby reducing the effective tunneling barrier. This effect can be visualized with the aid of the band diagram of the GaN/AlN/GaN tunnel junction versus that of the standard p++GaN/n++GaN tunnel junction shown in Fig. 1 below. The GaN/AlN/GaN heterojunction tunnel junction device was studied further by Simon *et al.* in the following years [22].

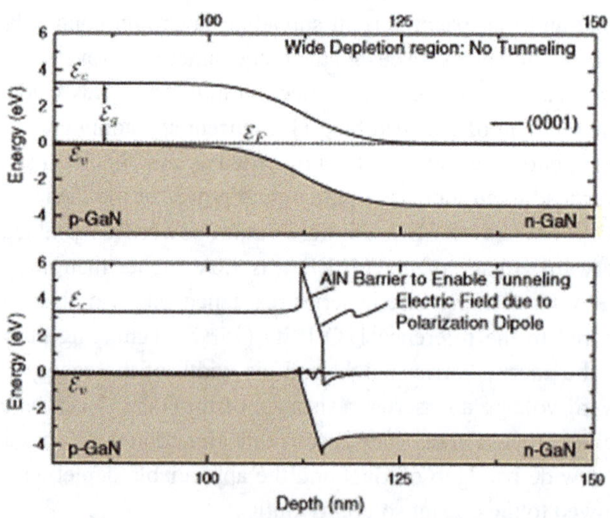

Fig. 1. The band diagram of the p++GaN/n++GaN homojunction tunnel junction (top), versus the band diagram of the p++GaN/AlN/n++GaN heterojunction tunnel junction (bottom) showing a greatly reduced tunneling barrier width due to the strong electric field within the heterojunction owed to polarization charges at the GaN/AlN interfaces [22].

Following the demonstration of the heterojunction enabled tunnel junctions theoretical works were completed which outlined the possible efficacy of utilizing such tunnel junction structures to overcome potential issues of hole injection in wide bandgap highly resistive p-AlGaN layers which are necessary components of UVLEDs [23, 24]. Krishnamoorthy *et al.* proposed a n++GaN/InGaN/p++GaN variant of the heterojunction tunnel junction device utilizing InGaN interlayers as grown on N-face GaN grown by MBE [25]. The GaN/InGaN/GaN tunnel junction design proposed by Krishnamoorthy *et al.* not only takes advantage of a reduced tunneling barrier width induced by the polarization charges at the GaN/InGaN interfaces but also reduced the energy barrier height of the tunneling barrier by utilizing a material with a narrow bandgap within the tunnel junction. Over the course of the next year Krishnamoorthy *et al.* continued to develop the GaN/InGaN/GaN tunnel junction design and by growing on N-face GaN templates they were able to grow high composition InGaN in the TJ layer, which resulted in the observation of NDR and record forward tunneling current density in a TJ/p-n junction (n-p-n) device structure [26, 27].

When designing polarization engineered tunnel junctions there are three important factors to consider, which will ensure that the polarity of the electric field caused by the polarization-induced charge is aligned with the polarity of the built in field of the heavily doped p and n layers of the tunnel junction. By increasing the overall electric field strength across the junction, aligning the polarity of the polarization-induced and built in fields enables more efficient tunneling within the heterojunction based tunnel junction. The first factor to consider is the polarity of the substrate (N-polar vs. Ga-polar). The polarity of the substrate will affect the sign of the polarization charge on either side of the heterojunction interface, which must be accounted for when designing the tunnel junction. Second, the orientation of the diode (p-side down vs. n-side down) must also be considered when designing the tunnel junction. The diode orientation determines the direction of the built in field of the heavily doped p and n layers of the tunnel junction. As discussed above the polarity of this built in field must be aligned with that of the polarization dipole created at the heterojunction interfaces. Finally, the material of the interlayer within the heterojunction tunnel junction, and what polarization charge it will induce when grown in between the heavily doped p and n layers of the tunnel junction must be considered. In the case of a GaN based tunnel junction grown on top of a p-side up LED (grown on Ga-face substrate) the addition of an $In_xGa_{1-x}N$ interlayer in the middle of the heavily doped n-GaN and p-GaN layers of the tunnel junction will create significant polarization charge at the heterojunction interfaces. In the described device structure, the electric field originating from the polarization charge will have the same polarity as the built in field of the heavily doped n-GaN and p-GaN layers of the tunnel junction. This can be seen in Fig. 2 below.

In 2013 Krishnamoorthy *et al.* continued to optimize their tunnel junction design and grew a TJ PN diode structure comprised of the GaN/InGaN/GaN heterojunction tunnel junction on top of an all MBE grown GaN PN junction. The device was shown to have a record low tunnel junction specific resistivity of 1.2×10^{-4} Ωcm^2 which at a driving current of 100 A/cm^2 meaning the tunnel junction consumed 0.012V in excess of the PN junction

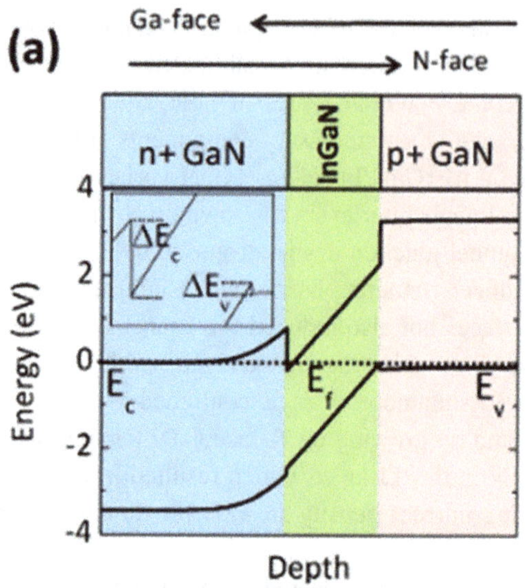

Fig. 2. The band diagram of a GaN based heterojunction tunnel junction with an $In_xGa_{1-x}N$ interlayer. The band diagram shows the increased strength of the electric field owed to the polarization charge generated at the heterojunction interface. The arrow points away from the substrate for the case of Ga-face versus N-face substrates, showing the orientation for the $GaN/In_xGa_{1-x}N/GaN$ tunnel junction which results in more efficient tunneling for both Ga-face and N-face substrates [28].

[28]. In achieving this low tunnel junction resistance the authors explained how a tunnel junction with such low resistance could enable cascaded LED active region as a means for addressing the efficiency droop issue. Later that year they also demonstrated another unique approach to reducing the operating voltage and resistance of III-nitride tunnel junctions, which was to insert midgap states inside the tunnel junction by growing GdN nanoislands at the interface of a p++GaN/n++GaN tunnel junction. The GdN nanoislands act as states in the bandgap which act as intermediate tunneling states, thus reducing the tunneling distance and increasing tunneling probability. Utilizing this novel tunnel junction approach the authors were able to achieve a low tunnel junction specific resistivity of $1.3\text{x}10^{-3}\ \Omega\text{cm}^2$ [29]. Utilizing these newly achieved low tunnel junction resistances Akyol *et al.* computationally demonstrated the viability of implementing such low resistance tunnel junction structures in a tunnel junction enabled cascaded LED structure. The efficacy of such an approach for overcoming the efficiency droop phenomenon in III-nitride visible LEDs by using the tunnel junctions as carrier regeneration centers was discussed. They also demonstrated cascaded PN junctions utilizing the low resistance tunnel junctions [30].

Meanwhile Kuwano *et al.* had demonstrated a method for realizing tunnel junction-based visible LEDs via metal organic chemical vapor deposition (MOCVD). This was especially tricky as the MOCVD growth of the tunnel junction on top of the LED structure would cause hydrogen passivation of the Mg acceptors in the p-GaN layer of the LED. By optimizing a lateral hydrogen diffusion process they were able to demonstrate TJLEDs grown entirely by MOCVD that could be activated from the sidewall after mesa isolation [31]. The lateral hydrogen diffusion process was confirmed via optical emission patterns after various activation annealing conditions and times. This was a huge breakthrough in the field of III-nitride tunnel junction based optoelectronics as MOCVD is the preferred commercial growth technology for III-nitride optoelectronic devices. Not only does MOCVD growth offer greatly enhanced throughput as compared to its MBE counterpart, it also boasts much higher IQE emitters.

It was also demonstrated that a heterojunction tunnel junction approach could help reduce the operating voltage and series resistance of MOCVD grown tunnel junctions as well. By utilizing a p++InGaN layer instead of a p++GaN layer within the tunnel junction in combination with high Si doping on the n++GaN side of the tunnel junction Kaga *et al.* were able to demonstrate an MOCVD TJLED with 0.68V excess forward voltage as compared to the ITO reference LED at 22A/cm^2 [32].

3. Recent Progress in III-Nitride Tunnel Junctions

After the successful demonstration of novel tunnel junction designs utilizing polarization induced charges to enhance the tunneling process as well as the development of an activation process for buried p-GaN layers the general interest of the III-nitride optoelectronics community has increased. In recent years there have been a large number of developments in the approaches being taken towards improving the III-nitride tunnel junctions. From a growth perspective there are three major approaches which have been explored thus far.

3.1. *III-nitride tunnel junctions grown by molecular beam epitaxy*

MBE growth is well suited for the growth of highly efficient tunnel junction devices. It allows for precise control over doping levels, offering sharp dopant profiles, it doesn't suffer from Mg memory effect [33], nor does it cause hydrogen induced re-passivation of underlying p-GaN layers in a tunnel junction optoelectronic device. One major drawback to MBE growth of tunnel junction-based optoelectronic devices is the poor IQE associated with MBE grown emitters. Another drawback is the low throughput of MBE growth which may have potential limitations on commercial application. With these factors considered MBE grown tunnel junctions offer a wide range of versatility in experimental options and are an excellent method for proof of concept and prototyping of III-nitride tunnel junction-based optoelectronics.

There have been many studies on III-nitride tunnel junctions via MBE growth. As mentioned earlier Krishnamoorthy *et al.* explored the use of polarization engineered GaN/InGaN/GaN tunnel junction structures grown by MBE and ultimately achieved a record low tunnel junction resistance of $1.2\text{x}10^{-4}$ Ωcm^2. Akyol *et al.* demonstrated MBE grown heavily doped GaN homojunction tunnel junction structures on p-n diodes with low TJ resistance values. The structure employed a n++GaN/p++GaN tunnel junction structure with doping concentrations of [Si]=$4\text{x}10^{20}$ cm^{-3} and [Mg]=$3\text{x}10^{20}$ cm^{-3}. The differential

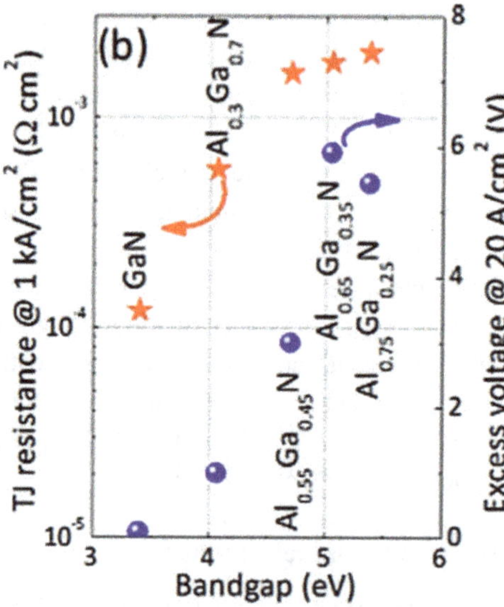

Fig. 3. The tunnel junction resistance at 1kA/cm^2 and excess voltage drop at 20A/cm^2 versus the bandgap energy of the tunnel junction for the tunnel injected UVLEDs reported by Zhang *et al.* [35].

resistance at current densities of 100A/cm^2, 1kA/cm^2, and 150kA/cm^2 were extracted as 2x10^{-3} Ωcm^2, 3.1x10^{-4} Ωcm^2, and 1.5x10^{-5} Ωcm^2, respectively [34]. Zhang *et al.* have demonstrated a range of efficient $Al_xGa_{1-x}N$ based wide bandgap tunnel junctions ranging from Al mole fraction x=0.3 through x=0.75 [9, 35, 36]. Even at such wide bandgaps efficient tunneling is signified by the low tunnel junction resistances obtained in the tunnel junction enabled UVLEDs as shown in Fig. 3. The tunnel junction resistance ranged from 5.6x10^{-4} Ωcm^2 for $Al_{0.3}Ga_{0.7}N$ to 1.9x10^{-3} Ωcm^2 for $Al_{0.75}Ga_{0.25}N$.

In order to achieve such efficient tunneling at these high bandgap energies Zhang *et al.* utilized advanced polarization engineering within the tunnel junction structure in order to compensate for the large tunneling barrier height associated with the wide band gap high composition $Al_xGa_{1-x}N$ materials. The issue of poor p-type doping in high composition $Al_xGa_{1-x}N$ was also addressed via polarization engineering. By utilizing graded $Al_xGa_{1-x}N$ layers on both sides of the $In_yGa_{1-y}N$ interlayer both 3D polarization charges and 2D polarization sheet charges were employed to reduce the large depletion widths caused by the high dopant ionization energies in high composition $Al_xGa_{1-x}N$ [37]. This can be visualized from the charge diagrams and energy band diagrams shown in Fig. 4. The advantages of polarization enhanced tunnel junctions have been explored elsewhere as well [38]. The efficacy of a metal/semiconductor tunnel junction was also demonstrated for UVLEDs [39].

Another new approach to achieving efficient GaN based tunnel junctions was recently explored. Here a metal modulated epitaxy (MME) approach was carried out by Clinton *et al.* via MBE growth to study highly doped GaN homojunction tunnel junctions. The tunnel junctions were grown on p-i-n diodes for study. A reference p-i-n diode was grown

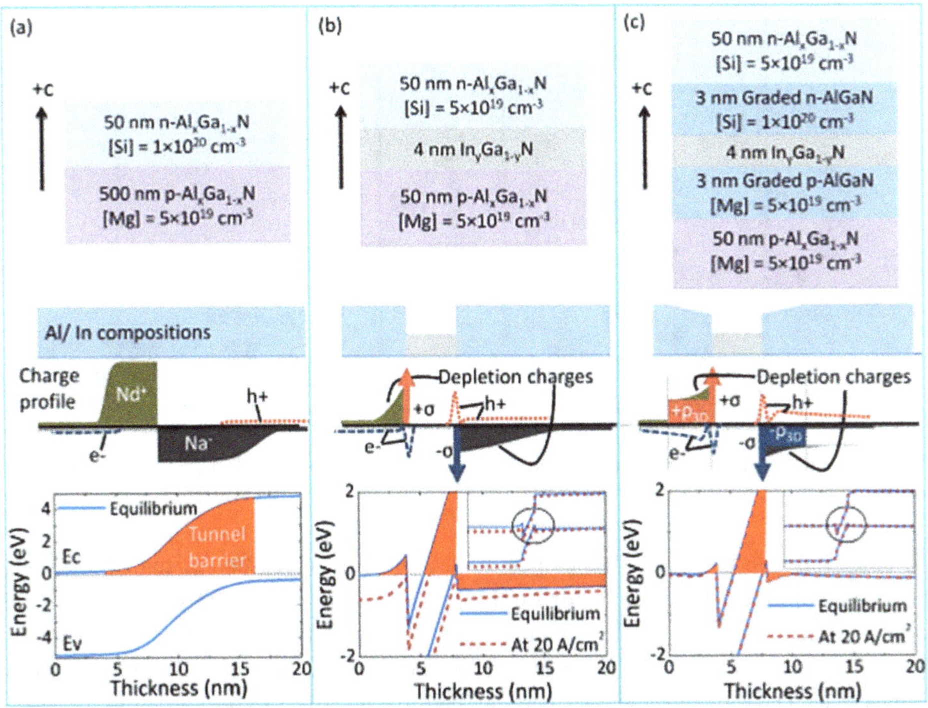

Fig. 4. A comparison of tunnel junction structures comprised of (a) heavily doped p-n homojunction, (b) heterojunction with thin InGaN layer inserted between p-n AlGaN layers, (c) heterojunction with graded Al composition layers. Al mole fraction is x=0.7 and In mole fraction is y=0.3 [37].

and processed with a p-contact as reference. The tunnel junctions had high doping concentrations of [Si]=$4.6\text{x}10^{20}$ cm^{-3} and [Mg]=$3.4\text{x}10^{20}$ cm^{-3}. At 1250 A/cm^2 the forward voltage of the TJ-p-i-n structure was 0.1V higher than that of the reference p-i-n structure. The specific on resistance of the TJ-p-i-n structure was $3.24\text{x}10^{-4}$ Ωcm^2 versus the reference p-i-n structure which showed $3.75\text{x}10^{-4}$ Ωcm^2 [40]. The lower on resistance of the tunnel junction device was attributed to the higher p-contact resistance on the p-i-n reference sample.

Tunnel junction enabled cascaded PN diodes and LED active regions have also been demonstrated by MBE in the past [8, 30]. The low tunnel junction resistances obtainable by MBE growth demonstrate the feasibility of using tunnel junctions as carrier regeneration centers between emitter active regions to increase output power at a given current density and thus circumvent efficiency droop. This effect can be visualized by looking at the calculated plot of IV characteristics for different numbers of cascaded LED active regions in Fig. 5 below. It is important to note that as signified by the stars on the plot, a constant input power can be obtained at much lower current density as the number of active regions is increased (and thus increasing the operating voltage).

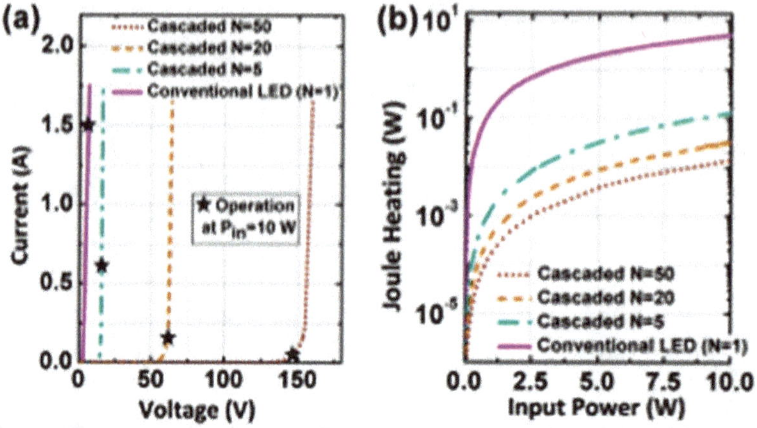

Fig. 5. (a) I-V characteristics for various numbers of cascaded LEDs and (b) the joule heating vs input power for various numbers of cascaded LEDs. Black stars on the I-V curves denote constant input power [30].

There has also been significant research into the implementation of tunnel junctions into nanowire UVLEDs grown by MBE. They have been shown to significantly reduce the forward voltage of the device by circumventing the necessity to grow on p-Si substrates, by utilizing tunnel injection of holes into the p-cladding layer of the LED through an n-Si contact [41]. The same concept was later applied to nanowire LEDs emitting in the UVC spectrum at 242nm. The tunnel junction LED showed a reduction in on resistance of one order of magnitude as compared to the p-contact reference nanowire LEDs, and the TJ nanowire UVLED showed two orders of magnitude improvement in light output power as compared to the reference device [42].

Recently Siekacz *et al.* have realized a tunnel junction enabled cascaded edge emitting laser diode stack containing two tunnel junctions and two laser diode active regions. The slope efficiency is 0.7W/A when only the first diode is lasing, and as the second diode begins lasing the slope efficiency doubles to 1.4W/A. The TJ laser diode has an excess forward voltage of 0.8V at 4kA/cm^2 driving current as compared to the reference laser diode with p-contact [43].

3.2. *III-nitride tunnel junctions grown by metal organic chemical vapor deposition*

Another such approach to the growth of III-nitride tunnel junction optoelectronics is monolithic MOCVD growth, which has obvious advantages from the perspective of commercialization, where MOCVD growth is the growth method of choice. One of the major challenges to this approach as discussed above is the necessity of a buried p-GaN activation process requiring mesa isolation before p-GaN activation. Another major challenge to this approach is the Mg memory effect which causes increased Mg levels present on the surface of the sample after Mg layers are grown in the MOCVD chamber [33]. The memory effect therefore makes abrupt doping profiles harder to achieve. Abrupt doping profiles are of significant importance in achieving a low resistance tunnel junction.

Even considering the range of challenges associated with this growth approach there has been significant progress made towards low resistance low operating voltage tunnel junctions grown monolithically by MOCVD.

One such approach to developing low resistance tunnel junctions grown monolithically by MOCVD has been to utilize Ge as a donor dopant instead of Si. This allows for a higher electron concentration at the n++GaN[Ge]/p++GaN[Mg] tunnel junction and resulted in a TJLED which had an extra 0.79V forward voltage as compared to the ITO reference LED at 2 A/cm^2 [44]. Another approach to the MOCVD growth of tunnel junctions was employed to realize VCSELs with tunnel junction contacts that were grown entirely by MOCVD in a two-step growth process. The VCSEL was grown and growth continued through to the p++ layer of the tunnel junction. At this point growth was stopped and the sample was removed from the chamber for a buffered HF dip which serves to remove excess Mg from the surface (thereby improving the abrupt nature of the doping profile which is crucial for a low voltage drop across the tunnel junction). Next the sample was returned to the chamber where an *in situ* activation process was completed. Finally the regrowth of the n++ tunnel junction and top contact layers were completed [45]. A similar two step MOCVD growth process was used to realize MOCVD grown micro-LEDs with tunnel junction contacts with the addition of Ozone treatment alongside a BHF dip in order to reduce both surface contaminant concentration and Mg concentration at the tunnel junction regrowth interface, respectively. The smallest sized micro-LED showed an excess voltage drop of 0.6V on the TJLED as compared to the reference ITO LED. The tunnel junction LED showed an 8.4% increase in EQE over the reference ITO LED and also showed about 40% less droop than the reference ITO LED [46].

Another method of improving the all MOCVD grown III-nitride tunnel junction is that demonstrated by Sohi *et al.* where low temperature Si doped n++ GaN layers were grown at 740C on commercially available LED wafers. The low temperature growth is suspected to inhibit re-passivation of the Mg doped p-GaN layers in the commercially available LEDs. InGaN interlayers were also utilized to reduce the tunnel junction resistance and lower the voltage dropped across the tunnel junction by taking advantage of polarization engineering. It was shown that the low temperature 5nm InGaN interlayer / n++ GaN TJ regrowth showed the best performance of all the tunnel junction structures grown, with an excess forward voltage of 0.09V at driving current of 1 A/cm^2 on the TJLED as compared with the reference p-contact LED. A low total differential resistance of 8.4×10^{-4} Ωcm^2 was obtained for the device [47]. It is worthy to note that usually the voltage penalty is reported at higher operating current densities where it is clear that the device is fully turned on, so it is hard to compare this result to other presented in tunnel junction literature. Sohi *et al.* themselves even reported their specific on resistance for the same device at significantly higher current densities. One metric that has been considered difficult to improve on in all MOCVD tunnel junctions has been the extra voltage drop across the tunnel junction. Table 1 below shows some of the most recent state-of-the-art results discussed above.

Table 1. A comparison of recent all MOCVD III-nitride tunnel junctions and their excess voltage dropped across the tunnel junction (V_{TJ}) as a function of driving current.

Reference	Excess TJ Voltage [V]	Current Density [A/cm^2]
P. Sohi (2018)	0.09	1
D. Hwang (2017)	0.6	20
S. Neugebauer (2017)	0.79	2

All MOCVD tunnel junctions have been implemented to demonstrate monolithically grown cascaded blue LEDs previously. The tunnel junctions took advantage of polarization engineering by inserting heterojunction interlayers in-between the p++GaN and n++GaN layers of the tunnel junctions. It was shown that EQE as high as 92.8% could be achieved in a tunnel junction enabled cascaded blue LED stack containing two LED active regions and two tunnel junctions, as compared to an EQE of 53.5% for the reference ITO LED [48, 49]. These improvements in optical performance can be seen below in Fig. 6. Two types of tunnel junctions were compared. One with only an InGaN interlayer between the heavily doped GaN p-n layers (n++GaN/2.5nm u-InGaN/p++GaN) and another with an extra AlGaN layer inserted in the middle of the InGaN layer (n++GaN/1.0nm u-InGaN/ 0.5nm u-AlGaN/1.0nm u-InGaN/ p++GaN). Ultimately the tunnel junction utilizing both InGaN and AlGaN showed a lower voltage drop across each tunnel junction (1.225V vs 1.42V) and a lower series resistance (1.95×10^{-3} Ωcm^2 vs 6.05×10^{-3} Ωcm^2) as predicted by Tsai *et al.* [50].

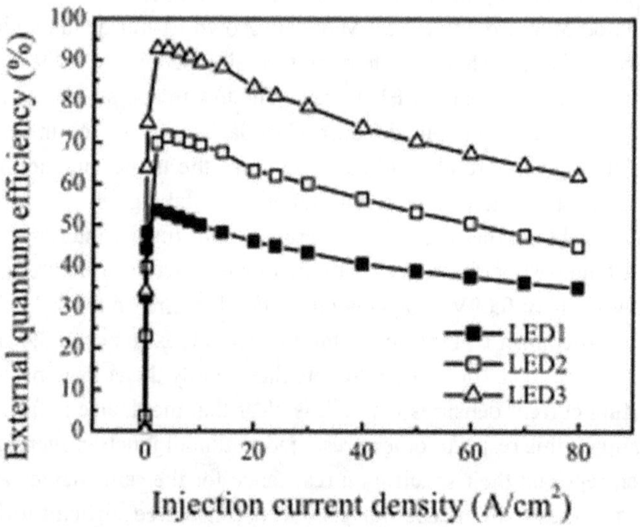

Fig. 6. External Quantum Efficiency versus injection current for the reference ITO LED (LED1), the tunnel junction cascaded LED with InGaN interlayer (LED2), and the tunnel junction cascaded LED with hybrid InGaN/AlGaN interlayer (LED3) [49].

3.3. *III-nitride tunnel junctions grown by hybrid MOCVD/MBE growth*

The final growth approach to realizing efficient III-nitride tunnel junctions is known as the hybrid tunnel junction. This approach was first demonstrated by Malinverni *et al.* in 2015. MOCVD was implemented to grow a standard LED active region capped by a thin, heavily doped [Mg] = $1x10^{20}$ cm^{-3} cap layer. Next NH_3 based MBE regrowth was carried out to grow a heavily Si doped n++GaN TJ layer with [Si] = $2x10^{20}$ cm^{-3} [51]. Since this initial demonstration of a hybrid tunnel junction there has been significant development of the hybrid tunnel junction technology and it has been used to realize a wide range of III-nitride optoelectronic devices. A violet non-polar VCSEL is grown with hybrid tunnel junction contact and compared to a standard ITO VCSEL. The threshold current density of the TJVCSEL (3.5 kA/cm^2) is significantly reduced as compared to the reference ITO VCSEL (8 kA/cm^2). However, the forward voltage of the TJ VCSEL is 1.5V higher than that of the reference ITO VCSEL [11].

The hybrid tunnel junction has also been explored as a contact to a standard blue LED emitting at 450nm. In this experiment Young *et al.* used NH_3 MBE to regrow n++GaN layers on top of MOCVD grown LEDs capped with a thin heavily Mg doped p-GaN layer. The effect of an *in situ* pre growth anneal on the hybrid TJ performance was conducted and it was found that pre-growth anneal degraded the quality of the hybrid tunnel junction by increasing the forward voltage of the TJLED as compared to the un-annealed case. The tunnel junction LED showed a 0.67V lower forward voltage than did the reference ITO LED at 20mA driving current. The total on-resistance for the hybrid TJLED including contact resistances was measured to be $1.5x10^{-4}$ Ωcm^2 [52].

A hybrid tunnel junction was then combined with a commercial grade LED, and flip chip packaging in order to achieve a LED with over 70% WPE. It was confirmed by SIMS measurement that acid treatment with HF prior to the MBE regrowth of the hybrid junction reduced the Magnesium concentration at the hybrid TJ interface from $3x10^{19}$ cm^{-3} to $1.3x10^{19}$ cm^{-3} which lead to a forward voltage reduction of 1.25V at 20 A/cm^2 [53]. A hybrid tunnel junction was recently shown to significantly reduce the efficiency droop in a green III-nitride LED. In this case a TJLED was compared to an ITO reference LED and although similar peak EQEs were measured in both cases the droop ratio was 23.7% in the TJLED and 34.2% in the reference ITO LED [54]. A collection of the lowest reported tunnel junction resistance values can be found in Fig. 7 below.

4. Conclusion

This paper has reviewed the recent progress made in III-nitride tunnel junction-based optoelectronic devices. The history of III-nitride tunnel junctions was discussed, including the initial limitations faced in realizing efficient tunnel junction-based devices. We have seen that polarization engineering and non-traditional growth approaches have been successful in reducing the excess voltage penalty dropped across the tunnel junctions significantly as well as showing appreciable reduction in tunnel junction resistances as well. It has been demonstrated that the excess voltage penalty of the tunnel junction has

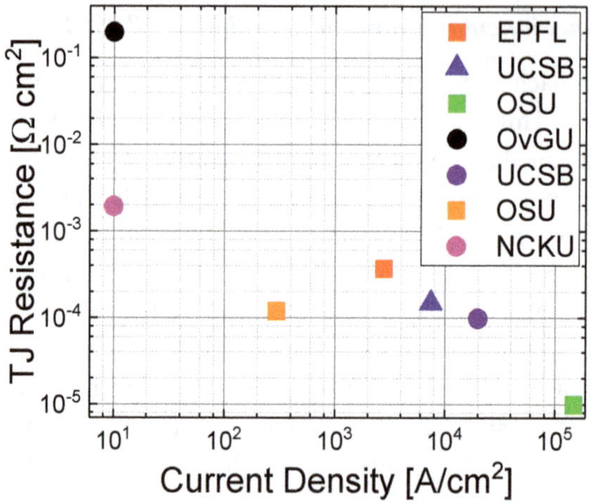

Fig. 7. A collection of some of the lowest reported tunnel junction resistance values. Squares represent MBE grown tunnel junctions, triangles represent hybrid tunnel junctions, and circles represent MOCVD grown tunnel junctions.

been reduced to 0.6V at 20A/cm^2 for the case of all MOCVD tunnel junctions, 0.012V at 100A/cm^2 for polarization engineered GaN/InGaN/GaN grown by MBE, and for the case of the hybrid tunnel junction LEDs the forward voltage was actually 0.67V lower than the ITO reference LED. Tunnel junction resistances have also been reduced significantly over the years as these different III-Nitride tunnel junction growth techniques have matured.

References

1. M. Razeghi, "III-Nitride Optoelectronic Devices: From Ultraviolet Toward Terahertz", *IEEE Photonics Journal*, **3** (2011) 263-267.
2. E. T. Yu, X. Z. Dang, P. M. Asbeck, S. S. Lau, and G. J. Sullivan, "Spontaneous and piezoelectric polarization effects in III–V nitride heterostructures", *Journal of Vacuum Science & Technology B*, **17** (1999) 1742-1749.
3. S. Brochen, J. Brault, S. Chenot, A. Dussaigne, M. Leroux, and B. Damilano, "Dependence of the Mg-related acceptor ionization energy with the acceptor concentration in p-type GaN layers grown by molecular beam epitaxy", *Appl. Phys. Lett.*, **103** (2013) 032102.
4. K. B. Nam, M. L. Nakarmi, J. Li, J. Y. Lin, and H. X. Jiang, "Mg acceptor level in AlN probed by deep ultraviolet photoluminescence", *Appl. Phys. Lett.*, **83** (2003) 878-880.
5. J. Piprek, "Efficiency droop in nitride-based light-emitting diodes", **207** (2010) 2217-2225.
6. M. J. Grundmann, and U. K. Mishra, "Multi-color light emitting diode using polarization-induced tunnel junctions", *Phys. Status Solidi C*, **4** (2007) 2830-2833.
7. I. Ozden, E. Makarona, A. V. Nurmikko, T. Takeuchi, and M. Krames, "A dual-wavelength indium gallium nitride quantum well light emitting diode", *Appl. Phys. Lett.*, **79** (2001) 2532-2534.
8. F. Akyol, S. Krishnamoorthy, Y. Zhang, and S. Rajan, "GaN-based three-junction cascaded light-emitting diode with low-resistance InGaN tunnel junctions", *Applied Physics Express*, **8** (2015) 082103.

9. Y. Zhang, S. Krishnamoorthy, J. M. Johnson, F. Akyol, A. Allerman, M. W. Moseley, A. Armstrong, J. Hwang, and S. Rajan, "Interband tunneling for hole injection in III-nitride ultraviolet emitters", *Appl. Phys. Lett.*, **106** (2015) 141103.
10. B. P. Yonkee, E. C. Young, C. Lee, J. T. Leonard, S. P. DenBaars, J. S. Speck, and S. Nakamura, "Demonstration of a III-nitride edge-emitting laser diode utilizing a GaN tunnel junction contact", *Opt. Express*, **24** (2016) 7816-7822.
11. J. Leonard, E. Young, B. P. Yonkee, D. A. Cohen, T. Margalith, S. Denbaars, J. Speck, and S. Nakamura, "Demonstration of a III-nitride vertical-cavity surface-emitting laser with a III-nitride tunnel junction intracavity contact", *Applied Physics Letters*, **107** (2015) 091105.
12. L. Yuan, H. Chen, and K. J. Chen, "Normally Off AlGaN/GaN Metal–2DEG Tunnel-Junction Field-Effect Transistors", *IEEE Electron Device Letters*, **32** (2011) 303-305.
13. L. Esaki, "New Phenomenon in Narrow Germanium $p\ensuremath{-}n$ Junctions", *Physical Review*, **109** (1958) 603-604.
14. D. L. Miller, S. W. Zehr, and J. S. H. Jr., "GaAs-AlGaAs tunnel junctions for multigap cascade solar cells", *Appl. Phys. Lett.*, **53** (1982) 744-748.
15. G. C. DeSalvo, "Ultrathin delta doped GaAs and AlAs tunnel junctions as interdevice ohmic contacts", *Appl. Phys. Lett.*, **74** (1993) 4207-4212.
16. J. K. Kim, S. Nakagawa, E. Hall, and L. A. Coldren, "Near-room-temperature continuous-wave operation of multiple-active-region 1.55 μm vertical-cavity lasers with high differential efficiency", *Appl. Phys. Lett.*, **77** (2000) 3137-3139.
17. X. Guo, X. Luan, W. Wang, C. Guo, and G. J. F. o. O. Shen, "Scalabilities of LEDs and VCSELs with tunnel-regenerated multi-active region structure", *Front. Optoelectron.*, **6** (2013) 97-101.
18. S.-R. Jeon, Y.-H. Song, H.-J. Jang, G. M. Yang, S. W. Hwang, and S. J. Son, "Lateral current spreading in GaN-based light-emitting diodes utilizing tunnel contact junctions", *Appl. Phys. Lett.*, **78** (2001) 3265-3267.
19. J. J. Sakurai, and J. Napolitano. (2017). *Modern Quantum Mechanics* (2 ed.). Cambridge: Cambridge University Press.
20. M. Diagne, Y. He, H. Zhou, E. Makarona, A. V. Nurmikko, J. Han, K. E. Waldrip, J. J. Figiel, T. Takeuchi, and M. Krames, "Vertical cavity violet light emitting diode incorporating an aluminum gallium nitride distributed Bragg mirror and a tunnel junction", *Applied Physics Letters*, **79** (2001) 3720-3722.
21. C. H. Chen, S. J. Chang, Y. K. Su, J. K. Sheu, J. F. Chen, C. H. Kuo, and Y. C. Lin, "Nitride-based cascade near white light-emitting diodes", *IEEE Photonics Technology Letters*, **14** (2002) 908-910.
22. J. Simon, Z. Zhang, K. Goodman, H. Xing, T. Kosel, P. Fay, and D. Jena, "Polarization-Induced Zener Tunnel Junctions in Wide-Band-Gap Heterostructures", *Physical Review Letters*, **103** (2009) 026801.
23. M. F. Schubert, "Polarization-charge tunnel junctions for ultraviolet light-emitters without p-type contact", **96** (2010) 031102.
24. M. F. Schubert, "Interband tunnel junctions for wurtzite III-nitride semiconductors based on heterointerface polarization charges", *Physical Review B*, **81** (2010) 035303.
25. S. Krishnamoorthy, D. N. Nath, F. Akyol, P. S. Park, M. Esposto, and S. Rajan, "Polarization-engineered GaN/InGaN/GaN tunnel diodes", *Appl. Phys. Lett.*, **97** (2010) 203502.
26. S. Krishnamoorthy, P. S. Park, and S. Rajan, "Demonstration of forward inter-band tunneling in GaN by polarization engineering", *Appl. Phys. Lett.*, **99** (2011) 233504.

27. S. Krishnamoorthy, P. S. Park, and S. Rajan. (2011, 20-22 June 2011). *III-nitride tunnel diodes with record forward tunnel current density.* Paper presented at the 69th Device Research Conference.
28. S. Krishnamoorthy, F. Akyol, P. S. Park, and S. Rajan, "Low resistance GaN/InGaN/GaN tunnel junctions", *Applied Physics Letters*, **102** (2013) 113503.
29. S. Krishnamoorthy, T. F. Kent, J. Yang, P. S. Park, R. C. Myers, and S. Rajan, "GdN Nanoisland-Based GaN Tunnel Junctions", *Nano Letters*, **13** (2013) 2570-2575.
30. F. Akyol, S. Krishnamoorthy, and S. Rajan, "Tunneling-based carrier regeneration in cascaded GaN light emitting diodes to overcome efficiency droop", *Appl. Phys. Lett.*, **103** (2013) 081107.
31. Y. Kuwano, M. Kaga, T. Morita, K. Yamashita, K. Yagi, M. Iwaya, T. Takeuchi, S. Kamiyama, and I. Akasaki, "Lateral Hydrogen Diffusion at p-GaN Layers in Nitride-Based Light Emitting Diodes with Tunnel Junctions", *Japanese Journal of Applied Physics*, **52** (2013) 08JK12.
32. M. Kaga, T. Morita, Y. Kuwano, K. Yamashita, K. Yagi, M. Iwaya, T. Takeuchi, S. Kamiyama, and I. Akasaki, "GaInN-Based Tunnel Junctions in n–p–n Light Emitting Diodes", *Japanese Journal of Applied Physics*, **52** (2013) 08JH06.
33. H. Xing, D. S. Green, H. Yu, T. Mates, P. Kozodoy, S. Keller, S. P. DenBaars, and U. K. Mishra, "Memory Effect and Redistribution of Mg into Sequentially Regrown GaN Layer by Metalorganic Chemical Vapor Deposition", *Japanese Journal of Applied Physics*, **42** (2003) 50-53.
34. F. Akyol, S. Krishnamoorthy, Y. Zhang, J. Johnson, J. Hwang, and S. Rajan, "Low-resistance GaN tunnel homojunctions with 150 kA/cm2 current and repeatable negative differential resistance", *Appl. Phys. Lett.*, **108** (2016) 131103.
35. Y. Zhang, Z. Jamal-Eddine, F. Akyol, S. Bajaj, J. M. Johnson, G. Calderon, A. A. Allerman, M. W. Moseley, A. M. Armstrong, J. Hwang, and S. Rajan, "Tunnel-injected sub 290 nm ultra-violet light emitting diodes with 2.8% external quantum efficiency", *Appl. Phys. Lett.*, **112** (2018) 071107.
36. Y. Zhang, S. Krishnamoorthy, F. Akyol, S. Bajaj, A. A. Allerman, M. W. Moseley, A. M. Armstrong, and S. Rajan, "Tunnel-injected sub-260 nm ultraviolet light emitting diodes", *Appl. Phys. Lett.*, **110** (2017) 201102.
37. Y. Zhang, S. Krishnamoorthy, F. Akyol, A. A. Allerman, M. W. Moseley, A. M. Armstrong, and S. Rajan, "Design and demonstration of ultra-wide bandgap AlGaN tunnel junctions", *Appl. Phys. Lett.*, **109** (2016) 121102.
38. X. Yan, W. Li, S. M. Islam, K. Pourang, H. Xing, P. Fay, and D. Jena, "Polarization-induced Zener tunnel diodes in GaN/InGaN/GaN heterojunctions", *Appl. Phys. Lett.*, **107** (2015) 163504.
39. Y. Zhang, S. Krishnamoorthy, F. Akyol, J. M. Johnson, A. A. Allerman, M. W. Moseley, A. M. Armstrong, J. Hwang, and S. Rajan, "Reflective metal/semiconductor tunnel junctions for hole injection in AlGaN UV LEDs", **111** (2017) 051104.
40. E. A. Clinton, E. Vadiee, S.-C. Shen, K. Mehta, P. D. Yoder, and W. A. Doolittle, "Negative differential resistance in GaN homojunction tunnel diodes and low voltage loss tunnel contacts", *Applied Physics Letters*, **112** (2018) 252103.
41. A. T. M. G. Sarwar, B. J. May, J. I. Deitz, T. J. Grassman, D. W. McComb, and R. C. Myers, "Tunnel junction enhanced nanowire ultraviolet light emitting diodes", *Appl. Phys. Lett.*, **107** (2015) 101103.
42. S. Zhao, S. M. Sadaf, S. Vanka, Y. Wang, R. Rashid, and Z. Mi, "Sub-milliwatt AlGaN nanowire tunnel junction deep ultraviolet light emitting diodes on silicon operating at 242 nm", *Appl. Phys. Lett.*, **109** (2016) 201106.

43. M. Siekacz, G. Muziol, M. Hajdel, M. Żak, K. Nowakowski-Szkudlarek, H. Turski, M. Sawicka, P. Wolny, A. Feduniewicz-Żmuda, S. Stanczyk, J. Moneta, and C. Skierbiszewski, "Stack of two III-nitride laser diodes interconnected by a tunnel junction", *Opt. Express*, **27** (2019) 5784-5791.
44. S. Neugebauer, M. P. Hoffmann, H. Witte, J. Bläsing, A. Dadgar, A. Strittmatter, T. Niermann, M. Narodovitch, and M. Lehmann, "All metalorganic chemical vapor phase epitaxy of p/n-GaN tunnel junction for blue light emitting diode applications", *Applied Physics Letters*, **110** (2017) 102-104.
45. S. Lee, C. A. Forman, C. Lee, J. Kearns, E. C. Young, J. T. Leonard, D. A. Cohen, J. S. Speck, S. Nakamura, and S. P. DenBaars, "GaN-based vertical-cavity surface-emitting lasers with tunnel junction contacts grown by metal-organic chemical vapor deposition", *Applied Physics Express*, **11** (2018) 062703.
46. D. Hwang, A. J. Mughal, M. S. Wong, A. I. Alhassan, S. Nakamura, and S. P. DenBaars, "Micro-light-emitting diodes with III–nitride tunnel junction contacts grown by metalorganic chemical vapor deposition", *Applied Physics Express*, **11** (2017) 012102.
47. P. Sohi, M. Mosca, Y. Chen, J.-F. Carlin, and N. Grandjean, "Low-temperature growth of n++-GaN by metalorganic chemical vapor deposition to achieve low-resistivity tunnel junctions on blue light emitting diodes", *Semiconductor Science and Technology*, **34** (2018).
48. S. Chang, W. Lin, and C. Yu, "GaN-Based Multiquantum Well Light-Emitting Diodes With Tunnel-Junction-Cascaded Active Regions", *IEEE Electron Device Letters*, **36** (2015) 366-368.
49. S. Chang, W. Lin, and W. Chen, "Cascaded GaN Light-Emitting Diodes With Hybrid Tunnel Junction Layers", *IEEE Journal of Quantum Electronics*, **51** (2015) 1-5.
50. M. Tsai, B. Leung, T. Hsu, and Y. Kuo, "Low Resistivity GaN-Based Polarization-Induced Tunnel Junctions", *Journal of Lightwave Technology*, **31** (2013) 3575-3581.
51. M. Malinverni, D. Martin, and N. Grandjean, "InGaN based micro light emitting diodes featuring a buried GaN tunnel junction", *Appl. Phys. Lett.*, **107** (2015) 051107.
52. E. C. Young, B. P. Yonkee, F. Wu, S. H. Oh, S. P. DenBaars, S. Nakamura, and J. S. Speck, "Hybrid tunnel junction contacts to III–nitride light-emitting diodes", *Applied Physics Express*, **9** (2016) 022102.
53. B. P. Yonkee, E. C. Young, S. P. DenBaars, S. Nakamura, and J. S. Speck, "Silver free III-nitride flip chip light-emitting-diode with wall plug efficiency over 70% utilizing a GaN tunnel junction", *Appl. Phys. Lett.*, **109** (2016) 191104.
54. A. I. Alhassan, E. C. Young, A. Y. Alyamani, A. Albadri, S. Nakamura, S. P. DenBaars, and J. S. Speck, "Reduced-droop green III–nitride light-emitting diodes utilizing GaN tunnel junction", *Applied Physics Express*, **11** (2018) 042101.

www.ingramcontent.com/pod-product-compliance
Lightning Source LLC
LaVergne TN
LVHW081930200926
848484LV00009B/2401